E.E. 422
MODULATION
TAKEN WINTER
OF 1975

B?
Rm
ALUMNI QUAD
Ph. 742-0426

Communication Systems

McGraw-Hill Electrical and Electronic Engineering Series

Frederick Emmons Terman, *Consulting Editor*
W. W. Harman and J. G. Truxal, *Associate Consulting Editors*

Communication Systems

An Introduction to Signals and Noise
in Electrical Communication

A. Bruce Carlson
Associate Professor of Systems Engineering
Rensselaer Polytechnic Institute

McGraw-Hill Book Company

New York, St. Louis, San Francisco, Toronto, London, Sydney

Communication Systems: An Introduction to Signals and Noise in Electrical Communication

Library of Congress Catalog Card Number 68–22761
07–009955–3

89101112 KPKP 7987654

To the memory of my father, Albin John Carlson

Preface

A book title containing the word *communication* is apt to be ambiguous or misinterpreted. (I well recall discovering Harman's "Principles of the Statistical Theory of Communication" shelved in the journalism section at a certain university library.) To clear the air immediately on that score, this text is an introduction to electrical communication systems written at a level appropriate for advanced undergraduates and first-year graduate students of electrical engineering. Because electrical communication involves many diverse considerations—from the details of applied electronics and electromagnetics to the more abstract notions of mathematical communication theory—I have attempted here to chart a middle path by adopting the systems engineering viewpoint. Correspondingly, the focus is upon those concepts, techniques, and problems which characterize information transfer via electric signals, given the inevitable limitations of physical systems.

The first chapter is a qualitative discussion of the general subject, intended to place what follows in perspective. The next three chapters review and develop the necessary working tools, namely: signal analysis and spectral interpretation using Fourier theory (Chap. 2); the transfer-function approach to signal transmission, processing, and distortion in linear systems (Chap. 3); and elementary probability and statistics applied to the description of random signals and noise (Chap. 4). (The treatment of electrical noise in terms of circuit models, noise figure, etc., is covered briefly as an appendix.) The heart of the study, Chaps. 5 to 7, deals with modulation-demodulation theory and practice, including sampling and multiplexing as related subjects. Particular attention is given to the more significant modulation types, their relative merits, instrumentation requirements, and performance in the presence of interference and noise. For the purpose of system evaluations and comparisons, and to shed further light on communication in the broader sense, the basics of modern information theory are introduced in Chap. 8. Finally, Chap. 9 surveys the special problems and techniques of digital signal transmission.

By way of background, I have assumed the student is reasonably adept at circuit analysis methods, including Fourier series, Fourier transforms, and Laplace transforms, and has some knowledge of electronics. Prior exposure to probability theory, while desirable, is not essential.

With these minimum prerequisites, most but not all of the material can be covered in one semester of three class hours per week. At Rensselaer Polytechnic Institute, for example, we omit Chap. 9 entirely and leave the Appendix as preparatory reading for concurrent laboratory work, a practice that, surprisingly, has proved quite successful. For a shorter course, Chaps. 8 and 9, the Appendix, and optional sections designated by the symbol ★ are suitable candidates for omission. Conversely, with the addition of extra lecture material as seen fit, a substantial year-long course can be constructed.

A number of homework problems have been supplied. They range from routine but necessary drills to discussion questions and design exercises, most being well within the capabilities of the diligent student. However, since the problems are listed in order of subject matter rather than difficulty, discretion is called for on the part of the instructor in selecting assignments.

The references are generally of two types, those cited in the text proper for further details on a particular item, and those given at the end of each chapter under the heading Selected Supplementary Reading. The latter, in fact, constitute brief annotated bibliographies listing other tutorial treatments at various levels, as well as the landmark papers. Unless otherwise indicated, referenced material can be profitably read by any student who has completed the chapter in question. All references have been screened for easy accessibility.

Like many textbooks, this work evolved over several years in the form of class notes, notes used for both the regular curriculum offering at Rensselaer and for the Long Lines Engineering Development Program of The American Telephone and Telegraph Company. Thus, I am especially indebted to my students who, in large measure, stimulated and encouraged the project. Credit must also be given to the administration of Rensselaer Polytechnic Institute for providing time and support; to my colleagues Drs. Dean N. Arden, Charles M. Close, Dean K. Frederick, and David G. Gisser for their criticisms and suggestions; to Miss Rosana Laviolette for typing the manuscript; to Mr. Malcolm Bailey for the illustrations; and to my wife Patricia who cheerfully endured the role of "author's widow."

A. Bruce Carlson

Contents

Communication Systems

1
Introduction

"Attention, the Universe! By kingdoms, right wheel!" This prophetic phrase is the first telegraph message on record; it was sent over a 10-mile line by Samuel F. B. Morse in 1838. Thus was born a new era in communication, the era of electrical communication.

Today, electrical communication systems are found wherever information is to be conveyed from one point to another—from man to man or machine to machine. Telephone, radio, and television have become integral parts of everyday life. Long-distance circuits span the globe, carrying text, voice, and images. Radar and telemetry systems play vital roles in navigation, defense, and scientific research. Computers talk to computers via transcontinental data links. Industrial operations are speeded and improved through the use of closed-loop control systems. The accomplishments are many, and the list is seemingly endless.

Certainly, great strides have been made since the days of Morse— witness the 1964–1965 Mariner IV mission, in which photographs of the planet Mars were transmitted back to earth, a distance in excess of

100 million miles! Equally certain, the next decade will see many new triumphs of communication engineering. Indeed, the potential applications of electrical communication technology are bounded only by man's needs, aspirations, and imagination.

In view of the myriad types of communication systems, a book like this, devoted to the subject of electrical communication, cannot possibly cover every application. The result would be no more than a catalog, and one that would soon be out of date. Nor can we discuss in detail the individual components, or hardware, which go to make up a specific type of system. A typical system consists of numerous and diverse parts, and understanding them draws on virtually all the specialties of electrical engineering: energy conversion, network theory, applied electronics, and electromagnetics, to name a few. Moreover, a part-by-part analysis would miss the essential point that a system is an integrated *whole*, a whole which truly exceeds the sum of its parts.

Instead, this book approaches the subject from a more general viewpoint. Recognizing that all communication systems have the same basic function, namely, *information transmission*, we shall seek out and isolate the principles and problems of transmitting information in electrical form. These will be examined in detail so as to provide the fundamental analysis and design techniques applicable to any type of electrical communication. In short, this book treats communication systems as *systems*.

The system-engineering approach is a powerful one; it is also somewhat abstract, relying heavily on mathematical tools and models to cut through to the heart of complex problems. However, it should be kept in mind that the mathematics is only a means to an end, that end being a basic understanding of electrical communication. To that end, the use of abstract mathematical tools and models must be tempered with physical reasoning and engineering judgment.

The purpose of this introductory chapter is twofold: (1) to give a general description of communication systems, enumerating the essential elements, their functions, and associated problems, and (2) to place in perspective the roles of the various disciplines in communication system engineering.

1.1 COMMUNICATION, MESSAGES, AND SIGNALS

To begin with, we define *communication* as the process whereby information is transferred from one point in space and time, called the *source*, to another point, the *destination*, or user. Usually, but not always, the destination is remote from the source in the sense of being separated by some distance.

A *communication system* is the totality of mechanisms which provides the information link between source and destination. An electrical communication system is one that achieves this function primarily, but not exclusively, through the use of electric devices and phenomena.

Clearly, the concept of information is central to communication. But what is *information?* Here we can run into difficulty, for information is a loaded word implying such semantic and philosophical notions as knowledge or meaning. These subjective aspects, fascinating though they may be, are largely irrelevant to the technical problems of communication. Eventually, in our discussion of *information theory*, a precise mathematical definition of information will be formulated, a definition which specifies information as a measurable quantity and thereby allows us to describe and compare communication systems in terms of their *information capacity*.

For the present, however, a slightly different avenue will be followed. Instead of coming to grips with information per se, we shall concentrate on the physical manifestation of the information as produced by the source, i.e., the *message*.

Because there are many kinds of information sources, including men and machines, messages appear in a variety of forms: (1) a sequence of discrete symbols or letters, e.g., words written on a telegraph blank, the holes punched in an IBM card; (2) a single time-varying quantity, e.g., the acoustic pressure produced by speech or music, the angular position of an aircraft gyro; (3) several functions of time and other variables, e.g., the light intensity and color of a television scene; (4) various combinations of the above. But whatever the message may be, the purpose of a communication system is to provide an acceptable replica of it at the destination.

In general, the message produced by a source is not electrical; hence, if an electrical system is used, an input *transducer* is required. This transducer converts the message to a *signal*, a time-varying electrical quantity such as voltage or current, which is better suited to further processing by the system. Similarly, another transducer at the destination converts the output signal to the appropriate message form.

Though transducer design is an important part of communication engineering, we shall limit our consideration to the strictly electric portion of the system, i.e., that portion where the message appears as an electric signal. Hereafter, the terms *signal* and *message* will be used interchangeably, since the signal, like the message, is a physical embodiment of the information.

With these preliminaries out of the way, let us turn our attention to the system, its parts, and its problems.

1.2 THE ELEMENTS OF A COMMUNICATION SYSTEM

Figure 1.1 shows the functional elements of a complete communication system, including input and output transducers. For convenience we have isolated them as distinct entities, though in actual systems the separation may not be so obvious. Also shown are some of the unwanted elements which inevitably enter the picture.

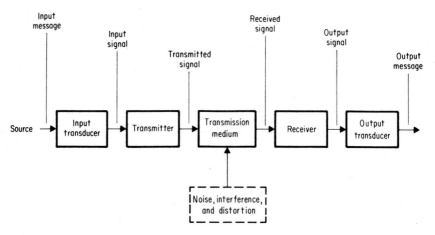

Fig. 1.1 *The elements of a communication system.*

Functional elements

Exclusive of transducers, there are three essential parts in an electrical communication system, *transmitter, transmission medium,* and *receiver.* Each has a characteristic function to perform.

Transmitter: The transmitter couples the message onto the transmission medium in the form of a transmitted signal. For purposes of effective and efficient transmission, several signal-processing operations may be performed. The commonest and most important of these operations is *modulation,* a process designed to match the transmitted signal to the properties of the medium through the use of a carrier wave. More will be said of this process shortly; later on, the subject will consume three chapters.

The items of hardware which go to make up a typical transmitter include the modulator and carrier circuitry, filters, amplifiers, and the transmitting antenna or equivalent coupling device.

Transmission medium: The transmission medium, sometimes called the *channel,* is the electrical connection between transmitter and receiver, bridging the distance from source to destination. It may be a pair of wires (transmission line), coaxial cable or waveguide, or a radiating electromagnetic wave such as radio waves and laser beams.

But regardless of type, all electrical transmission media are characterized by *attenuation*, the progressive decrease of signal power or power density with increasing distance.   The amount of attenuation can be very small or very large, ranging from a few percent on short runs of wire to a relative power loss exceeding 10^{20} (200 db) in deep-space transmission.   Usually, the transmission loss is large and therefore is a factor to be reckoned with.

Receiver: The function of the receiver is to extract the desired signal from the transmission medium and deliver it to the output transducer. Since received signals are often very feeble as a result of transmission loss, the receiver may have several stages of amplification.   However, the key operation performed by the receiver is *demodulation* (or detection), the reverse of the transmitter's modulation process, which restores the signal to its original form.

The components of a typical receiver include the receiving device, tuning circuits and filters, demodulator, and amplifiers.

Contaminations

In the course of electrical signal transmission, certain unwanted and undesirable effects take place.   One is attenuation, which reduces the signal *strength;* more serious, however, are distortion, interference, and noise, which appear as alterations of the signal *shape*.   While such contaminations are introduced throughout the system, it is a common and convenient practice to blame them all on the transmission medium, treating the transmitter and receiver as being ideal.   Figure 1.1 reflects this convention.

Broadly speaking, any unintended signal perturbation may be classified as "noise," and it is sometimes difficult to distinguish the various offenders in a contaminated signal.   Nonetheless, there are good reasons and an adequate basis for separating the three effects, as follows.

Distortion: Distortion is signal alteration due to imperfect response of the system to the desired signal itself.   Unlike noise and interference, distortion disappears when the signal is turned off.

Improved system design or compensating networks can reduce distortion.   Theoretically, perfect compensation is possible.   Practically, some distortion must be accepted, though the amount can be held within tolerable limits in all but extreme cases.

Interference: Interference is contamination by extraneous signals, usually man-made, of a form similar to the desired signal.   The problem is particularly common in broadcasting, where two or more signals may be picked up at the same time by the receiver. The cure for interference is obvious: eliminate, in one way or another, the interfering signal or its source. Again, a perfect solution is possible, if not always practical.

Noise: Finally, saving the worst for last, we come to noise. By noise we shall mean the random and unpredictable electric signals which come from natural causes, both internal and external to the system. When such random variations are added to an information-bearing signal, the information may be partially masked or totally obliterated. Of course the same can be said for interference and distortion; what makes noise unique is that it can never be completely eliminated, even in theory. As we shall see, noneliminable noise poses one of the basic problems of electrical communication.

1.3 MODULATION

Most input signals, as they come from the transducer, cannot be sent directly over the transmission medium. Instead, a *carrier wave*, whose properties are better suited to the medium in question, is modified to represent the message. *Modulation* is the systematic alteration of a carrier wave in accordance with the message (modulating signal) and may include *coding*.

It is interesting to note that many nonelectrical forms of communication also involve a modulation process, speech being a good example. When a person speaks, the movements of the mouth take place at rather low rates, on the order of 10 Hz,[1] and as such cannot effectively produce propagating acoustic waves. Transmission of voice through air is achieved by generating higher-frequency carrier tones in the vocal cords and modulating these tones with the muscular actions of the oral cavity. What the ear hears as speech is thus a modulated acoustic wave, similar in many respects to a modulated electric wave.

Types of modulation

To a large extent, the success of a communication system in a given mission depends on the modulation, so much so that the type of modulation is a pivotal decision in system design. Correspondingly, many different modulation techniques have evolved to suit various tasks and system requirements; and as new tasks arise, new techniques will be developed.

---

[1] The *hertz*, abbreviated Hz, is the generally accepted unit for frequency in cycles per second. It will be used throughout this book, often with prefixes as follows: $1 \text{ kHz} = 1 \text{ kilohertz} = 10^3 \text{ cps}$; $1 \text{ MHz} = 1 \text{ megahertz} = 10^6 \text{ cps}$; $1 \text{ GHz} = 1 \text{ gigahertz} = 10^9 \text{ cps}$.

Despite the multitude of varieties, it is possible to identify two basic types of modulation, according to the kind of carrier wave: (1) *continuous-wave* (c-w) *modulation,* for which the carrier is simply a sinusoidal waveform, and (2) *pulse modulation,* for which the carrier is a periodic train of pulses.

C-w modulation, being a *continuous* process, is obviously suited to signals that are continuously varying with time.   Usually, the sinusoidal carrier is at a frequency much higher than any of the frequency components contained in the modulating signal.   The modulation process is then characterized by *frequency translation;* i.e., the message spectrum (its frequency content) is shifted upward to a new and higher band of frequencies.

Pulse modulation is a discontinuous, or *discrete,* process, in the sense that the pulses are present only at certain distinct intervals of time. Hence, pulse modulation is best suited to messages that are discrete in nature.   Nonetheless, with the aid of *sampling,* continuously varying signals can be transmitted on pulsed carriers.   Often, as in telegraph and teletype, pulse modulation and coding go hand in hand.

As an alternative to the above classification it sometimes is preferable to speak of modulation as being *analog* or *coded* (digital).   This is particularly true of the more complex systems, which employ both c-w and pulsed techniques, making distinction by carrier type hazy.   The analog-versus-digital distinction is as follows.   In analog modulation, the modulated parameter varies in direct proportion to the modulating signal.   In coded modulation, a digital transformation takes place whereby the message is converted from one symbolic language to another. If the message is originally a continuous time function, it must be sampled and digitized (quantized) prior to encoding.

But regardless of type—c-w or pulsed, analog or coded—modulation must be a *reversible* process, so that the message can be retrieved at the receiver by the complementary operation of demodulation.

In Chaps. 5 to 7 some of the commoner modulation techniques will be covered in detail.   For the present, let us return to the purposes of modulation.

Why modulate?

In his book on modulation theory, Black (1953)[1] devotes an entire chapter to this question.   We have already given a concise answer, namely, that modulation is required to *match* the signal to the transmission medium. However, this match involves several considerations deserving further amplification.

[1] References are indicated in this fashion throughout.   Complete data are listed alphabetically by author in the References at the end of the book.

Modulation for ease of radiation: Efficient electromagnetic radiation requires radiating elements (antennas) whose physical dimensions are at least $\frac{1}{10}$ wavelength or so. But many signals, especially audio signals, have frequency components down to 100 Hz or lower, necessitating antennas some 200 miles long if radiated directly.

Utilizing the frequency-translation property of modulation, these signals can be impressed on a high-frequency carrier, thereby permitting substantial reduction of antenna size. For example, in the FM broadcast band, where carriers are in the 88- to 108-MHz range, antennas need be no more than a few feet across.

Modulation to reduce noise and interference: We have said that it is impossible to eliminate noise from the system. And though it is possible to eliminate interference, it may not be practical. Fortunately, certain types of modulation have the useful property of suppressing both noise and interference. The suppression, however, is not without a price; it generally requires a *transmission bandwidth* (frequency range) which is much larger than that of the original signal bandwidth, hence the designation *wideband noise reduction*. This trade-off of bandwidth for noise reduction is one of the most interesting and sometimes frustrating aspects of communication system design.

Modulation for channel assignment: The owner of a broadcast receiver, e.g., a radio or television set, has the option of selecting one of several stations even when all stations are broadcasting similar program material in the same transmission medium. The selection and separation of any one station is possible because each has a different assigned carrier frequency, or channel.

Were it not for modulation, only one station could operate in a given area. Two or more stations transmitting directly in the same medium, without modulation, would produce a hopeless jumble of interfering signals.

Modulation for multiplexing: Often it is desired to send many signals simultaneously between the same two points. *Multiplexing* techniques, inherently forms of modulation, permit multiple-signal transmission on one system such that each signal can be picked out at the receiving end.

Applications of multiplexing include data telemetry, FM stereophonic broadcasting, and long-distance telephone. It is quite common, for instance, to have as many as 1,860 intercity telephone conversations multiplexed for transmission on a coaxial cable less than $\frac{1}{2}$ in. in diameter.

Modulation to overcome equipment limitations: The design of a system is usually constrained by available equipment, equipment whose performance is often contingent upon the frequencies involved. Modulation can be used to place a signal in that portion of the frequency spectrum where equipment limitations are minimum or where design requirements are more easily met.

For this purpose, modulation devices are found in receivers as well as transmitters. The mixer, or frequency converter, in a superheterodyne receiver—e.g., a common home radio—is actually a modulator producing downward frequency translation.

1.4 FUNDAMENTAL LIMITATIONS IN ELECTRICAL COMMUNICATION

In the design of a communication system, or any system for that matter, the engineer is faced with two general kinds of constraints. On the one hand are the *technological problems*, the engineering facts of life. On the other hand are the *fundamental physical limitations* imposed by the system itself, the laws of nature as they pertain to the task.

Since engineering is, or should be, the art of the possible, both kinds of constraints must be recognized in system design. Nonetheless, there is a difference. Technological problems are problems of feasibility, including such diverse considerations as equipment availability, interaction with existing systems, economic factors, etc., problems that can be solved in theory, though the solution may not be practical. But the fundamental physical limitations are just that; when they are encountered head on, there is no recourse, even in theory. Technological questions notwithstanding, it is these limitations which ultimately dictate what can or cannot be accomplished.

Our purpose here is to state the fundamental limitations of information transmission by electrical means and give an indication of how they affect system design. The limitations, as currently understood, are two in number, *bandwidth* and *noise*.

The bandwidth limitation

Although not explicitly shown in Fig. 1.1, the time element is an integral part of communication systems. Frequently, a message must be sent in *real time*, the output signal keeping pace with the input as it comes from the source. In such cases the source rate, and hence the *signaling speed*, is beyond the control of the engineer; the system must be designed accordingly. But even if the source rate is adjustable, efficient system utilization calls for minimizing transmission time, i.e., sending the most information in the least time.

Rapid information transmission is achieved by using signals which change rapidly with time.   To decrease transmission time we would like to speed up the signals, like speeding up a tape recorder on playback. But we are dealing with an electrical system, a system which always includes energy storage; and it is a well-known physical law that for all but lossless systems, a change in stored energy requires a definite amount of time.   Thus, we cannot arbitrarily increase signaling speed, for eventually the system will cease to respond to the signal changes.

A convenient measure of signal speed is its bandwidth, the width of the signal spectrum.   Similarly, the rate at which a system can change stored energy is reflected by its usable frequency response, measured in terms of the *system bandwidth*.   Transmitting a large amount of information in a small amount of time requires wideband signals to represent the information and wideband systems to accommodate the signals. Bandwidth therefore emerges as a fundamental limitation that should be recognized.   When real-time transmission is required, the design must provide for adequate system bandwidth.   If the bandwidth is insufficient, it may be necessary to decrease signaling speed and thereby increase transmission time.

Along these same lines it may be remarked that equipment design is not so much a question of *absolute* bandwidth as of *fractional* bandwidth, i.e., absolute bandwidth divided by the center frequency.   Modulating a wideband signal onto a high-frequency carrier can reduce the fractional bandwidth, even if the modulation process increases absolute bandwidth, and thereby simplifies equipment design.   This is one reason why TV signals, having a bandwidth of about 5 MHz, are sent on much higher carriers than AM radio, where the bandwidth is about 10 kHz.

Likewise, given a fractional bandwidth dictated by equipment considerations, the absolute bandwidth can be increased almost indefinitely by going to higher carrier frequencies.   A 5-GHz microwave system can accommodate 10,000 times as much information in a given period as a 500-kHz radio-frequency carrier, while a laser beam of frequency 5×10^{14} Hz has a theoretical information capacity exceeding that of the microwave system by a factor of 10^5, or roughly equivalent to 10 million TV channels.[1] Thus it is that communication engineers are continually seeking new and usable high-frequency carrier sources to compensate for the bandwidth factor.

The noise limitation

A measuring instrument having 1 percent resolution can yield more information than an instrument having 10 percent resolution; the difference is one of *accuracy*. Similarly, successful electrical communication

---

[1] The absolute bandwidth would be $10^7 \times 5$ MHz $= 5 \times 10^{13}$ Hz, which is 10 percent of the carrier frequency.

depends on how accurately the receiver can determine which signal was actually sent, as distinguished from signals that might have been sent. Perfect signal identification might be possible in the absence of noise and other contaminations, but noise is always present in electrical systems, and superimposed noise perturbations limit our ability to correctly identify the intended signal and therefore limit information transmission.

Why is noise inevitable? Curiously, the answer comes from kinetic theory. Any particle at a temperature other than absolute zero has *thermal energy*, manifested as random motion or thermal agitation. If the particle happens to be an electron, its random motion constitutes a random current. If the random current takes place in a conducting medium, a random voltage, known as *thermal noise* or *resistance noise*, is produced. While resistance noise is only one of the possible sources in a system, most others are related in one way or another to random electron motion. Moreover, as might be expected from the wave-particle duality, there is thermal noise associated with electromagnetic radiation. Hence, just as we cannot have electrical communication without electrons or electromagnetic waves, we cannot have electrical communication without noise.

Typically, noise variations are quite small, on the order of microvolts. If the signal variations are substantially greater, say several volts peak to peak, then the noise may be all but ignored. Indeed, in ordinary systems under ordinary conditions, the *signal-to-noise ratio* is large enough for noise to go unnoticed. But in long-range or minimum-power systems, the received signal may be as small as the noise or smaller. When this happens, the noise limitation becomes very real.

It is important to note that if the signal strength is insufficient, adding more stages of amplification at the receiver is to no avail; the noise will be amplified along with the signal, leaving the signal-to-noise ratio unimproved. Increasing the transmitted power will help, but power cannot be increased indefinitely because of technological problems. (One of the early transatlantic cables was apparently destroyed by high-voltage rupture in an effort to obtain a usable received signal.) Alternately, as mentioned earlier, we can exchange bandwidth for signal-to-noise ratio via modulation and coding techniques. For this purpose, some techniques are more effective than others, the most effective being the most difficult and costly to instrument, which is not very surprising. Also note that swapping bandwidth for signal-to-noise ratio may take us from one limitation to the other.

In the final analysis, given a system of fixed bandwidth and signal-to-noise ratio, there is a definite upper limit on the rate at which information can be transmitted by that system. This upper limit is called the information *capacity* and is one of the central concepts of information theory.   Because the capacity is finite, it can truly be said that communication system design is a matter of compromise; a compromise between transmission time, transmitted power, bandwidth, and signal-to-noise ratio; a compromise further constrained by the technological problems.

1.5 A CHRONOLOGY OF ELECTRICAL COMMUNICATION

The organization of this text is dictated by pedagogical considerations and does not necessarily reflect the evolutionary order.   But the history of electrical communication is both interesting and informative, if only to provide some feeling for the significant events of the past, when they occurred, and the names associated with them.   A complete history is impossible here, of course, for it would require a book unto itself.   As an alternate, the following selected chronology is presented, listing the more important inventions, discoveries, and papers.   For additional historical background, see Black (1953, chap. 1), Albert (1940, chap. 1), or Still (1946).

Year	Event
1800–1837	**Preliminary developments** Volta discovers the primary battery; the mathematical treatises of Fourier, Cauchy, and Laplace; experiments on electricity and magnetism by Oersted, Ampere, Faraday, and Henry; Ohm's law (1826); early telegraph systems by Gauss and Weber and by Wheatstone and Cooke.
1838–1866	**The birth of telegraphy** Morse perfects his system with the help of Gale, Henry, and Vail; Steinheil finds that the earth can be used for a current path; commercial service initiated (1844); multiplexing techniques devised; William Thomson (Lord Kelvin) calculates the pulse response of a telegraph line (1855); transatlantic cables installed by Cyrus Field and associates.
1845	Kirchhoff's circuit laws enunciated.
1864	"A Dynamical Theory of the Electromagnetic Field," by James Clerk Maxwell, predicts electromagnetic radiation.
1876–1899	**The birth of telephony** Acoustic transducer perfected by Alexander Graham Bell, after earlier attempts by Reis; first telephone exchange, in New Haven, Conn., with eight lines (1878); Edison's carbon-button transducer; cable circuits introduced; Strowger devises automatic step-by-step switching (1887); the theory of cable loading by Heaviside, Pupin, and Campbell.
1887–1907	**Wireless telegraphy** Heinrich Hertz verifies Maxwell's theory; demonstrations by Marconi and Popov; Marconi patents a complete wireless telegraph system (1897); the theory of tuning circuits developed by Sir Oliver Lodge; commercial service begins, including ship-to-shore and transatlantic systems.

Year	*Event*
1892–1899	Oliver Heaviside's publications on operational calculus, circuits, and electromagnetics.
1904–1920	**Electronics applied to radio and telephone** Lee De Forest invents the Audion (triode) based on Fleming's diode; basic filter types devised by G. A. Campbell and others; experiments with AM radio broadcasting; transcontinental telephone line with electronic repeaters completed by the Bell System (1915); multiplexed carrier telephony introduced; E. H. Armstrong perfects the superheterodyne radio receiver (1918); first broadcasting station, KDKA, Pittsburgh.
1920–1922	J. R. Carson applies sampling theory to communication, and publishes "Notes on the Theory of Modulation."
1924	"Certain Factors Affecting Telegraph Speed," by H. Nyquist.
1928	J. B. Johnson investigates resistance noise.
1928	"Transmission of Information," by R. V. L. Hartley.
1923–1938	**The birth of television** Mechanical image-formation systems demonstrated by Baird and Jenkins; theoretical analysis of bandwidth requirements by Gray, Horton, and Mathes; Farnsworth and Zworykin propose electronic systems; vacuum cathode-ray tubes perfected by DuMont and others; field tests and experimental broadcasting begin.
1931	Teletypewriter service initiated.
1934	H. S. Black develops the negative-feedback amplifier.
1936	"A Method of Reducing Disturbances in Radio Signaling by a System of Frequency Modulation," by Armstrong, states the case for FM radio.
1937	Alec Reeves conceives pulse-code modulation.
1938–1945	**World War II** Radar and microwave systems developed; FM used extensively for military communications; improved electronics, hardware, and theory in all areas; Wiener and Kolmogoroff apply statistical methods to signal-detection problems.
1948	"A Mathematical Theory of Communication," by Claude Shannon.
1948–1951	Transistor devices invented by Bardeen, Brattain, and Shockley.
1950	Time-division multiplexing applied to telephony.
1955	J. R. Pierce proposes satellite communication systems.
1956	First transoceanic telephone cable.
1958	Long-distance data transmission systems developed for military purposes.
1960–1967	**Recent developments** Maiman demonstrates the first laser; commercial data transmission service; Telstar I launched as first communication satellite (1962); experimental pulse-code modulation systems for telephone; Mariner IV; experimental laser communication systems.

The above chronology, being a mere listing of events, gives little indication of the relative importance and interrelationships, but further comments are withheld till later chapters where additional discussion will be more meaningful.

1.6 PROSPECTUS

This study of communication systems begins with the fundamental physical limitations, their description, and analysis.   In Chaps. 2 and 3 the mathematical tools of signal theory and linear systems are reviewed, emphasizing the frequency-domain approach leading to the concepts of spectrum and bandwidth.   By and large this will be familiar territory, based on elementary transform methods and circuit theory.

Chapter 4 is a very short course on probability and statistics as applied to particular random signals of interest in communication.   The material is germane not only to electrical noise but to information-bearing signals as well, since, at the receiving end, the desired signal is random in the sense that it cannot be predicted in advance.   However, our principal task will be the description of undesired noise, particularly from the spectral viewpoint.   A discussion of noise sources and circuit noise, being somewhat of a digression, is relegated to the Appendix.

Having developed the necessary tools, we then turn to the stuff of communication engineering, namely, modulation theory and practice. Continuous-wave modulation is dealt with in Chaps. 5 and 6, while Chap. 7 treats sampling theory and pulse modulation.   Throughout these chapters, the problem of demodulation in the presence of noise and interference will be considered.   Multiplexing techniques, wideband noise reduction, and threshold effects are examined where applicable.   For the most part, system elements and instrumentation will be described as "black boxes" having certain terminal properties, but on occasion the lid is lifted to see what goes on inside and how the various functions are achieved in practice.

Information theory and its implications for electrical communication are surveyed in Chap. 8.   Of particular importance is the Hartley-Shannon equation, which expresses in quantitative form the bandwidth and noise limitations on information transmission.   In the light of this theory it is possible to draw some conclusions about the relative merits of conventional systems and also get some hints as to how better systems can be designed.

The text concludes, in Chap. 9, with a brief consideration of digital data transmission, the most rapidly expanding area in communication engineering.   This topic not only involves all the previous material but also gives added meaning to the mathematical theory of communication.

2
Signals and Spectra

Electrical communication signals are time-varying quantities, such as voltage or current. The usual description of a signal $x(t)$ is in the *time domain*, where the independent variable is t. But for communications work, it is often more convenient to describe signals in the *frequency domain*, where the independent variable is f. Roughly speaking, we think of the time function as being composed of a number of frequency components, each with appropriate amplitude and phase. Thus, while the signal physically *exists* in the time domain, we can say that it *consists* of those components in its frequency-domain description, called the *spectrum*.

Spectral analysis, the study of signals and their spectra, is a powerful tool in communication engineering. It is based primarily on the Fourier series and transform. This chapter therefore is devoted to a review and elaboration of Fourier techniques, with emphasis on spectral interpretation. But before getting into the details, some additional comments are in order.

There are two major reasons why the frequency-domain approach, via Fourier transforms, is of more use in communications than other techniques, such as time-domain analysis or Laplace transforms. First, the frequency domain is essentially a steady-state viewpoint; and for many purposes it is reasonable to restrict attention to the steady-state behavior of a communication system. Indeed, considering the multitude of possible signals that a system may handle, detailed transient solutions for each would be an impossible task. Second, and more important, the spectral approach allows us to treat entire *classes* of signals which have similar properties in the frequency domain. This not only gives insight for analysis but is invaluable for design. It is quite unlikely, for example, that such a significant technique as single-sideband modulation could have been developed without the aid of spectral concepts.

However, despite its many successes, the spectral approach should not be thought of as the only tool in communications engineering. There are some problems where it cannot be applied directly, e.g., nonlinear-device analysis, and some problems where other techniques are more convenient, e.g., time-domain analysis in pulse-modulation and pole-zero techniques in filter design. Thus, each new problem must be approached with an open mind and a good set of analytic tools.

As the first step in much of our work we shall write equations for signals as a function of time, but one must bear in mind that such equations are only mathematical *models* of physical signals, usually imperfect models. In fact, a completely faithful description of the simplest signal would be prohibitively complex in mathematical form and consequently useless for engineering purposes. Hence the models we seek are those which represent, with minimum complexity, the properties of the signal which are pertinent to the problem at hand. This leads to constructing several models for the same signal, according to need. Then, given a particular problem, the choice of which model to use is based on understanding the physical phenomena involved and the limitations of the mathematics; in short, it is engineering.

2.1 FROM PHASORS TO SPECTRA

It is well known that the steady-state solution to an ac circuit problem is found most simply using the time function $e^{j\omega t}$. (This relates to the fact that the characteristic functions, or eigenfunctions, of any time-invariant linear system are the set of complex exponentials $\{e^{st}\}$.) Since communication systems inevitably involve electric circuits, it is convenient to represent the signals in exponential form. Thus, given $x(t) =$

$A \cos(\omega_1 t + \phi)$, we can invoke *Euler's theorem*[1]

$$e^{\pm j\theta} = \cos\theta \pm j\sin\theta \tag{2.1}$$

and write $x(t) = \text{Re}\,[Ae^{j(\omega_1 t + \phi)}]$. The bracketed term then is thought of as a rotating vector, or *phasor*, in a complex plane (Fig. 2.1) such that its projection on the real axis is $x(t)$. The phasor has amplitude (length) A, rotates at a rate of $f_1 = \omega_1/2\pi$ revolutions per second, and at $t = 0$ makes an angle of $+\phi$ with respect to the real axis.

Three parameters are needed to specify a phasor: amplitude, relative phase, and rotational frequency. To describe the same phasor in the *frequency domain* we first note that it is defined only for the particular frequency f_1. With this frequency we must associate the corresponding

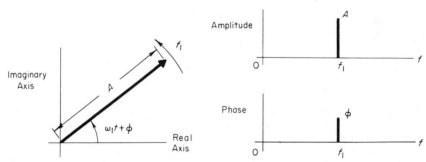

Fig. 2.1 *The phasor $Ae^{j(\omega_1 t + \phi)}$.*

Fig. 2.2 *The line spectrum of the expression $A\cos(2\pi f_1 t + \phi)$.*

amplitude and phase. Hence, a suitable frequency-domain description would be the *line spectrum* of Fig. 2.2.

In Fig. 2.2, as in all spectra to follow, the independent variable is *cyclic frequency f*, measured in hertz. Cyclic frequency has the advantage of being the direct reciprocal of the period. However, radian frequency, $\omega = 2\pi f$, will often be used for notational convenience.

Carrying the idea of line spectrum further, the signal $y(t) = 2 + 6\cos(2\pi 10t + \pi/6) + 3\sin 2\pi 30t - 4\cos 2\pi 35t$ can be written as the sum of four phasors, one having zero frequency, and the line spectrum will be as in Fig. 2.3. Notice the conversion from sine to cosine through the use of $\sin\theta = \cos(\theta - \pi/2)$ and the negative amplitude treated as a phase shift of π rad (180°). The latter is required since, by convention, amplitude is usually considered to be a positive quantity.

Figures 2.2 and 2.3 are called *positive-frequency line spectra* and can

---

[1] Euler's theorem is the most versatile of the trigonometric identities. A short table of other useful relations is given in Table B at the end of the book.

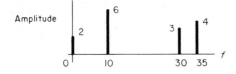

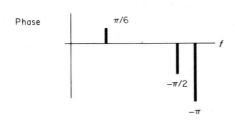

Fig. 2.3 *The line spectrum of 2 + 6 cos (2π10t + π/6) + 3 sin 2π30t − 4 cos 2π35t.*

be generated for any signal which is a sum of sinusoids. There is, however, another type of line spectrum which is only slightly more complicated, and much more useful. In our first example we represented $x(t)$ as the real part of a phasor. We could just as well have written $x(t) = \frac{1}{2}Ae^{j(\omega_1 t + \phi)} + \frac{1}{2}Ae^{j(-\omega_1 t - \phi)}$ and in doing so dispensed with the need for taking real parts. In this formulation, $x(t)$ is the sum of two phasors which have equal amplitudes and opposite phase and rotate in opposite directions but at the same speed. They are called *conjugate phasors*, for obvious reasons.

A line spectrum taken from the conjugate-phasor expression must include negative frequencies to allow for the two rotational directions. Thus in the *two-sided line spectrum*, half of the original amplitude is associated with each of the two frequencies. Figure 2.4 shows the two-sided version of Fig. 2.3. The rules for constructing such spectra are quite simple; a little thought will show that the amplitude lines always have even symmetry in f, while the phase lines always have odd symmetry. The symmetry is a direct consequence of $x(t)$'s being a real (noncomplex) function of time.

There is one significant difference in interpreting the two-sided line spectrum compared to the positive-frequency spectrum. A single line in the latter represents a cosine wave, via Re $[e^{j\omega t}]$. But in the two-sided case, one line by itself has no meaning; the conjugate term is required to get a real-time function. Thus, whenever we speak of some frequency interval in a two-sided spectrum, such as f_1 to f_2, we must include the corresponding negative-frequency interval, $-f_1$ to $-f_2$. A simple notation specifying both intervals is $f_1 \leq |f| \leq f_2$.

It should be emphasized that these line spectra are just shorthand ways of representing certain signals in terms of the form $e^{j\omega t}$. The effect on a signal by a network thus reduces to a steady-state ac problem and

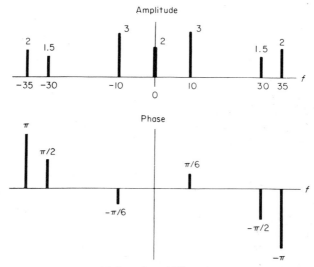

Fig. 2.4   *The two-sided version of Fig. 2.3.*

hence is a simple problem conceptually. (The details are a little more involved, as seen in Chap. 3.) This is one reason for the use of spectra in communications work; there are many others. The specific advantages of the two-sided spectrum, compared to the positive-frequency spectrum, will become apparent as we go along.

The next section develops line spectra for periodic signals through use of the Fourier series.

2.2 PERIODIC SIGNALS—FOURIER SERIES

A signal $x(t)$ is said to be periodic, with period T_0, if $x(t + T_0) = x(t)$. This definition implies that the signal exists for all time, $-\infty < t < \infty$. Clearly, no physical signal is strictly periodic; it must have a beginning and an end. Nonetheless, a periodic signal of finite duration can be treated as strictly periodic under certain conditions. In particular, if the signal has been on long enough for the initial transients to have died away, and if many periods will elapse before the signal is turned off, then in this steady-state interval the periodic assumption is both useful and meaningful. Hereafter these assumptions are implicit, and no distinction will be made between the physical signal and its periodic model.

The spectrum of a periodic signal is a two-sided line spectrum obtained from its *exponential Fourier series*. Letting $\omega_0 = 2\pi f_0 = 2\pi/T_0$, the series expansion is

$$x(t) = \sum_{n=-\infty}^{\infty} c_n e^{jn\omega_0 t} \tag{2.2}$$

where c_n is the nth Fourier coefficient

$$c_n = \frac{1}{T_0} \int_{-T_0/2}^{T_0/2} x(t) e^{-jn\omega_0 t} \, dt \tag{2.3}$$

and n is any integer, including zero. Equation (2.2) states that $x(t)$ can be expressed as a weighted sum of harmonically related phasors, the weighting factors being given by (2.3). The spectral interpretation is discussed below, after a brief digression on time averages and convergence.

The integral in (2.3) is similar to a time average or mean value, and integrals of this form will occur quite often. To simplify notation we here define the *time average* of any function $u(t)$, periodic in T_0, as

$$\langle u(t) \rangle = \frac{1}{T_0} \int_{T_0} u(t) \, dt \tag{2.4}$$

where the brackets denote time average and $\int_{T_0}$ is symbolic for $\int_{-T_0/2}^{T_0/2}$. Note that time averaging is a *linear operation* in the sense that $\langle a_1 u_1(t) + a_2 u_2(t) \rangle = a_1 \langle u_1(t) \rangle + a_2 \langle u_2(t) \rangle$, where a_1 and a_2 are constants. In this notation (2.3) becomes

$$c_n = \langle x(t) e^{-jn\omega_0 t} \rangle \tag{2.5}$$

The usual mathematical conditions on $x(t)$ for the Fourier series expansion are known as the *strong Dirichlet conditions*. Specifically, if $x(t)$ is *absolutely integrable* in T_0, that is, $\int_{T_0} |x(t)| \, dt < \infty$, then the coefficients c_n will exist. For the series to converge[1] uniformly, $x(t)$ must be finite and have a finite number of discontinuities in T_0. These conditions are sufficient but not strictly necessary. An alternate condition is that $x(t)$ be *integrable square* in T_0, so

$$\int_{T_0} |x(t)|^2 \, dt < \infty$$

When $x(t)$ is integrable square, the series converges to $x(t)$ in the sense that the mean-square difference between $x(t)$ and its Fourier series expansion vanishes as more terms are included. This is called *convergence in the mean*, written as

$$\lim_{N \to \infty} \int_{T_0} \left| x(t) - \sum_{n=-N}^{N} c_n e^{jn\omega_0 t} \right|^2 \, dt = 0$$

[1] A simple proof of the convergence is given by Stuart (1966, chap. 1).

For our purposes, we shall treat the series as being *identical* to $x(t)$ when either condition is satisfied. Furthermore, the integrable-square condition is not a severe restriction for physical signals since it implies that $x(t)$ has finite average power, a point we shall shortly pursue.

Turning to the spectral interpretation, we see from (2.2) that a periodic signal contains only those frequency components which are harmonically related to the *fundamental frequency* $f_0 = 1/T_0$. The Fourier coefficients are in general complex, so we can write $c_n = |c_n|e^{j\phi_n}$, and each term of the series has the form $|c_n| \exp [j(n\omega_0 t + \phi_n)]$. The amplitudes of the frequency components are thus described by $|c_n|$, while ϕ_n gives their relative phase. The two-sided line spectrum then has $|c_n|$ and ϕ_n as dependent variables, defined only for the discrete frequencies nf_0, where $n = 0, \pm1, \pm2, \dots$.

Of special interest is the zeroth coefficient, $c_0 = \langle x(t) \rangle$, representing the *average value* of $x(t)$ in time and the *dc component* in frequency. A word of warning here: setting $n = 0$ *after* integrating (2.3) may yield an indeterminate result for c_0, particularly if $c_0 = 0$, and conventional limiting operations are invalid because n is discrete. Therefore, one should either set $n = 0$ *before* integration or determine c_0 by inspection when possible.

Symmetry plays an important role in spectral analysis. This is brought out by writing (2.5) in the form

$$c_n = \langle x(t) \cos n\omega_0 t \rangle - j\langle x(t) \sin n\omega_0 t \rangle \qquad (2.6)$$

so, assuming $x(t)$ is real,

$$|c_n|^2 = \langle x(t) \cos n\omega_0 t \rangle^2 + \langle x(t) \sin n\omega_0 t \rangle^2$$
$$\phi_n = -\arctan \frac{\langle x(t) \sin n\omega_0 t \rangle}{\langle x(t) \cos n\omega_0 t \rangle} \qquad (2.7)$$

Replacing n by $-n$ then shows that

$$|c_{-n}| = |c_n| \qquad \phi_{-n} = -\phi_n$$

or

$$c_{-n} = |c_n|e^{-j\phi_n} = c_n^* \qquad (2.8)$$

Thus for real signals (the usual case) amplitude lines will have even symmetry in frequency, and phase lines will have odd symmetry. This is the same property observed for conjugate phasors; and it should be the same, since the exponential Fourier series *is* a sum of conjugate phasors when $x(t)$ is real.

Using (2.8), we can regroup the exponential series by pairs of the form

$$c_n e^{jn\omega_0 t} + c_{-n}e^{-jn\omega_0 t} = 2|c_n| \cos (n\omega_0 t + \phi_n)$$

and rewrite (2.2) as

$$x(t) = c_0 + \sum_{n=1}^{\infty} 2|c_n| \cos (n\omega_0 t + \phi_n) \tag{2.9}$$

By this process we have arrived at a trigonometric Fourier series, and in a sense have come full circle, for $x(t)$ is now described as a sum of sinusoids rather than conjugate phasors. Indeed, (2.9) is less useful than (2.2) and certainly lacks the attractive symmetry of the exponential series. We shall, however, have some occasions for its use and for the corresponding positive-frequency line spectrum.

Spectral symmetry is a consequence of the signal's being real. If the signal itself has symmetry in *time*, other simplifications are possible. For example, if $x(t)$ is an even time function, as well as real, then the second term in (2.6) is zero, and $c_n = \langle x(t) \cos n\omega_0 t \rangle$. Under these conditions the phase has only three possible values, 0 or $\pm\pi$, the latter to account for negative amplitudes. The reader can consider the implications of $x(t)$'s being an odd function of time. Of course signal symmetry depends on the location of the time origin, so it is wise to select the origin with symmetry in mind. Usually one can choose this location freely, since the zero of time is not physically unique. (On the other hand, the zero of frequency has a definite physical meaning related to the average value, as mentioned above.) It will be shown later that displacing the time origin affects only the phase of the spectrum.

Line spectra and average power

We have said that signals are usually time-varying voltage or current. If $x(t)$ is then applied to a 1-ohm resistance, the instantaneous power dissipated is $x^2(t)$. Likewise, the *mean-square* value $\langle x^2(t) \rangle$ is the average power. Even though the resistance may not be unity, we shall define $P = \langle x^2(t) \rangle$ as average power measured on the basis of mean-square volts or amperes. A multiplying constant can be used if actual power in watts is required.

An important theorem, called *Parseval's theorem*, relates average power to the Fourier series. The key to its development is replacing $x(t)$ by its Fourier series at the appropriate time. Thus $x^2(t) = x(t)(\Sigma c_n e^{jn\omega_0 t})$, and

$$P = \langle x^2(t) \rangle = \frac{1}{T_0} \int_{T_0} x(t) \left(\sum c_n e^{jn\omega_0 t} \right) dt$$

Interchanging the order of summation and integration yields

$$\sum_{n=-\infty}^{\infty} c_n \left[\frac{1}{T_0} \int_{T_0} x(t) e^{-j(-n)\omega_0 t} \, dt \right]$$

But the term in brackets is just c_{-n} [see (2.3)]; hence

$$P = \sum_{n=-\infty}^{\infty} c_n c_{-n} = \sum_{n=-\infty}^{\infty} |c_n|^2 \tag{2.10}$$

which is Parseval's theorem. Alternately, the *root-mean-square* (rms) value of $x(t)$ is, by definition, the square root of the mean square, so

$$x_{\mathrm{rms}} = \sqrt{\sum_{n=-\infty}^{\infty} |c_n|^2} \tag{2.11}$$

Insofar as power is concerned, we can immediately see the merit of the two-sided spectrum: the average power is just the sum of the squares of the amplitude lines. Alternately, for positive-frequency line spectra, Eq. (2.10) can be rearranged in the form

$$P = c_0^2 + \sum_{n=1}^{\infty} \tfrac{1}{2}|2c_n|^2 \tag{2.12}$$

Here we have written $2 \sum_{n=1}^{\infty} |c_n|^2 = \sum_{n=1}^{\infty} \tfrac{1}{2}|2c_n|^2$ to emphasize the fact that a sinusoid of amplitude $A = |2c_n|$ has the average power $P = \tfrac{1}{2}A^2 = \tfrac{1}{2}|2c_n|^2$. Comparing (2.12) with (2.9) makes an interesting point, namely, that superposition of average power applies for harmonically related sinusoidal waves. Superposition does not apply for *instantaneous* power.

The sinc function

A few functions occur time and time again in spectral analysis. One of these is of the form $(\sin x)/x$, which some authors call the *sampling function*. Related to this, and more useful to us, is the function $(\sin \pi u)/\pi u$, which is designated as the *sinc function*

$$\mathrm{sinc}\, u = \frac{\sin \pi u}{\pi u} \tag{2.13}$$

By inspection, sinc u is seen to be an even function of u and is plotted to scale in Fig. 2.5. Note that including the factor of π in the definition causes the zero crossings to occur at integral values of u. The special case sinc $0 = 1$ is found by simple limiting.

The sinc function is related to the time average of exponential and cosine functions, as seen by the following:

$$\frac{1}{u} \int_{-u/2}^{u/2} e^{\pm j2\pi v}\, dv = \frac{1}{u} \int_{-u/2}^{u/2} \cos 2\pi v\, dv = \mathrm{sinc}\, u \tag{2.14}$$

To illustrate, consider the average of $x(t) = \cos 2\pi ft$ over $|t| \leq \tau/2$, where τ is not necessarily the period of $x(t)$. Setting $v = ft$ gives immediately

$$\frac{1}{\tau} \int_{-\tau/2}^{\tau/2} \cos 2\pi ft \, dt = \frac{1}{f\tau} \int_{-f\tau/2}^{f\tau/2} \cos 2\pi v \, dv = \text{sinc } f\tau$$

Because such integrations are common in our later work, (2.14) is a particularly handy result; its proof is left to the reader.

Numerical values of sinc u and sinc^2 u are given in Table C.

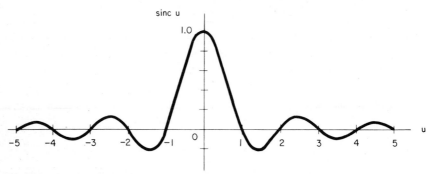

Fig. 2.5   *The sinc function.*

Example 2.1 Rectangular pulse train

Without unduly belaboring the point, let us apply these ideas to the periodic pulse train shown in Fig. 2.6a. Taking advantage of symmetry, we have

$$c_n = \frac{1}{T_0} \int_{-\tau/2}^{\tau/2} A \cos n\omega_0 t \, dt = \frac{A}{T_0} \frac{2 \sin (n\omega_0 \tau/2)}{n\omega_0}$$

$$= \frac{A\tau}{T_0} \text{sinc } nf_0\tau \qquad f_0 = \frac{\omega_0}{2\pi} = \frac{1}{T_0} \tag{2.15}$$

and

$$x(t) = \frac{A\tau}{T_0} \sum_{n=-\infty}^{\infty} \text{sinc } nf_0\tau \, e^{jn\omega_0 t} \tag{2.16}$$

The line spectrum, drawn from (2.15), is shown in Fig. 2.6b, where $\tau/T_0 = \frac{1}{4}$ for illustration. Observe that the continuous function $A\tau/T_0$ sinc $f\tau$ is the *envelope* of the spectral lines, shown dashed. In this figure we are able to show both the amplitude and phase of c_n because $x(t)$ is an even time function. In general, separate plots of $|c_n|$ and ϕ_n would be required. Also, with T_0 being an integral multiple of τ, certain spectral lines are missing because they fall at the zero values of the envelope; again, not true in general.

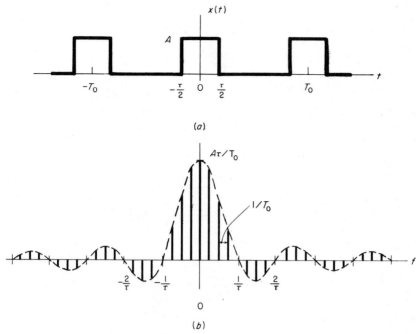

Fig. 2.6   *A rectangular pulse train and its line spectrum.*

It is interesting to note that $c_0 = A\tau/T_0$, which we certainly expect from inspection of $x(t)$.   The ratio τ/T_0 is the ratio of "on" time to period, called the *duty cycle* in pulse electronics and designated by d.   Thus $c_n = dA \operatorname{sinc} nd$.

Considering average power, we might think of applying Parseval's theorem, which, in terms of the duty cycle, becomes

$$P = \sum_{n=-\infty}^{\infty} (dA)^2 \operatorname{sinc}^2 nd$$

This actually would be rather foolish for the problem at hand, since the infinite summation is much harder than the time integration.   The latter gives $P = dA^2$ almost immediately.   However, consider this question: What percentage of the total power is contained in the spectral lines for $|f| < 1/\tau$?   (Anticipating Chap. 3, this might be pertinent to the effect of a lowpass filter.)   Again taking $T_0 = 4\tau$, we have $d = \frac{1}{4}$ and $f_0 = \frac{1}{4}\tau$, so there are seven lines contained in $|f| < 1/\tau$.   Squaring and adding the seven c_n's, with the aid of Table C, gives

$$P' = \sum_{n=-3}^{3} \left(\frac{A}{4}\right)^2 \operatorname{sinc}^2 \frac{n}{4} = \frac{A^2}{16} \times 3.612$$

whereas $P = A^2/4$. Hence $P'/P = 0.903$, and we can conclude that over 90 percent of the average power is in the designated frequency region.

Example 2.2 RF pulse train

As a second example, consider the train of sinusoidal pulses shown in Fig. 2.7a, called a *radio-frequency* (RF) pulse train when f_c falls in the

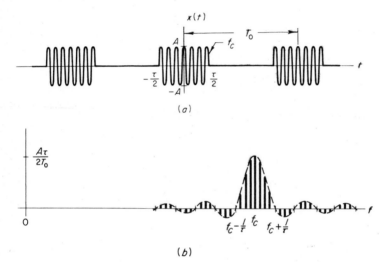

Fig. 2.7 *An RF pulse train.* (a) *Waveform;* (b) *spectrum (negative-frequency portion omitted).*

radio portion of the electromagnetic spectrum. Again invoking the signal symmetry,

$$c_n = \frac{1}{T_0} \int_{-\tau/2}^{\tau/2} A \cos \omega_c t \cos n\omega_0 t \, dt$$

$$= \frac{A\tau}{2T_0} [\text{sinc } (f_c - nf_0)\tau + \text{sinc } (f_c + nf_0)\tau]$$

providing that f_c is an integral multiple of $f_0 = 1/T_0$ so that the periodicity is preserved. The corresponding line spectrum is sketched in Fig. 2.7b for $f_c \gg 1/\tau$.

A comparison of this example with the previous one is both informative and interesting. We first note that the RF pulse train can be generated from a rectangular pulse train by *multiplying* by $\cos \omega_c t$. But the time-domain multiplication becomes *translation* in the frequency domain;

the spectrum of Fig. 2.7*b* consists of Fig. 2.6*b* shifted up (and down) in frequency by f_c Hz. It should come as no surprise that the resulting spectrum is concentrated about $\pm f_c$.

We could go even further and state a frequency-translation theorem for periodic signals of the form $x(t) \cos \omega_c t$. However, as demonstrated shortly, the Fourier series is a special case of the Fourier transform, at least from the spectral viewpoint, and all the theorems and most of the properties of Fourier transforms will apply to the series coefficients and the line spectra obtained therefrom. Let us therefore proceed directly to the study of Fourier transforms.

2.3 NONPERIODIC SIGNALS—FOURIER TRANSFORMS

We have described a periodic signal as one with a repeating characteristic applied for a long time interval, theoretically infinite. Now we consider nonperiodic signals whose effects are concentrated over a brief period of time. Such signals may be *strictly timelimited*, so $x(t)$ is identically zero outside of a specified interval, or *asymptotically timelimited*, so $x(t) \rightarrow 0$ as $t \rightarrow \pm \infty$.

In either case, if $x(t)$ is integrable square over all time, that is,

$$\lim_{T \to \infty} \int_T |x(t)|^2 \, dt < \infty$$

then the frequency-domain description is provided by the Fourier *transform* $X(f) = \mathfrak{F}[x(t)]$, defined as the integral

$$X(f) = \int_{-\infty}^{\infty} x(t) e^{-j\omega t} \, dt \tag{2.17}$$

Conversion back to the time domain is accomplished via the *inverse transform*

$$\mathfrak{F}^{-1}[X(f)] = \int_{-\infty}^{\infty} X(f) e^{j\omega t} \, df$$

We shall equate the inverse transform with the signal itself, that is,

$$x(t) = \int_{-\infty}^{\infty} X(f) e^{j\omega t} \, df \tag{2.18}$$

though it is more precise to say that $\mathfrak{F}^{-1}[X(f)]$ approaches $x(t)$ in the sense of convergence in the mean.

There are a number of ways of writing (2.17) and (2.18), the major differences being in the use of ω or f for the integration variable in (2.18). The form displayed here has the advantages of symmetry and simplicity, but it must be kept in mind that $\omega = 2\pi f$ when carrying out frequeney integrations.

The Fourier transform pair can be thought of as an extension of the exponential Fourier series by the following argument. Suppose $x(t)$ is nonperiodic but integrable square in $|t| < T_0/2$. From Eqs. (2.2) and (2.3) we can construct a Fourier series representation of $x(t)$ over this interval, namely,

$$x(t) = \sum_{n=-\infty}^{\infty} \left[\frac{1}{T_0} \int_{-T_0/2}^{T_0/2} x(t')e^{-j2\pi nf_0 t'}\,dt' \right] e^{j2\pi nf_0 t} \qquad |t| < \frac{T_0}{2}$$

where $f_0 = 1/T_0$ and the dummy variable t' is introduced for clarity. To represent $x(t)$ for *all* time we simply let $T_0 \to \infty$ such that $nf_0 = n/T_0$ becomes the *continuous* variable f, $1/T_0$ becomes the differential df, and the infinite sum becomes an integral, that is,

$$x(t) = \int_{-\infty}^{\infty} \left[\int_{-\infty}^{\infty} x(t')e^{-j2\pi ft'}\,dt' \right] e^{j2\pi ft}\,df$$

Now the inside integral is the same as $X(f)$ as defined in (2.17), so

$$x(t) = \int_{-\infty}^{\infty} X(f)e^{j\omega t}\,df$$

Furthermore, comparing (2.17) and (2.3) in the light of this limiting procedure reveals that

$$X(f) = \lim_{T_0 \to \infty} T_0 c_n$$

As an important illustration of this point, recall that $c_0 = \langle x(t) \rangle$ is the area under $x(t)$ for one period divided by the length of the period. Now $n = 0$ corresponds to $f = 0$, so $X(0) = \lim_{T_0 \to \infty} T_0 c_0$ is the *total area* under $x(t)$ for all time, that is,

$$X(0) = \int_{-\infty}^{\infty} x(t)\,dt \tag{2.19}$$

which also follows from (2.17) by setting $f = 0$.

To interpret the above limiting operation in the frequency domain, suppose that $x(t)$ is periodic in T_0. Its frequency-domain representation is then a line spectrum where the lines are spaced by $1/T_0$, located at nf_0, and proportioned according to c_n. As $T_0 \to \infty$, $1/T_0 \to df$, so the lines merge into a *continuum*. But $c_n \to 0$ as $T_0 \to \infty$, so the continuum must be described by $T_0 c_n$ to prevent it from vanishing. Thus $X(f) = \lim_{T_0 \to \infty} T_0 c_n$ is a complex function giving both the amplitude and phase of the frequency-domain description.

Because of the similarity between the Fourier transform and the Fourier coefficients we expect that, for real functions of time,

$$X(-f) = X^*(f) \tag{2.20}$$

and hence

$$|X(-f)| = |X(f)| \qquad \arg X(-f) = -\arg X(f)$$

so the spectral amplitude is an even function of frequency while the phase is odd. Complex functions satisfying (2.20) are said to be *hermitian;* proof that $X(f)$ is hermitian follows directly from (2.17) when written as $X(f) = \int_{-\infty}^{\infty} x(t) \cos \omega t \, dt - j \int_{-\infty}^{\infty} x(t) \sin \omega t \, dt$. The consequences of symmetry in time can also be seen from this formulation.

Energy and energy spectral density

In dealing with nonperiodic functions the definition of time average must be modified as $\langle u(t) \rangle = \lim\limits_{T \to \infty} \dfrac{1}{T} \int_{T} u(t) \, dt$. Hence average power becomes

$$P = \langle x^2(t) \rangle = \lim_{T \to \infty} \frac{1}{T} \int_{T} x^2(t) \, dt$$

But here we run into a snag; for if $x(t)$ is integrable square, the above integral remains finite, and $P \to 0$ as $T \to \infty$. This is not too surprising a turn since timelimited signals must have zero averages when averaged over an *infinite* time. Average power is therefore not a useful concept here, so we turn instead to energy.

By definition, total energy E is the integral of instantaneous power. Again assuming $x(t)$ applied to a 1-ohm resistance,

$$E = \int_{-\infty}^{\infty} x^2(t) \, dt \tag{2.21}$$

Relating energy to the spectrum, we proceed as in deriving Parseval's theorem.

$$E = \int_{-\infty}^{\infty} x(t) \left[\int_{-\infty}^{\infty} X(f) e^{j\omega t} \, df \right] dt$$
$$= \int_{-\infty}^{\infty} X(f) \left[\int_{-\infty}^{\infty} x(t) e^{-j(-\omega)t} \, dt \right] df$$

Now the bracketed term is just $X(-f)$, and hence

$$E = \int_{-\infty}^{\infty} X(f) X(-f) \, df = \int_{-\infty}^{\infty} |X(f)|^2 \, df \tag{2.22}$$

This is *Rayleigh's energy theorem* and compares directly with Parseval's power theorem for Fourier series (2.10).

Consider now the interpretation of $|X(f)|^2$. If $x(t)$ is a voltage waveform, then $X(f)$ has dimensions of voltage per unit frequency and describes the distribution, or *density,* of the signal voltage in frequency.

By like reasoning, $|X(f)|^2$ is the density of energy in the frequency domain. This conceptual viewpoint is both physically meaningful and useful; its importance is underscored by defining the *energy spectral density* $S(f)$ for nonperiodic signals

$$S(f) = |X(f)|^2 \qquad (2.23)$$

Obviously, from (2.23), $S(f)$ is always *positive* and *real*. (The reader should consider why this must be true from a physical viewpoint.) Moreover, if $x(t)$ is real, then $X(f)$ is hermitian and $S(f)$ is an *even* function of frequency. The total energy is therefore

$$E = \int_{-\infty}^{\infty} S(f) \, df = 2 \int_{0}^{\infty} S(f) \, df$$

Line spectra versus continuous spectra

We are now in position to compare the two types of spectra. For this purpose consider a very narrow frequency interval of width Δf centered at f_1, that is, $f_1 - \frac{1}{2}\Delta f < |f| < f_1 + \frac{1}{2}\Delta f$, and suppose this interval includes the mth harmonic of a periodic signal, so $f_1 = mf_0$. The frequency component of the periodic signal contained within the interval is

$$c_m e^{j\omega_1 t} + c_{-m} e^{j(-\omega_1)t} = 2|c_m| \cos (\omega_1 t + \phi_m)$$

in the sense that the time function is found by *summing* terms of the above form. Note that this particular component has an average power of $2|c_m|^2$.

For a nonperiodic signal, the frequency component represented by the interval is approximately

$$X(f_1)e^{j\omega_1 t} \, \Delta f + X(-f_1)e^{j(-\omega_1)t} \, \Delta f = 2|X(f_1)| \, \Delta f \cos [\omega_1 t + \arg X(f_1)]$$

in the sense that the time function is found by *integrating*, over all frequency, terms of the above form. Likewise, this interval contains an energy of approximately $2|X(f_1)|^2 \, \Delta f$. Observe that these expressions are proportional to the width of the frequency interval and vanish as $\Delta f \to 0$.

To summarize, the difference between line spectra and continuous spectra can be stated as follows. A line spectrum represents a signal that can be constructed from a *sum* of discrete frequency components (sinusoids); the signal power is concentrated at specific frequencies. On the other hand, a continuous spectrum represents a signal that is constructed by *integrating* over a continuum of frequency components (they can be thought of as sinusoids of infinitesimal amplitude); similarly, the signal energy is distributed continuously in frequency.

Example 2.3 Rectangular pulse

The timelimited restriction placed on nonperiodic signals implies that we are speaking primarily of *pulses*. As an example of a strictly time-limited signal we consider the single *rectangular pulse* of Fig. 2.8. This is a common signal model in communications, and for simplicity in writing let us adopt the notation

$$
\Pi\left(\frac{t - t_0}{\tau}\right) =
\begin{cases}
1 & |t - t_0| < \dfrac{\tau}{2} \\[2mm]
0 & |t - t_0| > \dfrac{\tau}{2}
\end{cases}
\tag{2.24}
$$

In words, $\Pi[(t - t_0)/\tau]$ is a rectangular pulse with unit amplitude and duration τ and is centered at $t = t_0$. For Fig. 2.8, $x(t) = A\Pi(t/\tau)$.

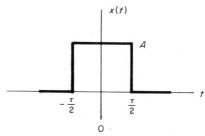

Fig. 2.8   *The rectangular pulse $x(t) = A\Pi(t/\tau)$.*

The spectrum of $x(t)$ can be found from direct integration or by taking the limiting process to the results of Example 2.1. Both give

$$
X(f) = A\tau \, \text{sinc} \, f\tau
\tag{2.25}
$$

which should be verified by the reader, using both methods. The spectrum is plotted in Fig. 2.9 along with the energy spectral density $S(f) = (A\tau)^2 \, \text{sinc}^2 f\tau$. We again have a case where magnitude and phase of $X(f)$ can be shown on one plot. Had the pulse been centered at $t_0 = \tau/2$, destroying the time symmetry, the spectrum would be $X(f) = A\tau$ sinc $f\tau \, e^{-j\pi f\tau}$, which is illustrated in Fig. 2.10. Note the discontinuities in the phase, which account for sign changes of the sinc function.

It is apparent from $X(f)$ and $S(f)$ that the significant portion of the spectrum is in the range $|f| < 1/\tau$. We can therefore take $1/\tau$ as a measure of the spectral "width." Now if the pulse duration is reduced (small τ), the frequency width is increased, and vice versa. Thus, short pulses have broad spectra, long pulses have narrow spectra. This phenomenon is called *reciprocal spreading* and is a general property of all signals, pulses or not.

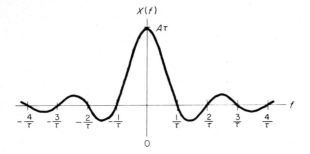

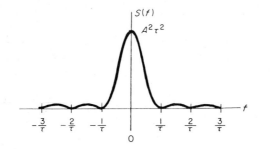

Fig. 2.9 *The spectrum and energy spectral density of a rectangular pulse.*

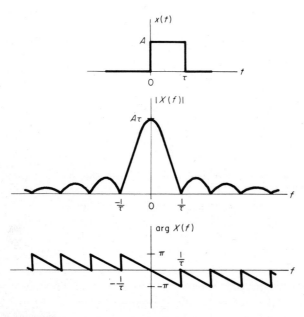

Fig. 2.10

Paralleling Example 2.1 we can ask: What percentage of the total energy is contained in $|f| < 1/\tau$? Using energy spectral density,

$$E' = 2 \int_0^{1/\tau} S(f) \, df = 2(A\tau)^2 \int_0^{1/\tau} \text{sinc}^2 f\tau \, df$$

Unfortunately, the integral is nonelementary and requires numerical or graphical solution. The result is $E' \approx 0.92A^2\tau$, whereas $E = A^2\tau$ by inspection of $x(t)$. Thus, over 90 percent of the signal energy is contained in $|f| < 1/\tau$, supporting our qualitative conclusion about spectral width.

We shall use the rectangular pulse quite often as a signal model, despite the fact that no physical signal can take this exact form. The simplicity and ease of manipulation more than compensate for its non-realizability, but in certain types of analysis there are serious mathematical objections to this function because of the discontinuities. (Consider, for example, its first derivative.) An alternate model that meets the objections is the gaussian pulse of our next example.

Example 2.4 Gaussian pulse

The *gaussian pulse* (Fig. 2.11a) is written as

$$x(t) = Ae^{-\pi(t/\tau)^2} \tag{2.26}$$

This signal, being asymptotically timelimited, does not have a well-defined duration, but, as indicated in the figure, the parameter τ can be taken as a measure. The specific advantage of this model is the continuity of the function and all its derivatives; the disadvantage is that $x(t)$ exists for all time.

The spectrum is readily found to be[1]

$$X(f) = \int_{-\infty}^{\infty} Ae^{-\pi(t/\tau)^2}e^{-j\omega t} \, dt = 2A \int_0^{\infty} e^{-\pi(t/\tau)^2} \cos \omega t \, dt$$

$$= A\tau e^{-\pi(f\tau)^2} \tag{2.27}$$

which is a gaussian "pulse" in frequency. Comparing (2.26) and (2.27) shows that the transform of a gaussian function is a gaussian function.

The gaussian spectrum is sketched in Fig. 2.11b, which shows that the spectral width is measured approximately by $1/\tau$. Again we have the reciprocal-spreading effect. A comparison of Fig. 2.11b with Fig. 2.9 reveals that the rectangular pulse has appreciably more spectral content at frequencies above $1/\tau$. Physically, these high-frequency components are demanded by the discontinuities of the rectangular shape. The gaussian pulse, having a smoother time variation, requires relatively little high-frequency content.

[1] Certain of the commoner integrals we shall encounter are listed in Table B.

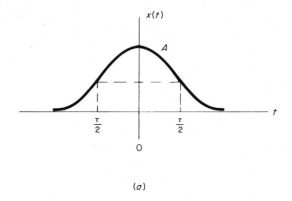

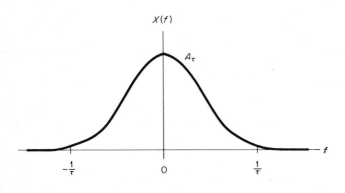

Fig. 2.11 *A gaussian pulse and its spectrum.*

2.4 TRANSFORM THEOREMS

Given below are a few of the many theorems associated with Fourier transforms. They are included not just as exercises in manipulation but for two very practical reasons: (1) using theorems, one can build an extensive catalog of transform pairs with a minimum of integration problems (see Table A), and such a catalog will be useful as we seek new signal models; (2) the theorems are invaluable when interpreting spectra, for they express the relationship between time-domain and frequency-domain operations.

In stating the theorems, we indicate a signal and its transform (spectrum) by lowercase and uppercase letters, so $X(f) = \mathfrak{F}[x(t)]$ and $x(t) = \mathfrak{F}^{-1}[X(f)]$. This is denoted more compactly by $x(t) \leftrightarrow X(f)$.

Linearity (superposition)

Theorem: For the constants a_1 and a_2

$$a_1 x_1(t) + a_2 x_2(t) \leftrightarrow a_1 X_1(f) + a_2 X_2(f) \tag{2.28}$$

This theorem simply states that linear combinations in the time domain become linear combinations in the frequency domain. Although proof of the theorem is trivial and left to the reader, the importance of the theorem cannot be overemphasized. From a practical viewpoint (2.28) greatly facilitates spectral analysis when the signal in question is a linear combination of functions whose individual spectra are known. From a theoretical viewpoint it underscores the applicability of the Fourier transform for the study of linear systems.

Time delay

Theorem: If a signal $x(t)$ is delayed in time by t_0 sec, producing the new signal $x(t - t_0)$, the spectrum is modified by a linear phase shift of slope $-2\pi t_0$, that is,

$$x(t - t_0) \leftrightarrow X(f)e^{-j\omega t_0} \tag{2.29}$$

Translation of a signal in time thus changes the spectral phase but not the amplitude. Note that if t_0 is a negative number, the signal is *advanced* in time, and the phase shift has *positive* slope. Since time advancement is a physical impossibility, we conclude that in actual signal processing the spectral phase will have negative slope, though not necessarily linear. More of this in Chap. 3.

Proof of this theorem is accomplished by change of variable in the transform integral. [Observe that time is indeed a dummy variable in the direct transform (2.17), just as frequency is a dummy variable in the inverse transform.] The change-of-variable technique is basic to the proof of most transform theorems, and is demonstrated here.

Proof: $\mathfrak{F}[x(t - t_0)] = \int_{-\infty}^{\infty} x(t - t_0)e^{-j\omega t} \, dt$

Setting $u = t - t_0$, so $t = u + t_0$ and $dt = du$,

$$\mathfrak{F}[x(t - t_0)] = \int_{-\infty}^{\infty} x(u)e^{-j\omega(u + t_0)} \, du$$

$$= \left[\int_{-\infty}^{\infty} x(u)e^{-j\omega u} \, du \right] e^{-j\omega t_0}$$

But the integral in brackets is just $X(f)$, so

$$\mathfrak{F}[x(t - t_0)] = X(f)e^{-j\omega t_0}$$

To illustrate the use of this and the previous theorem, consider the signal

$$y(t) = \tfrac{1}{2}x(t - t_0) - \tfrac{1}{2}x(t - 3t_0)$$

which is a linear combination of delayed signals. Its spectrum follows immediately from (2.28) and (2.29) as

$$\begin{aligned} Y(f) &= \tfrac{1}{2}X(f)e^{-j\omega t_0} - \tfrac{1}{2}X(f)e^{-j\omega 3t_0} \\ &= j\sin \omega t_0\, e^{-j\omega 2t_0}X(f) \end{aligned}$$

Scale change

Time delay is equivalent to translation of the time origin. Another geometric operation is scale change, in which the time axis is expanded, compressed, or reversed. Thus, $x(at)$ is a compressed version of $x(t)$ when a is positive and greater than 1. Similarly, if a is negative and less than 1, $x(at)$ is the expanded image of $x(t)$ reversed in time. Such operations may occur in playback of recorded signals, for example.

Theorem: $\quad x(at) \leftrightarrow \dfrac{1}{|a|}\, X\!\left(\dfrac{f}{a}\right)$ (2.30)

The theorem formally expresses the property of reciprocal spreading encountered in Examples 2.3 and 2.4; for if the signal is compressed in time by the factor a, its spectrum is expanded in frequency by $1/a$. On the other hand, if the signal is reversed in time ($a = -1$), its spectrum is $X(-f)$, which means that only the phase is altered; specifically, the phase is reversed.

The theorem is proved by change of variables, considering positive and negative values of a separately.

Duality

The powerful concept of duality is well known in circuit analysis. In spectral analysis there is also duality, a duality between the time and frequency domains which stems from the similarity of the Fourier transform integrals. To state the theorem, we consider $x(\ )$ and $X(\ )$ to be *functions* whose independent variables are not designated.

Theorem: If $x(t) \leftrightarrow X(f)$, then

$$X(t) \leftrightarrow x(-f)$$ (2.31)

Proof: $\mathfrak{F}[X(t)] = \int_{-\infty}^{\infty} X(t)e^{-j\omega t}\, dt = \int_{-\infty}^{\infty} X(u)e^{j2\pi u(-f)}\, du$

Now

$x(t) = \int_{-\infty}^{\infty} X(f)e^{j\omega t}\, df = \int_{-\infty}^{\infty} X(u)e^{j2\pi u t}\, du$

so

$x(-t) = \int_{-\infty}^{\infty} X(u)e^{j2\pi u(-t)}\, du$

Comparing integrals, we have

$\mathfrak{F}[X(t)] = x(-f)$

In the form of (2.31) this theorem is rather abstract, and it is difficult to visualize its use in generating new transform pairs. The following example should help to clarify the procedure.

Example 2.5 Sinc pulse

Consider the signal $y(t) = A \operatorname{sinc} 2Wt$, a sinc function in time. (While the idea of a sinc pulse may seem strange at first, it plays a major role in the study of digital data transmission.) Recalling the transform pair of Example 2.3, $A\Pi(t/\tau) \leftrightarrow A\tau \operatorname{sinc} f\tau$, we apply duality by writing $y(t)$ in the form

$$y(t) = X(t) = \frac{A}{2W} 2W \operatorname{sinc} t2W$$

so

$$Y(f) = x(-f) = \frac{A}{2W} \Pi\left(\frac{-f}{2W}\right)$$

Because the rectangular function has even symmetry, $\Pi(-f/2W) = \Pi(f/2W)$, and we have derived the transform pair

$$A \operatorname{sinc} 2Wt \leftrightarrow \frac{A}{2W} \Pi\left(\frac{f}{2W}\right) \qquad (2.32)$$

which is shown in Fig. 2.12. As can be seen, the spectrum of a sinc pulse has clearly defined spectral width W. In fact, the spectrum is zero for

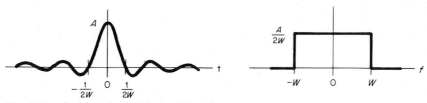

Fig. 2.12 *A sinc pulse and its bandlimited spectrum.*

$|f| \geq W$, and the signal is said to be *bandlimited* in W. This is our first example of a signal which is strictly bandlimited in frequency. Note that the signal itself is only asymptotically limited in time.

Frequency translation (modulation)

Duality can be used to generate transform theorems as well as transform pairs. For example, a dual of the time-delay theorem (2.29) is

Theorem: $x(t)e^{j\omega_c t} \leftrightarrow X(f - f_c)$ $\hspace{3cm}$ (2.33)

We designate this as *frequency translation* or *complex modulation*, since multiplying a time function by $e^{j\omega_c t}$ causes its spectrum to be translated in frequency by $+f_c$.

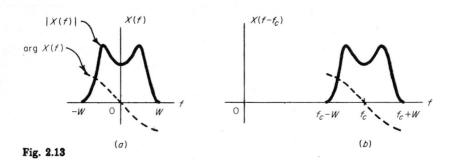

Fig. 2.13

To see the effects of frequency translation, let $x(t)$ have the band-limited spectrum of Fig. 2.13a. Inspection of the translated spectrum $X(f - f_c)$ (Fig. 2.13b) reveals the following:

1. The significant components are concentrated around the frequency f_c.
2. Though $X(f)$ was bandlimited in W, $X(f - f_c)$ has a spectral width of $2W$. Translation has therefore doubled spectral width. Stated another way, the negative-frequency portion of $X(f)$ now appears at positive frequencies.
3. $X(f - f_c)$ is not hermitian but does have symmetry with respect to translated origin at $f = f_c$.

These considerations may appear somewhat academic in view of the fact that $x(t)e^{j\omega_c t}$ is not a real time function and cannot occur as a communication signal. However, signals of the form $x(t) \cos \omega_c t = \frac{1}{2}x(t)(e^{j\omega_c t} + e^{-j\omega_c t})$ are common in communications—in fact, they are the basis of carrier modulation—and by direct extension of (2.33) we have the following *modulation theorem:*

⋮ Theorem: $x(t)\,\cos\,\omega_c t \leftrightarrow \tfrac{1}{2}X(f - f_c) + \tfrac{1}{2}X(f + f_c)$ (2.34)

    In words, multiplying a signal by $\cos\,\omega_c t$ translates its spectrum *up and down* in frequency by f_c. (Recall the RF pulse train of Example 2.2.) All the comments about complex modulation also apply here. In addition, the resulting spectrum is hermitian, which it must be if $x(t)\,\cos\,\omega_c t$ is a real function of time. The theorem is easily proved with the aid of Euler's theorem and (2.33).

Example 2.6 RF pulse

Figure 2.14a is the single RF pulse, written as

$$x(t) \;=\; A\Pi\left(\frac{t}{\tau}\right)\,\cos\,\omega_c t$$

From the modulation theorem and Example 2.3 we have immediately

$$X(f) \;=\; \tfrac{1}{2}A\tau\,\text{sinc}\,(f - f_c)\tau + \tfrac{1}{2}A\tau\,\text{sinc}\,(f + f_c)\tau$$

The spectrum is sketched in Fig. 2.14b and can be compared with the RF pulse-train spectrum (Fig. 2.7b).

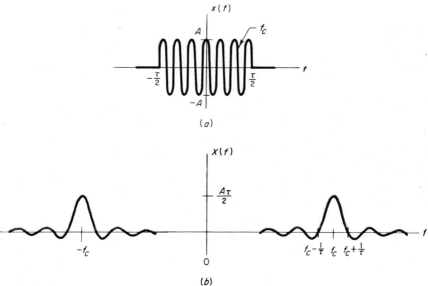

(a)

(b)

Fig. 2.14 *An RF pulse and its spectrum.*

Differentiation and integration

Certain processing techniques involve the time derivative or integral of a signal. The effects of these operations on the spectrum are indicated by the following theorems:

Theorem: If $dx(t)/dt$ is integrable square, then

$$\frac{d}{dt} x(t) \leftrightarrow j\omega X(f) = (j2\pi f) X(f) \tag{2.35}$$

Theorem: If $\int_{-\infty}^{t} x(t')\, dt'$ is integrable square, then

$$\int_{-\infty}^{t} x(t')\, dt' \leftrightarrow \frac{1}{j\omega} X(f) = (j2\pi f)^{-1} X(f) \tag{2.36}$$

where the dummy variable t' is required for clarity.

Note that the theorems do not guarantee the existence of these transforms; but if the transforms exist, they are given by (2.35) and (2.36).

Inspecting the theorems, it can be said that differentiation enhances the high-frequency components of a signal while integration suppresses high-frequency components. Spectral interpretation thus agrees with the time-domain viewpoint that differentiation accentuates time variations while integration smoothes them out.

The proof of (2.35) is given below; proof of (2.36) follows similar lines.

Proof: $\dfrac{d}{dt} x(t) = \dfrac{d}{dt} \displaystyle\int_{-\infty}^{\infty} X(f) e^{j\omega t}\, df = \displaystyle\int_{-\infty}^{\infty} X(f) \dfrac{d}{dt} e^{j\omega t}\, df$

$$= \int_{-\infty}^{\infty} j\omega X(f) e^{j\omega t}\, df$$

Comparing the above with (2.18) gives

$$\mathfrak{F}\left[\frac{d}{dt} x(t)\right] = j\omega X(f)$$

By iteration one can also show that

$$\frac{d^n}{dt^n} x(t) \leftrightarrow (j2\pi f)^n X(f) \tag{2.37}$$

$$\int_{-\infty}^{t} \cdots \int_{-\infty}^{t_3} \int_{-\infty}^{t_2} x(t_1)\, dt_1\, dt_2 \cdots dt_n \leftrightarrow (j2\pi f)^{-n} X(f) \tag{2.38}$$

These relations are particularly valuable for the analysis of linear systems.

Convolution

To conclude this section we consider *convolution* in the time and frequency domains. Let x, y, and z be three functions of time related by

$$z(t) = \int_{-\infty}^{\infty} x(t') y(t - t')\, dt' \tag{2.39}$$

where t' is a dummy variable of integration.   The integral is known as the *convolution integral*, and z is said to be the convolution of x and y.   A convenient notation for this is

$$z(t) = x(t) * y(t)$$

It is a simple matter to show, using change of variables, that convolution operates symbolically like ordinary multiplication; specifically, it is commutative, distributive, and associative, so $x * y = y * x$, $z * (x + y) = (z * x) + (z * y)$, and $z * (x * y) = (z * x) * y$.

In Chap. 3 a physical interpretation of convolution is discussed, with reference to linear systems.   For the moment, however, we shall

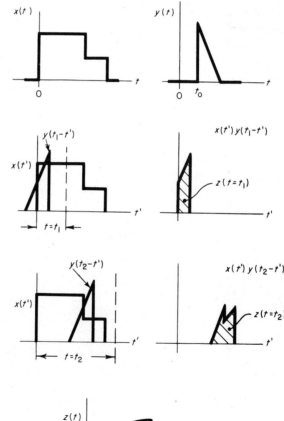

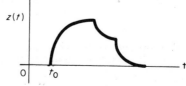

Fig. 2.15 *The graphical in-terpretation of*

$$z(t) = x(t) * y(t)$$
$$= \int_{-\infty}^{\infty} x(t')y(t - t') \, dt'$$

deal with the operation mathematically and graphically as a tool in spectral analysis. The graphical interpretation of convolution is particularly helpful when dealing with simple functions, as illustrated by Fig. 2.15. Note that in the domain of integration t', $y(t - t')$ is $y(t)$ folded backward about $t = 0$ and slid past $x(t')$. Then $z(t)$ is the area of the product $x(t')y(t - t')$, and its independent variable t is measured in the t' domain by the spacing between the original origins of x and y. As a simple exercise in graphical technique, the reader should verify that the convolution of two rectangular pulses is a trapezoidal pulse (Fig. 2.16).

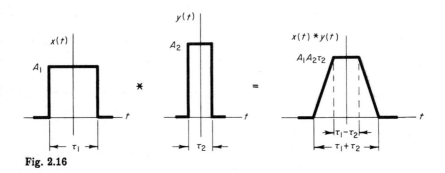

Fig. 2.16

With these preliminaries, we now give the *convolution theorems*, which are two in number.

Theorem: $x(t) * y(t) \leftrightarrow X(f)Y(f)$ (2.40)

Theorem: $x(t)y(t) \leftrightarrow X(f) * Y(f)$ (2.41)

These theorems state that convolution in the time domain corresponds to multiplication in the frequency domain, and vice versa. Proof of (2.40) follows.

Proof: $\mathcal{F}[x(t) * y(t)] = \int_{-\infty}^{\infty} \left[\int_{-\infty}^{\infty} x(t')y(t - t') \, dt' \right] e^{-j\omega t} \, dt$

$= \int_{-\infty}^{\infty} x(t') \left[\int_{-\infty}^{\infty} y(t - t')e^{-j\omega t} \, dt \right] dt'$

$= \int_{-\infty}^{\infty} x(t') Y(f)e^{-j\omega t'} \, dt'$

$= \left[\int_{-\infty}^{\infty} x(t')e^{-j\omega t'} \, dt' \right] Y(f) = X(f)Y(f)$

Example 2.7 Triangular pulse

As a special case of Fig. 2.16, the convolution of a rectangular pulse with itself is a triangular pulse of base 2τ and height $B = A^2\tau$ (Fig. 2.17a).

Using the results of Example 2.3 and the convolution theorem (2.40), the spectrum of the triangular pulse is

$$(A\tau \text{ sinc } f\tau)(A\tau \text{ sinc } f\tau) = A^2\tau^2 \text{ sinc}^2 f\tau = B\tau \text{ sinc}^2 f\tau$$

as shown in Fig. 2.17b.

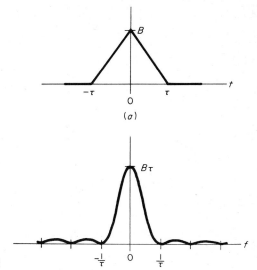

Fig. 2.17 *A triangular pulse and its spectrum.*

Introducing the symbolic notation

$$\Lambda\left(\frac{t}{\tau}\right) = \begin{cases} 1 - \dfrac{|t|}{\tau} & |t| < \tau \\ 0 & |t| > \tau \end{cases}$$

we have the transform pair

$$A\Lambda\left(\frac{t}{\tau}\right) \leftrightarrow A\tau \text{ sinc}^2 f\tau$$

Example 2.8 The spectrum of $x^2(t)$

Suppose $x(t)$ is bandlimited in W, with spectrum as shown in Fig. 2.18a. What then is the spectrum of $x^2(t)$?   From (2.41) we must convolve $X(f)$ with itself, the result being something like Fig. 2.18b.   Without any further specific knowledge of $x(t)$ we reach this important conclusion: when $x(t)$ is bandlimited in W, $x^2(t)$ is bandlimited in $2W$.   (Note the difference between this operation and the modulation theorem, both of which double spectral width.)   The process may be iterated for $x^3(t)$, etc., with predictable conclusions.

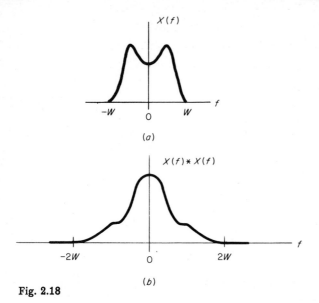

Fig. 2.18

2.5 IMPULSES AND TRANSFORMS IN THE LIMIT

We have so far discussed two rather restricted classes of signals, periodic signals with finite average power and timelimited nonperiodic signals (primarily pulses) of finite total energy. The power and energy restrictions are equivalent to the integrable-square condition of Fourier theory. But strict adherence to this condition precludes spectral description of many other signal models which are useful in the study of communications. Furthermore, if a signal consists of both periodic and nonperiodic time functions, its frequency-domain representation is quite awkward, requiring a line spectrum for the periodic component and a continuous spectrum for the nonperiodic one.

In many cases these limitations can be overcome by extending Fourier transform theory to include signals which are not integrable square but do have finite power when averaged over all time, i.e., when $x(t)$ satisfies

$$\lim_{T \to \infty} \frac{1}{T} \int_T |x(t)|^2 \, dt < \infty$$

Thus any signal having a line spectrum, in the previous sense of Sec. 2.1, can also be described by a continuous spectrum. These continuous spectra are obtained from limiting operations on signals whose transforms are known to exist, and the derived transforms are called *transforms in the limit*. In general, the limiting process leads to impulses, whose properties we now review.

Impulses

The unit impulse (also called Dirac impulse or delta function) is sometimes defined as a function with unit area satisfying

$$\delta(u) = \begin{cases} 0 & u \neq 0 \\ \infty & u = 0 \end{cases}$$

More precisely it should be stated that $\delta(u)$ has the property

$$\int_{-\infty}^{\infty} g(u)\delta(u - u_0)\, du = g(u_0) \tag{2.42}$$

where $g(u)$ is any regular function continuous at u_0. As a special case of (2.42) we have the more familiar relation

$$\int_{-\infty}^{\infty} \delta(u)\, du = \int_{0-}^{0+} \delta(u)\, du = 1 \tag{2.43}$$

In words, an impulse has unit area or *weight* concentrated at the point where its argument is zero and no *net* area elsewhere. For most purposes we can also say that $\delta(u - u_0)$ is located at $u = u_0$ and is zero everywhere else. Thus $A\delta(u - u_0)$, an impulse of weight A, is graphically represented by Fig. 2.19.

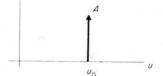

Fig. 2.19   *The graphical representation of $A\delta(u - u_0)$.*

Strictly speaking, impulses are not *functions* in the usual sense. Consequently, (2.42) and (2.43)—or any expressions containing impulses —require a standard for interpretation. The usual convention is to replace $\delta(u)$ by a unit-area pulse of finite amplitude and nonzero duration, the pulse shape being relatively unimportant. Operations involving $\delta(u)$ are then carried out with the finite pulse, after which we consider the limit as the duration approaches zero.

Of the many possible shapes which become impulses in the limit, we shall have use for the *gaussian, rectangular, sinc,* and *sinc*^{2} pulses. Letting the duration measure be $1/a$, we have

$$\delta(u) = \lim_{a \to \infty} \begin{cases} ae^{-\pi(au)^2} \\ a\Pi(au) \\ a \operatorname{sinc} au \\ a \operatorname{sinc}^2 au \end{cases} \tag{2.44}$$

bearing in mind that the limit is to be taken last. The reader can verify that each of these satisfies (2.42) and (2.43). Further detailed treatment of impulses and the associated concept of *generalized functions* can be found in the literature, e.g., Bracewell (1965, chap. 5) and Lighthill (1958).

As to operations with impulses, (2.42) expresses the *sampling*, or *sifting*, property since $\int_{-\infty}^{\infty} g(u)\delta(u - u_0)\, du$ "sifts out" the sample value $g(u_0)$. Related to sampling is *convolution* with impulses, specifically,

$$g(u) * \delta(u - u_0) = \int_{-\infty}^{\infty} g(u')\delta(u - u_0 - u')\, du' = g(u - u_0) \quad (2.45)$$

so convolving a regular function with $\delta(u - u_0)$ merely reproduces the entire function displaced by u_0. Note also that with $u_0 = 0$

$$g(u) * \delta(u) = g(u)$$

The difference between (2.42) and (2.45) should be clearly understood: sampling picks out a particular value; convolving repeats the function completely. Finally we observe that multiplication is equivalent to weighting the impulse with a sample value, that is,

$$g(u)\delta(u - u_0) = g(u_0)\delta(u - u_0)$$

since, *in the limit*, $\delta(u - u_0) = 0$ for $u \neq u_0$.

Before leaving this subject, it should be mentioned that certain improper integrals can be evaluated in the limiting sense of the impulse. For example,

$$\int_{-\infty}^{\infty} e^{\pm j2\pi uv}\, dv = \lim_{a \to \infty} \int_{-a/2}^{a/2} e^{\pm j2\pi uv}\, dv$$
$$= \lim_{a \to \infty} a \operatorname{sinc} au = \delta(u) \quad (2.46)$$

a result which, when coupled with (2.42), plays a pivotal role in transform theory. To illustrate, let us now prove the Fourier integral theorem

$$x(t) = \mathcal{F}^{-1}[X(f)] = \int_{-\infty}^{\infty} \left[\int_{-\infty}^{\infty} x(t')e^{-j2\pi ft'}\, dt \right] e^{j2\pi ft}\, df$$

Consider the integral $I_a(t) = \int_{-a/2}^{a/2} df \int_{-\infty}^{\infty} dt'\, x(t')e^{-j2\pi f(t'-t)}$; we wish to show that $\lim_{a \to \infty} I_a(t) = x(t)$. Reversing the order of integration,

$$I_a(t) = \int_{-\infty}^{\infty} dt'\, x(t') \int_{-a/2}^{a/2} e^{-j2\pi f(t'-t)}\, df$$
$$= \int_{-\infty}^{\infty} x(t')\, a \operatorname{sinc} a(t' - t)\, dt'$$

Thus, invoking (2.46),

$$\lim_{a \to \infty} I_a(t) = \int_{-\infty}^{\infty} x(t')\delta(t' - t)\, dt'$$
$$= x(t)$$

providing $x(t')$ is continuous at $t' = t$.

Impulses in frequency

To begin our discussion of transforms in the limit, consider the signal $x(t) = A$, a constant for all time. This signal does not have a Fourier series, since the period is ambiguous. Nor does it have a Fourier transform in the strict sense, for the total energy is infinite. But the power averaged over all time is finite, specifically, $P = A^2$, and intuitively we expect the line spectrum to consist of a line of height A at zero frequency.

To put these notions on firmer mathematical ground we turn to the continuous spectrum and a transform in the limit, for which we seek another signal whose transform exists and which can be made to approach $x(t)$. One such signal is the gaussian pulse of Example 2.4, $y(t) = Ae^{-\pi(t/\tau)^2}$. Thus

$$x(t) = \lim_{\tau \to \infty} y(t) = \lim_{\tau \to \infty} Ae^{-\pi(t/\tau)^2} = A$$

and

$$X(f) = \lim_{\tau \to \infty} Y(f) = \lim_{\tau \to \infty} A\tau e^{-\pi(\tau f)^2} = A\delta(f)$$

Therefore

$$A \leftrightarrow A\delta(f) \tag{2.47}$$

As a check on (2.47) we take the inverse transform

$$\mathcal{F}^{-1}[A\delta(f)] = \int_{-\infty}^{\infty} A\delta(f)e^{j\omega t}\, df = Ae^{j2\pi ft}\Big|_{f=0} = A$$

We could have taken another route to this result by starting with a rectangular pulse and letting $\tau \to \infty$. The spectrum would then be $\lim_{\tau \to \infty} A\tau \operatorname{sinc} f\tau = A\delta(f)$. The variety of approaches to such problems points out why there are many limiting forms for the impulse and emphasizes that their *shape* is not critical.

The continuous spectrum of the constant A is thus an *impulse* of *weight* A at zero frequency, whereas the intuitive line spectrum is a *line* of *height* A at zero frequency. The impulsive form is required to indicate that a nonzero amount of power is concentrated at a specific point in the continuous spectrum.

Impulses in frequency are given further meaning by considering the continuous spectrum of $A \cos \omega_c t$, which does have a Fourier series. Applying the modulation theorem (2.34) to (2.47) gives

$$A \cos \omega_c t \leftrightarrow \frac{A}{2} \delta(f - f_c) + \frac{A}{2} \delta(f + f_c) \tag{2.48}$$

(The formality of transforms in the limit can be maintained by taking the limit $\tau \to \infty$ to the RF pulse of Example 2.6, which yields the same result.)

The continuous spectrum of a cosine wave is thus a pair of impulses (Fig. 2.20). Moreover, since $A \cos \omega_c t = (A/2)e^{j\omega_c t} + (A/2)e^{-j\omega_c t}$, we

Fig. 2.20 *The spectrum of A cos $\omega_c t$.*

can interpret the impulse in frequency as representing a *phasor* whose amplitude is equal to the weight of the impulse and whose frequency is indicated by the impulse location. This interpretation is justified by

$$\mathfrak{F}[e^{j\omega_c t}] = \int_{-\infty}^{\infty} e^{-j(\omega - \omega_c)t}\, dt = \delta(f - f_c) \tag{2.49}$$

which follows directly from (2.46).

Invoking superposition, we can further state that any signal which is a *sum* of conjugate phasors (or sinusoids) has a continuous spectrum consisting of impulse pairs. Likewise, any two-sided line spectrum can be converted to a continuous spectrum by changing the spectral lines to impulses whose weights are equal to the line heights.

Continuous spectra for periodic signals

A summation of sinusoids is not necessarily a periodic time function (all the frequencies must be harmonically related to make the sum periodic); but if the signal is periodic and has a Fourier series, its continuous spectrum is found by inspection of the exponential Fourier series expansion. Formally, the signal

$$x(t) = \sum_{n=-\infty}^{\infty} c_n e^{jn\omega_0 t}$$

has as its transform in the limit

$$X(f) = \sum_{n=-\infty}^{\infty} c_n \delta(f - nf_0) \qquad (2.50)$$

This is easily derived using (2.49) and superposition, to wit,

$$X(f) = \int_{-\infty}^{\infty} x(t)e^{-j\omega t}\,dt = \int_{-\infty}^{\infty} \left(\sum c_n e^{jn\omega_0 t}\right)e^{-j\omega t}\,dt$$
$$= \sum \left(c_n \int_{-\infty}^{\infty} e^{-j(\omega - n\omega_0)t}\,dt\right) = \sum c_n\,\delta(f - nf_0)$$

The impulse weights c_n, being complex, give both the spectral amplitude and relative phase.

With the aid of transforms in the limit, we can represent both periodic and nonperiodic signals by continuous spectra. In addition, the transform theorems developed in Sec. 2.4 can now be applied to periodic signals. That strange beast the impulse function thus emerges as a key to unifying spectral analysis.

But one may well ask: What is the difference between the line spectrum and the "continuous" spectrum of a periodic signal? Obviously there can be no physical difference; the difference lies in the mathematical conventions. To return to the time domain from the line spectrum, we sum the phasors which the lines represent. To return to the time domain from the continuous spectrum, we *integrate* $X(f)$, weighted by $e^{j\omega t}$, over all frequency.

Impulses in time

As a signal model, the impulse in time may seem farfetched, but in Chap. 3 we shall find conditions for which it is a reasonable model. Even more important is the convenience of the time impulse as an analytic tool.

The impulse in time $\delta(t)$ can be constructed using the gaussian form of (2.44), with $\tau = 1/a$. Thus

$$\delta(t) = \lim_{\tau \to 0} \frac{1}{\tau} e^{-\pi(t/\tau)^2}$$

whose transform in the limit is

$$\mathfrak{F}[\delta(t)] = \lim_{\tau \to 0} e^{-\pi(\tau f)^2} = 1$$

Therefore

$$A\,\delta(t) \leftrightarrow A \qquad (2.51)$$

and, with the time-delay theorem,

$$A\,\delta(t - t_0) \leftrightarrow A e^{-j\omega t_0} \qquad (2.52)$$

The transform of the time impulse has *constant amplitude,* so the spectrum contains all frequencies in equal proportion. This property is used to advantage in experimental electronics, where very short pulses are generated to test the frequency response of amplifiers, etc. In communications, the time impulse has special importance for the theory of sampling, as seen in Chap. 7.

The reader may have observed that (2.51) is the dual of $A \leftrightarrow A\delta(f)$, Eq. (2.47). This dual relationship has its roots in reciprocal spreading, (2.47) and (2.51) being the two extremes; for a constant signal of infinite duration has "zero" spectral width, whereas an impulse in time has "zero" duration and infinite spectral width.

Impulses and transform calculations

The transform pair $A\delta(t - t_0) \leftrightarrow Ae^{-j\omega t_0}$ coupled with the iterated differentiation theorem (2.37) often expedites transform calculations, particularly when the signal in question is a piecewise-linear time function. Since, from (2.37),

$$\frac{d^n}{dt^n} x(t) \leftrightarrow (j2\pi f)^n X(f)$$

it follows that

$$X(f) = (j2\pi f)^{-n} \mathfrak{F}\left[\frac{d^n}{dt^n} x(t)\right] \tag{2.53}$$

Therefore we can repeatedly differentiate $x(t)$ until we obtain a time function $d^n x(t)/dt^n$ whose transform is known. Then dividing by $(j2\pi f)^n$ yields $X(f)$.

To illustrate this technique, consider the triangular pulse of Fig. 2.21*a*. The first and second derivatives are easily found, as sketched in Fig. 2.21*b*. (Note the importance of such sketches in determining discontinuities, etc.) Thus

$$\frac{d^2 x}{dt^2} = \frac{A}{\tau} \delta(t + \tau) - \frac{2A}{\tau} \delta(t) + \frac{A}{\tau} \delta(t - \tau)$$

and

$$\mathfrak{F}\left[\frac{d^2 x}{dt^2}\right] = \frac{A}{\tau} e^{+j\omega\tau} - \frac{2A}{\tau} + \frac{A}{\tau} e^{-j\omega\tau}$$

$$= -\frac{2A}{\tau} (1 - \cos 2\pi f\tau)$$

$$= -\frac{4A}{\tau} \sin^2 \pi f\tau$$

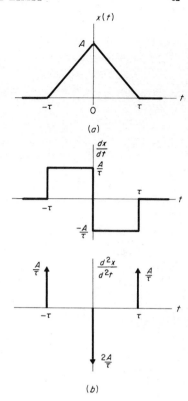

Fig. 2.21 *A triangular pulse and its first and second derivatives.*

Therefore

$$X(f) = \frac{-4A/\tau \sin^2 \pi f\tau}{-4\pi^2 f^2} = A\tau \operatorname{sinc}^2 f\tau$$

in agreement with Example 2.7, where the result was obtained by the convolution theorem.

The ideal sampling wave ★

By way of groundwork for the study of sampling, we here examine a very peculiar waveform called the *ideal sampling wave* (Fig. 2.22a). This

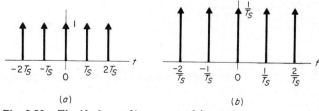

Fig. 2.22 *The ideal sampling wave and its spectrum.*

waveform is a periodic string of impulses spaced by T_s and looks like a picket fence. Mathematically, we are dealing with an exceedingly improper function. It has an infinite number of infinite discontinuities and violates all the usual conditions for existence of a Fourier transform. Nonetheless, a transform in the limit can be found, and has great analytic value.

Formally, the ideal sampling wave is written as

$$\sum_{m=-\infty}^{\infty} \delta(t - mT_s)$$

which, being periodic, has a Fourier series representation, at least in a limiting sense. Thus

$$\sum_{m=-\infty}^{\infty} \delta(t - mT_s) = \sum_{n=-\infty}^{\infty} c_n e^{jn\omega_s t} \qquad \omega_s = \frac{2\pi}{T_s}$$

where

$$c_n = \frac{1}{T_s} \int_{T_s} \delta(t) e^{-jn\omega_s t} \, dt = \frac{1}{T_s} = f_s$$

Therefore

$$\sum_{m=-\infty}^{\infty} \delta(t - mT_s) = f_s \sum_{n=-\infty}^{\infty} e^{jn\omega_s t}$$

In the frequency domain, it follows from (2.50) that

$$\mathfrak{F}\left[\sum_{m=-\infty}^{\infty} \delta(t - mT_s) \right] = \sum_{n} c_n \delta(f - nf_s) = f_s \sum_{n=-\infty}^{\infty} \delta(f - nf_s)$$

or symbolically

$$\sum_{m=-\infty}^{\infty} \delta(t - mT_s) \leftrightarrow f_s \sum_{n=-\infty}^{\infty} \delta(f - nf_s) \tag{2.54}$$

Thus, like a gaussian pulse, the spectrum of the ideal sampling wave has the same form as its time function.

While (2.54) is correct, the method we have used does not strictly adhere to the formalisms of transforms in the limit. A more rigorous derivation is obtained by starting with the rectangular pulse train of Example 2.1 and taking suitable limits. The reader is encouraged to try this for himself.

2.6 POWER SPECTRA AND CORRELATION

Looking back upon Secs. 2.3 and 2.5, the following observations can be made:

1. Signals of finite total energy $E = \int_{-\infty}^{\infty} x^2(t) \, dt$ are represented in the frequency domain by Fourier transforms, which are continuous functions of frequency and free of impulses. The distribution of signal energy in frequency is described by the energy spectral density $S(f) = |X(f)|^2$. Since the energy is finite, the power averaged over all time is zero.

2. Signals of finite average power $P = \lim_{T \to \infty} \frac{1}{T} \int_T x^2(t) \, dt$ are represented in the frequency domain by transforms in the limit, which are impulsive functions of frequency. Since the power is nonzero, the total energy is infinite.

Clearly, the concept of energy spectral density is of little use in the second case. But it may be meaningful to speak of the distribution of average power in frequency, as described by a *power spectral density*, or *power spectrum* for short.

Analogous to $S(f)$, the power spectrum $G(f)$ should be a real, even, and positive function of frequency satisfying

$$P = \int_{-\infty}^{\infty} G(f) \, df = 2 \int_0^{\infty} G(f) \, df = \langle x^2(t) \rangle \tag{2.55}$$

But, in contrast to $S(f)$, which is a continuous function of frequency, we suspect that $G(f)$ will be *impulsive*. This conclusion stems from the observation that signals of nonzero average power discussed so far have been *sums of sinusoids*, and therefore their power is concentrated at discrete frequencies.

The purpose of this section is to develop more fully the concept and properties of power spectra. Our method is first to derive $G(f)$ for signals having transforms in the limit. We then generalize our results by introducing the *correlation function* $R(\tau)$ and the *Wiener-Kinchine theorem*, which states that $G(f)$ and $R(\tau)$ are a Fourier transform pair. As we shall see in Chap. 4, the Wiener-Kinchine approach is invaluable to the frequency-domain representation of *random* signals and noise. For the present it puts the definition of power spectrum on firmer mathematical ground.

Power spectra for sums of sinusoids

Consider a signal $x(t)$, not necessarily periodic, which contains only discrete frequency components, i.e., a summation of sinusoidal waves. If the average power is not zero, the total energy is infinite, and $X(f)$ exists only as a transform in the limit. But suppose we restrict our attention to the time interval $|t| < T/2$ such that the *truncated* signal $x_T(t) = x(t)\Pi(t/T)$ has a proper transform

$$X_T(f) = \int_{-T/2}^{T/2} x(t)e^{-j\omega t}\, dt \tag{2.56}$$

and finite energy $E_T = \int_{-\infty}^{\infty} |X_T(f)|^2\, df$. Then as $T \to \infty$, $x_T(t) \to x(t)$ and $X_T(f) \to X(f)$ in the limiting sense. The signal power in $x(t)$, averaged over all time, can thus be written as

$$P = \lim_{T \to \infty} \frac{E_T}{T} = \lim_{T \to \infty} \frac{1}{T} \int_{-\infty}^{\infty} |X_T(f)|^2\, df$$

Now, from (2.55), we require that the power spectral density $G(f)$ be such that $\int_{-\infty}^{\infty} G(f)\, df = P$. Comparing with the above suggests that we might interchange the order of limiting and integration so

$$P = \int_{-\infty}^{\infty} \left[\lim_{T \to \infty} \frac{1}{T} |X_T(f)|^2 \right] df = \int_{-\infty}^{\infty} G(f)\, df$$

and hence

$$G(f) = \lim_{T \to \infty} \frac{1}{T} |X_T(f)|^2 \tag{2.57}$$

In general, the interchange of order leading to (2.57) is *invalid*, and all that can be said with assurance is that for any finite frequency range, say f_1 to f_2, the *integrals* are equal, that is,

$$\int_{f_1}^{f_2} G(f)\, df = \lim_{T \to \infty} \int_{f_1}^{f_2} \frac{1}{T} |X_T(f)|^2\, df$$

where the order of operations cannot be interchanged.

However, for the special class of signals under consideration, namely sums of sinusoids, Eq. (2.57) is a plausible definition of $G(f)$. Moreover, it leads to power spectra which agree with intuitive reasoning, as demonstrated shortly.

Before giving a sample calculation, we should point out that power spectra contain no *phase* information. This is apparent from (2.57) since $|X_T(f)|^2$ depends only on the magnitude of $X_T(f)$. Thus, unlike the conventional (voltage) spectrum, $G(f)$ does not uniquely describe $x(t)$, and it is not possible to go from $G(f)$ back to $x(t)$. In other words, power spectra is a one-way transformation, and two signals differing in spectral phase can have the same power spectra. But then loss of phase information is to be expected when dealing with *averages*.

To illustrate the use of (2.56) and (2.57), let $x(t) = A \cos \omega_c t$, whose average power is $P = A^2/2$. Physical considerations indicate that the power is concentrated at a single frequency, f_c. Hence the two-sided power spectrum should consist of a *pair* of impulses at $\pm f_c$. To check this conclusion with the formal expression we first note that the truncated signal $x_T(t)$ is simply a rectangular RF pulse of duration T. Then, from Example 2.6, we have

$$|X_T(f)|^2 = \left(\frac{AT}{2}\right)^2 [\text{sinc}^2 (f - f_c)T + \text{sinc}^2 (f + f_c)T$$
$$+ 2 \text{ sinc} (f - f_c)T \text{ sinc} (f + f_c)T]$$

Inserting in (2.57) and regrouping yields

$$G(f) = \frac{A^2}{4} \lim_{T \to \infty} \left\{ T \text{ sinc}^2 (f - f_c)T + T \text{ sinc}^2 (f + f_c)T \right.$$
$$\left. + \frac{2}{T} [T \text{ sinc} (f - f_c)T][T \text{ sinc} (f + f_c)T] \right\}$$

The first two terms are recognized as approaching impulses at $\pm f_c$ [see (2.44)], while the last term becomes the null product $(2/T)\delta(f - f_c)\delta(f + f_c) = 0$. Hence

$$G(f) = \frac{A^2}{4} \delta(f - f_c) + \frac{A^2}{4} \delta(f + f_c)$$

which shows that the power is equally divided between the positive- and negative-frequency components. Clearly, $\int_{-\infty}^{\infty} G(f) \, df = A^2/2 = P$.

Extending the above result to sums of sinusoids is intuitively obvious, though mathematically tedious. In general, if

$$x(t) = \sum_m A_m \cos (\omega_m t + \phi_m)$$

then

$$G(f) = \sum_m \frac{A_m^2}{4} [\delta(f - f_m) + \delta(f + f_m)] \tag{2.58}$$

whether or not the frequencies are harmonically related. (Again, note the loss of phase information.) Similarly, for a periodic signal having Fourier series coefficients c_n

$$G(f) = \sum_{n=-\infty}^{\infty} |c_n|^2 \delta(f - nf_0) \tag{2.59}$$

and integrating over all frequency yields Parseval's theorem

$$P = \int_{-\infty}^{\infty} \left[\sum_n |c_n|^2 \delta(f - nf_0) \right] df = \sum_{n=-\infty}^{\infty} |c_n|^2$$

as it should.

Equations (2.58) and (2.59) show that any *two-sided line spectrum* can be converted to a power spectrum by changing the lines to impulses with weights equal to the magnitude squared of the line heights. Thus, hereafter we need not bother with the formalism of (2.57) if the line spectrum is known.

But Eq. (2.57) or the line-spectrum technique fails when we come to signals which are not sums of sinusoids but still have nonzero power averaged over all time. Many *random* signals satisfy this description, and one can also find nonrandom (deterministic) signals of this type as well. (Consider, for example, a constant-amplitude cosine wave whose frequency changes from f_1 to f_2 at $t = 0$.) A more general approach to power spectra is necessary to cope with such signals, so we turn to correlation functions and the Wiener-Kinchine theorem.

Correlation functions

The correlation function $R(\tau)$ is a special time average defined by

$$
\begin{aligned}
R(\tau) &= \langle x(t)x(t+\tau) \rangle \\
&= \lim_{T \to \infty} \frac{1}{T} \int_T x(t)x(t+\tau)\, dt
\end{aligned}
\tag{2.60}
$$

or, if $x(t)$ is periodic in T_0,

$$
R(\tau) = \frac{1}{T_0} \int_{T_0} x(t)x(t+\tau)\, dt
$$

Thus, the correlation function is found by taking a signal, multiplying it by the same signal displaced τ units in time, and averaging the product over all time. Although averages are usually numbers, rather than functions, $R(\tau)$ is a function whose independent variable is the displacement τ. Note that since $R(\tau)$ is an average, correlation functions, like power spectra, are nonunique signal descriptions.

There are two qualifications to the above definition. First, it has been assumed that $x(t)$ is real. In the case of complex time functions we use $R(\tau) = \langle x^*(t)x(t+\tau) \rangle$. Second, $R(\tau)$ is more precisely designated as the *autocorrelation* function, the correlation of $x(t)$ with itself, as distinguished from a *crosscorrelation* such as $R_{xy}(\tau) = \langle x(t)y(t+\tau) \rangle$, the correlation of $x(t)$ with $y(t)$. We shall deal almost exclusively with autocorrelation and call it simply *correlation*, with little danger of confusion.

One reason for introducing correlation at this particular point is evident by inspection of (2.60): when $\tau = 0$, $R(0) = \langle x^2(t) \rangle$, which is the average power of $x(t)$, and hence $R(0) = P = \int_{-\infty}^{\infty} G(f)\, df$. Other properties of $R(\tau)$ are:

1. $R(0) = \langle x^2(t) \rangle \geq |R(\tau)|$ (2.61a)
2. $R(-\tau) = \langle x(t)x(t - \tau) \rangle = R(\tau)$ (2.61b)
3. If $x(t)$ is periodic in T_0, then $R(\tau + T_0) = R(\tau)$ (2.61c)

In words, the correlation function is *even* in τ with a maximum at $\tau = 0$ equal to the signal power. If the signal is periodic, the correlation function is periodic with the same period.

      Proving (2.61b) and (2.61c) is a simple matter of changing variables in the definition. The inequality of (2.61a) comes from the observation that $\langle [x(t) \pm x(t + \tau)]^2 \rangle$ must be positive if $x(t)$ is real. Expanding the square and recalling that time averaging is a linear operation yields

$$\langle x^2(t) \rangle \pm 2\langle x(t)x(t + \tau) \rangle + \langle x^2(t + \tau) \rangle \geq 0$$

But $\langle x^2(t + \tau) \rangle = \lim_{T \to \infty} \dfrac{1}{T} \int_T x^2(t + \tau)\, dt$ can be recognized as $R(0)$, the displacement τ having no effect when averaging over all time. Hence

$$R(0) \pm 2R(\tau) + R(0) \geq 0$$

or

$$R(0) \geq |R(\tau)|$$

      Several characteristics of $R(\tau)$ are illustrated by the following example.

Example 2.9   Autocorrelation of a sinusoidal wave

Consider $x(t) = A \cos (\omega_c t + \phi)$, which is periodic in $T_0 = 2\pi/\omega_c$. Its correlation function is

$$R(\tau) = \frac{1}{T_0} \int_{T_0} A \cos (\omega_c t + \phi) A \cos (\omega_c t + \omega_c \tau + \phi)\, dt$$

Upon expanding the second term $R(\tau)$ becomes

$$A^2[\cos \omega_c \tau \langle \cos^2 (\omega_c t + \phi) \rangle - \sin \omega_c \tau \langle \cos (\omega_c t + \phi) \sin (\omega_c t + \phi) \rangle]$$

But $\langle \cos^2 (\omega_c t + \phi) \rangle = \frac{1}{2}$, while

$$\langle \cos (\omega_c t + \phi) \sin (\omega_c t + \phi) \rangle = \tfrac{1}{2} \langle \sin 2(\omega_c t + \phi) \rangle = 0$$

and hence

$$R(\tau) = \frac{A^2}{2} \cos \omega_c \tau \tag{2.62}$$

independent of ϕ. Clearly (2.62) has all the properties of (2.61).

      In this example there are values of τ such that $R(\tau) = +R(0)$, $-R(0)$, and 0. As an aid to the physical interpretation of correlation, the reader should compare $x(t + \tau)$ with $x(t)$ when τ satisfies these conditions.

Correlation and incoherence

Further meaning can be given to correlation by considering a sum of two signals, say $z(t) = x(t) + y(t)$. The correlation function of $z(t)$ is, after expansion,

$$R_z(\tau) = \langle x(t)x(t + \tau)\rangle + \langle y(t)y(t + \tau)\rangle + \langle x(t)y(t + \tau)\rangle$$
$$+ \langle y(t)x(t + \tau)\rangle$$
$$= R_x(\tau) + R_y(\tau) + R_{xy}(\tau) + R_{yx}(\tau)$$

where the last two terms are the crosscorrelations of x with y and y with x, respectively, which may be different.

Two signals are said to be *uncorrelated* if $R_{xy}(\tau) = R_{yx}(\tau) = \langle x(t)\rangle\langle y(t)\rangle$; that is, the crosscorrelation is the product of the means (dc values). This is the *definition* of uncorrelated signals. Suppose then that $x(t)$ and $y(t)$ are uncorrelated and at least one has no dc component, so $R_{xy}(\tau) = R_{yx}(\tau) = 0$. With these conditions

$$R_z(\tau) = R_x(\tau) + R_y(\tau)$$

Setting $\tau = 0$ and recalling that $R_z(0) = \langle z^2(t)\rangle = P_z$, etc., we have

$$P_z = P_x + P_y$$

Signals for which the average power of the sum equals the sum of the average powers are said to be *incoherent*.

Inasmuch as communication systems generally include blocking capacitors or transformers which remove dc components from the signals, we can view *uncorrelated* and *incoherent* as virtually synonymous terms and state the following generalization: if a signal is a sum of mutually uncorrelated signals, its correlation function is the sum of the autocorrelations, and its average power is the sum of the average powers. This conclusion is sufficiently important, particularly when extended to random signals, to formalize as a theorem.

Theorem: If $z(t) = \sum\limits_{p} x_p(t)$ and $R_{x_p x_q}(\tau) = 0$ for $p \neq q$, then

$$R_z(\tau) = \sum_{p} R_{x_p}(\tau) \tag{2.63a}$$

and

$$P_z = \sum_{p} P_{x_p} \tag{2.63b}$$

In this notation p and q are subscripts identifying the various signals which are summed to form $z(t)$.

But this is not the first time we have encountered superposition of average power; Parseval's theorem for periodic signals says essentially the same thing. And since the terms which form the Fourier series of a periodic signal are harmonically related sinusoids, we suspect that such time functions are uncorrelated.

To verify this suspicion consider the two signals

$$x_p(t) = A_p \cos (\omega_p t + \phi_p) \qquad x_q(t) = A_q \cos (\omega_q t + \phi_q)$$

where f_p and f_q are not necessarily harmonics of the same fundamental. Their crosscorrelation function is

$$
\begin{aligned}
R_{x_p x_q}(\tau) &= \langle x_p(t) x_q(t + \tau) \rangle \\
&= A_p A_q \langle (\cos \omega_p t \cos \phi_p - \sin \omega_p t \sin \phi_p) \\
&\qquad [\cos \omega_q t \cos (\omega_q \tau + \phi_q) - \sin \omega_q t \sin (\omega_q \tau + \phi_2)] \rangle
\end{aligned}
$$

Expanding and noting that

$$
\begin{aligned}
\langle \sin \omega_p t \cos \omega_q t \rangle &= 0 \\
\langle \sin \omega_p t \sin \omega_q t \rangle &= \langle \cos \omega_p t \cos \omega_q t \rangle = \begin{cases} 0 & f_p \neq f_q \\ \tfrac{1}{2} & f_p = f_q \neq 0 \end{cases}
\end{aligned}
$$

yields

$$
R_{x_p x_q}(\tau) = \begin{cases} 0 & f_p \neq f_q \\ \tfrac{1}{2} A_p A_q \cos (\omega_p t + \phi_q - \phi_p) & f_p = f_q \neq 0 \end{cases}
$$

so the signals are incoherent unless $f_p = f_q$.

The above calculation leads us to a very simple yet significant conclusion: time functions consisting only of discrete frequency components (and no dc component) are uncorrelated and incoherent unless one or more of the frequencies are *exactly equal*. This is a most useful rule, for suppose two sinusoidal generators are set at the same frequency but there is no physical synchronization between them. Then, no matter how carefully the generators are adjusted, the frequencies will be slightly different, so the outputs will be incoherent, and the average powers add. Stated another way, signals from physically independent sources are generally uncorrelated, and physical independence is a sufficient condition for superposition of average power. (It is not, however, a necessary condition.)

Superposition of average power from independent sources applies to nonelectric power as well and is a familiar physical effect. For example, consider 10 violinists playing in unison; we expect the total acoustical power to be about 10 times that of a single violinist. If one could perfectly synchronize the musicians, the total power would be 100 times that of the individuals.

As a final comment here, the mathematically inclined reader may have observed a similarity between *incoherence* and *orthogonality*. Indeed, functions which are orthogonal over all time such that

$$\int_{-\infty}^{\infty} x_p(t)x_q(t)\, dt = 0 \qquad p \neq q$$

are also incoherent if one or the other has zero mean.

Example 2.10 Autocorrelation of periodic signals

With the aid of (2.63a) and (2.62), the autocorrelation of a periodic signal follows from its Fourier series. Thus if

$$x(t) = \sum_{n=-\infty}^{\infty} c_n e^{jn\omega_0 t} = c_0 + \sum_{n=1}^{\infty} 2|c_n| \cos\left(n\omega_0 t + \phi_n\right)$$

we have a sum of mutually uncorrelated time functions, and

$$R(\tau) = \sum_{n=-\infty}^{\infty} |c_n|^2 e^{jn\omega_0 \tau} \tag{2.64}$$

Setting $\tau = 0$ again yields Parseval's theorem.

The Wiener-Kinchine theorem

However interesting the study of correlation may be, we have strayed from our primary consideration—power spectrum. Without further delay, we now *define* $G(f)$ as the Fourier transform of $R(\tau)$, that is,

$$G(f) = \mathcal{F}[R(\tau)] = \int_{-\infty}^{\infty} R(\tau)e^{-j\omega\tau}\, d\tau \tag{2.65a}$$

Then, from the theory of Fourier integrals

$$R(\tau) = \mathcal{F}^{-1}[G(f)] = \int_{-\infty}^{\infty} G(f)e^{j\omega\tau}\, df \tag{2.65b}$$

This transform pair is the *Wiener-Kinchine theorem* and says simply that the power spectrum of a signal is the Fourier transform of its correlation function, $R(\tau) \leftrightarrow G(f)$.

It was previously shown that $R(\tau)$ is an even function. Consequently, $G(f)$ is real and even, and the theorem can be written as

$$G(f) = 2 \int_0^{\infty} R(\tau) \cos 2\pi f\tau\, d\tau$$
$$R(\tau) = 2 \int_0^{\infty} G(f) \cos 2\pi f\tau\, df \tag{2.66}$$

which emphasizes the symmetry aspects. But in using (2.66), special care must be taken if $G(f)$ or $R(\tau)$ has an impulse at the origin.

Earlier we *derived* an expression for power spectrum, namely, Eq. (2.57); here we are *defining* $G(f)$ as the transform of $R(\tau)$, a procedure that may seem strange and inconsistent. However, one must recall that (2.57) applies only to a special class of signals and, in fact, fails completely for signals which are not sums of sinusoids. The Wiener-Kinchine theorem circumvents these difficulties and provides a more general and mathematically acceptable definition of power spectrum. Moreover, for signals which are sums of sinusoids, (2.65a) and (2.57) are in agreement. Indeed, the Wiener-Kinchine theorem can be "proved" from $G(f) = \lim\limits_{T\to\infty} (1/T)|X_T(f)|^2$ as follows.

Attacking the proof in reverse order, we wish to show that

$$\mathcal{F}^{-1}[G(f)] = R(\tau) = \lim_{T\to\infty} \frac{1}{T} \int_T x(t)x(t+\tau)\, dt$$

where $G(f)$ is given by (2.57). Transforming (2.57) yields

$$\mathcal{F}^{-1}\left[\lim_{T\to\infty}\frac{1}{T}|X_T(f)|^2\right] = \int_{-\infty}^{\infty}\left[\lim_{T\to\infty}\frac{1}{T}X_T(f)X_T(-f)\right]e^{j\omega\tau}\,df$$

We then substitute $X_T(-f) = \int_{-T/2}^{T/2} x(t')e^{+j\omega t'}\,dt'$, etc., giving

$$\lim_{T\to\infty}\frac{1}{T}\int_{-\infty}^{\infty}df\int_{-T/2}^{T/2}dt\int_{-T/2}^{T/2}dt'\,x(t)e^{-j\omega t}x(t')e^{+j\omega t'}e^{j\omega\tau}$$

$$= \lim_{T\to\infty}\frac{1}{T}\int_{-T/2}^{T/2}dt\int_{-T/2}^{T/2}dt'\,x(t)x(t')\int_{-\infty}^{\infty}e^{+j2\pi f(t'-t+\tau)}\,df$$

Now

$$\int_{-\infty}^{\infty}e^{+j2\pi f(t'-t+\tau)}\,df = \delta(t'-t+\tau)$$

and integrating $x(t')\delta(t'-t+\tau)$ over t' sifts out the value of $x(t')$ at $t' = t - \tau$. Therefore

$$\mathcal{F}^{-1}\left[\lim_{T\to\infty}\frac{1}{T}|X_T(f)|^2\right]$$

$$= \lim_{T\to\infty}\frac{1}{T}\int_{-T/2}^{T/2}x(t)x(t-\tau)\,dt = R(-\tau) = R(\tau)$$

Hence $\mathcal{F}^{-1}[G(f)] = R(\tau)$, and conversely $G(f) = \mathcal{F}[R(\tau)]$.

Of course the proof is subject to the same objections as Eq. (2.57). Therefore, the wisest course is to accept the Wiener-Kinchine theorem as the only valid definition of power spectrum, noting that it is consistent with all cases for which $G(f)$ is intuitively obvious. As an exercise, the reader may wish to show that (2.58), for instance, can be obtained from (2.65a) with the help of Example 2.9 and power superposition.

Power-spectra theorems

To conclude this chapter we briefly state some of the important theorems relating to power spectra. Since $R(\tau)$ and $G(f)$ form a Fourier transform pair, the theorems are simply modifications of Sec. 2.4. They will take on added meaning when applied to random signals.

Superposition

Theorem: If $R_z(\tau) = \sum_p R_{x_p}(\tau)$, then

$$G_z(f) = \sum_p G_{x_p}(f) \tag{2.67}$$

This theorem is a combination of (2.63a) and the linearity of Fourier transforms. It states that if a signal is the sum of mutually uncorrelated signals, the power spectrum is the sum of the individual power spectra.

Frequency translation

Theorem: If $x(t)$ is bandlimited in $W < f_c$ and $y(t) = x(t) \cos(\omega_c t + \phi)$, then

$$R_y(\tau) = \frac{1}{2} R_x(\tau) \cos \omega_c \tau \tag{2.68a}$$

and

$$G_y(f) = \frac{1}{4} G_x(f - f_c) + \frac{1}{4} G_x(f + f_c) \tag{2.68b}$$

Equation (2.68b) is analogous to (2.34); i.e., power spectra are translated in frequency in the same fashion as voltage spectra. However, one must note the condition $W < f_c$, which implies that $G_x(f - f_c)$ and $G_x(f + f_c)$ are nonoverlapping. It then follows that $P_y = \int_{-\infty}^{\infty} G_y(f)\, df = \frac{1}{4}\int_{-\infty}^{\infty} G_x(f - f_c)\, df + \frac{1}{4}\int_{-\infty}^{\infty} G_x(f + f_c)\, df = \frac{1}{2}\int_{-\infty}^{\infty} G_x(f)\, df$, hence $P_y = \frac{1}{2}P_x$.

To prove (2.68a) we first write $R_y(\tau) = \langle x(t) \cos(\omega_c t + \phi) x(t + \tau) \cos(\omega_c t + \omega_c \tau + \phi)\rangle$. Expanding the product of cosines and rearranging yields

$$\begin{aligned}
R_y(\tau) = {} & \frac{1}{2} \cos \omega_c \tau \langle x(t)x(t + \tau)\rangle \\
& + \frac{1}{2} \cos(2\omega_c \tau + 2\phi)\langle x(t)x(t + \tau) \cos 2\omega_c t\rangle \\
& - \frac{1}{2} \sin(2\omega_c \tau + 2\phi)\langle x(t)x(t + \tau) \sin 2\omega_c t\rangle
\end{aligned}$$

The first term is clearly $\frac{1}{2} \cos \omega_c \tau \, R_x(\tau)$; as for the remaining terms, they cannot be further simplified without knowing $x(t)$ explicitly, unless we invoke the condition that $x(t)$ is bandlimited in $W < f_c$. With this condition, $x(t)x(t + \tau)$ is bandlimited in $2W$, and therefore neither $x(t)x(t + \tau) \cos 2\omega_c t$ nor $x(t)x(t + \tau) \sin 2\omega_c t$ contains a dc component. Thus, their time averages are zero, and $R_y(\tau) = \frac{1}{2}R_x(\tau) \cos \omega_c \tau$. $G_y(f)$ then follows directly by transforming $R_y(\tau)$.

Differentiation and integration

Theorem: If $y(t) = dx(t)/dt$, then

$$G_y(f) = (2\pi f)^2 G_x(f) \tag{2.69}$$

The differentiation theorem reflects the fact that, from (2.35), $Y(f) = j2\pi f X(f)$, providing $X(f)$ exists. Proof of the theorem requires using the basic definition of differentiation, that is, $y(t) = \lim_{\epsilon \to 0} (1/\epsilon)$ $[x(t + \epsilon) - x(t)]$, from which one can obtain $R_y(\tau) = -d^2 R_x(\tau)/d\tau^2$. Equation (2.69) then follows by transformation.

The integration theorem is easily obtained from (2.69) by interchanging the roles of x and y. Specifically,

Theorem: If $y(t) = \int_{-\infty}^{t} x(t') \, dt'$, then

$$G_y(f) = (2\pi f)^{-2} G_x(f) \tag{2.70}$$

Duality

Theorem: If $R(\tau) \leftrightarrow G(f)$, then

$$G(\tau) \leftrightarrow R(f) \tag{2.71}$$

The duality theorem is simplicity itself because both $R(\tau)$ and $G(f)$ are even functions.

PROBLEMS

2.1 With the aid of Euler's theorem, write exponential Fourier series expansions for $\cos \omega_c t$, $\sin \omega_c t$, $(\sin \omega_c t)(\cos \omega_c t)$, and $\cos^2 \omega_c t$.

2.2 Find the Fourier series coefficients c_n and sketch the line spectra for a half-rectified and full-rectified cosine wave (Fig. P 2.1).

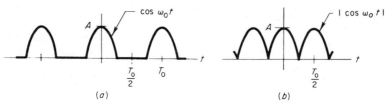

(a)   (b)

Fig. P 2.1

2.3 Find and compare c_n for the triangular wave and square wave of Fig. P 2.2. In particular, write the ratio of the c_n's. What is the time-domain relationship between these two waves?

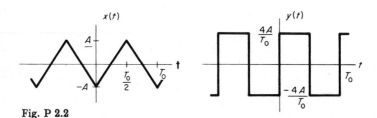

Fig. P 2.2

2.4 Referring to Example 2.1, what percentage of the total power is contained in $|f| < 1/\tau$ if the duty cycle is $d = \frac{1}{2}$ or $d = \frac{1}{10}$?

2.5 Sketch the line spectrum for the half-rectified cosine wave (Fig. P 2.1a) by starting with the results of Example 2.2 and judiciously selecting the parameter values.

2.6 Sketch, roughly to scale, the spectrum of $x(t) = \Pi(t/\tau)$ for $\tau = 1$, 2, and 4. Explain physically why the spectrum peaks up at $f = 0$ as τ is increased.

2.7 Find the spectrum of the *cosine pulse* and *raised-cosine pulse* (Fig. P 2.3), expressing your answer in terms of the sinc function. In each case sketch the spectrum and estimate the spectral width.

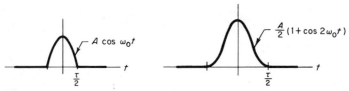

Fig. P 2.3

2.8 Using numerical or graphical integration, determine what portion of the total energy of a gaussian pulse (Example 2.4) is contained in $|f| < 1/2\tau$. Compare with Example 2.3.

2.9   Find the Fourier transforms of the exponential time functions

$$x(t) = Ae^{-|at|}$$

$$y(t) = \begin{cases} Ae^{-|at|} & t \ge 0 \\ 0 & t < 0 \end{cases}$$

2.10 Using the time-delay theorem, find and sketch the spectrum of the signals shown in Fig. P 2.4, taking $\tau \ll T$.

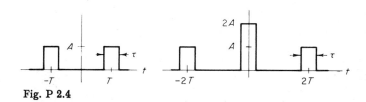

Fig. P 2.4

2.11 Prove the scale-change theorem, Eq. (2.30).

2.12 A finite-duration cosine wave of frequency f_c has precisely N cycles. Sketch its spectrum. How large must N be so that the significant spectral width is less than 1 percent of f_c?

2.13 Find and sketch the spectrum of $x(t) = A \cos \omega_c t \exp \{-\pi[(t - t_0)/\tau]^2\}$, taking $f_c \gg 1/\tau$.

2.14 Apply duality to Eq. (2.40) to obtain Eq. (2.41).   *Hint:* Write $z(t) = x(t) * y(t)$ and consider the condition on $z(t)$ so that $Z(f)$ is a real function.

2.15 Applying the convolution theorem, find

$$\mathfrak{F}[Ae^{-|at|} \text{ sinc } 2Wt]$$

2.16 Invoking any necessary theorems, find $X(f)$ for $x(t) = A \text{ sinc}^2 2Wt$. What is the spectral width?

2.17 Show mathematically that $\delta(u) = \lim_{a \to \infty} a\Pi(au)$ satisfies Eq. (2.42).   *Hint:* Expand $g(u)$ in a Taylor's series about $u = u_0$.

2.18 The transform pair $A \leftrightarrow A\delta(f)$ was derived by taking the limit of a gaussian pulse. Verify this result by taking limits to at least two other pulse shapes.

2.19 Use the convolution theorem to find the spectrum of $x(t) = A \cos^2 \omega_c t$. Check by finding the transform in the limit.

2.20 Generate the waveform shown in Fig. P 2.5 by convolving a rectangular pulse with two impulses. Find and sketch the corresponding spectrum.

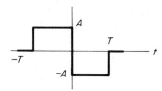

Fig. P 2.5

2.21 With the aid of Eq. (2.53), find the transform of the *trapezoidal pulse* shown in Fig. P 2.6.

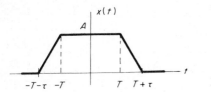

Fig. P 2.6

2.22 Using any technique you wish, determine the spectrum of the *parabolic pulse* $x(t) = A[1 - (2t/\tau)^2]\Pi(t/\tau)$.

2.23 Show that correlation can be written in terms of convolution as

$$R(\tau) = \lim_{T \to \infty} \frac{1}{T} [x(t) * x(-t)]$$

2.24 Using the above expression and the graphical approach to convolution, find the correlation function of a rectangular pulse train having $d < \frac{1}{2}$.

2.25 Prove the differentiation theorem for power spectra, Eq. (2.69), by first showing that $R_y(\tau) = -d^2 R_x(\tau)/d\tau^2$.

SELECTED SUPPLEMENTARY READING

Many fine texts dealing in whole or in part with signal representation and Fourier techniques have been written specifically for students of electrical engineering. Some of the more recent include Javid and Brenner (1963), Craig (1964), Lathi (1965), Marshall (1965), and Stuart (1966). Lathi gives extensive treatment of convolution and correlation. Craig, Marshall, and Stuart are pitched at a slightly lower level than the present volume but have useful and informative discussions. Goldman (1948) is also strongly recommended, despite its age, for the numerous examples of spectral analysis in the context of radio engineering.

Two books devoted exclusively to Fourier transforms and applications are Bracewell (1965) and Papoulis (1962). The former features imaginative notation and illustrations plus a pictorial dictionary of transform pairs. The latter strikes a nice balance between rigor and lucidity. Both present further consideration of impulses in time and frequency. Either work is an excellent supplement to the material of this chapter and should be given due attention by the serious student.

For the mathematically inclined, Churchill (1963) and Guillemin (1949) cover such topics as convergence and orthogonal functions. More advanced material, not for the casual reader, can be found in Lighthill (1958), for example.

3
Signals and Linear Networks

Much of the processing and transmission of signals in communication systems is accomplished by linear time-invariant two-port networks. A *two-port network* has two pairs of terminals, one pair (or port) for the input signal, the other pair for the output. The *linearity* property means that if the input $x_1(t)$ produces the output $y_1(t)$ and $x_2(t)$ produces $y_2(t)$, then $a_1x_1(t) + a_2x_2(t)$ produces $a_1y_1(t) + a_2y_2(t)$, where a_1 and a_2 are constants; in brief, superposition applies. If in addition the network is *time-invariant*, then the delayed input $x_1(t - t_0)$ produces the delayed output $y_1(t - t_0)$.

This chapter develops the relationship between input and output signals in terms of the properties of linear time-invariant two-ports. The spectral approach and the notion of bandwidth will be emphasized. Distortion due to system imperfections, including unwanted nonlinearities, are examined. Nonlinear and time-varying networks used intentionally in communications, primarily for modulation and detection, will be discussed later.

3.1 TRANSFER FUNCTION AND IMPULSE RESPONSE

As intimated previously, the effects of networks on signals are often best described in the frequency domain. In fact, the concept of a network *transfer function* permits the direct relationship of input and output spectra. In terms of the transfer function we can readily talk about distortionless transmission, bandwidth requirements, and filtering. Nonetheless, time-domain analysis is sometimes superior, and for that purpose we can describe a network in terms of its *impulse response*. Both descriptions are explored in the following.

Figure 3.1 shows an arbitrary two-port network with input signal $x(t)$ and output signal $y(t)$. The spectra of the input and output are $X(f)$

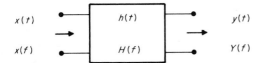

Input   Output   **Fig. 3.1** *A two-port network.*

and $Y(f)$, respectively. $H(f)$ is the two-port's transfer function, and $h(t)$ its impulse response. Normally $x(t)$ and $y(t)$ will be the same type of electrical quantity, usually voltage. At times, however, it may be more convenient to let $y(t)$ be a current responding to the voltage $x(t)$, etc. In such cases, the idea of a transfer function is equally valid, though $H(f)$ will not be dimensionless. (Similarly, in an electromechanical system, the transfer function might relate angular position of a rotor to the control current.) Finally, the two-port need not be passive; for our purposes, amplifiers and attenuators both can be described by transfer functions.

The fundamental property of a linear time-invariant two-port (or any linear system, for that matter) is that the input and output are related by a linear integrodifferential equation with constant coefficients. Typically we have

$$\cdots + a_{-1} \int_{-\infty}^{t} x(t)\, dt + a_0 x(t) + a_1 \frac{d}{dt} x(t) + \cdots$$

$$= \cdots + b_{-1} \int_{-\infty}^{t} y(t)\, dt + b_0 y(t) + b_1 \frac{d}{dt} y(t) + \cdots \quad (3.1a)$$

where the a's and b's are *constants*. Defining the *operator* p by

$$p^m[x(t)] = \frac{d^m}{dt^m} x(t)$$

$$p^{-m}[x(t)] = \int_{-\infty}^{t} \cdots \int_{-\infty}^{t_3} \int_{-\infty}^{t_2} x(t_1)\, dt_1\, dt_2 \cdots dt_m$$

the integrodifferential equation takes the form

$$\sum_m a_m p^m[x(t)] = \sum_n b_n p^n[y(t)] \tag{3.1b}$$

where m and n are integers.   The sums in (3.1a) and (3.1b) have a finite number of terms if the two-port consists of lumped elements.

We now take the Fourier transform of (3.1b) term by term.   This is easily accomplished since, from the differentiation and integration theorems,

$$p^m[x(t)] \leftrightarrow (j\omega)^m X(f) \qquad p^{-m}[x(t)] \leftrightarrow (j\omega)^{-m} X(f)$$

The Fourier transform of (3.1b) then becomes

$$\left[\sum_m a_m(j\omega)^m\right] X(f) = \left[\sum_n b_n(j\omega)^n\right] Y(f) \tag{3.2}$$

Transformation has reduced the system equation from *operator* form to an *algebraic* equation, which is precisely the purpose of transformation.   And while (3.1) could not be rearranged, (3.2) can be written as

$$\frac{Y(f)}{X(f)} = \frac{\displaystyle\sum_m a_m(j\omega)^m}{\displaystyle\sum_n b_n(j\omega)^n}$$

the right-hand side of which depends only on the elements of the two-port and *uniquely* describes its effects in the frequency domain.   We therefore define the transfer function by

$$H(f) = \frac{\displaystyle\sum_m a_m(j\omega)^m}{\displaystyle\sum_n b_n(j\omega)^n} \tag{3.3}$$

Having made this definition, we also see that

$$\frac{Y(f)}{X(f)} = H(f)$$

Thus, given the transfer function and the spectrum of the input signal, the spectrum of the output signal is simply

$$Y(f) = H(f)X(f) \tag{3.4}$$

so the output time function is

$$y(t) = \mathcal{F}^{-1}[Y(f)] = \int_{-\infty}^{\infty} H(f)X(f)e^{j\omega t}\, df \tag{3.5}$$

For considerations of energy and energy spectral density, taking the magnitude squared of (3.4) gives

$$S_y(f) = |Y(f)|^2 = |H(f)|^2|X(f)|^2$$
$$= |H(f)|^2 S_x(f) \tag{3.6}$$

and

$$E_y = \int_{-\infty}^{\infty} S_y(f)\, df = \int_{-\infty}^{\infty} |H(f)|^2 S_x(f)\, df \tag{3.7}$$

Comparing (3.5) with (3.4), it appears that the output *spectrum* is more easily obtained than the output *time function*, which indeed is the case if $H(f)$ and $X(f)$ are known. The power of frequency-domain analysis in signal processing and transmission is based on the simple relationship of input and output spectra. Furthermore, an experienced communications engineer can often infer all he needs to know from the spectrum. (Much of the previous chapter was pointed toward making such inferences.) On the other hand, if *specific details* of the time function are to be investigated, we must return to the time domain, for which an alternate to (3.5) involving the network's impulse response may be useful. More of this later; for the present we give further attention to the transfer function.

Properties of transfer functions

In general, $H(f)$ is a complex function of frequency usually expressed in the polar form

$$H(f) = |H(f)|e^{j\theta(f)} \tag{3.8}$$

where $|H(f)|$ is called the *amplitude response* of the network and $\theta(f)$ is its *phase shift*. Combining (3.8) with (3.4) yields

$$|Y(f)| = |H(f)|\, |X(f)| \qquad \arg Y(f) = \theta(f) + \arg X(f)$$

which explains the nomenclature.

Except in a very special sense, the transfer function is not the same thing as a signal spectrum. However, it is equal to the ratio of two spectra. Since the spectra are hermitian, $H(f)$ will be hermitian and

$$H(-f) = H^*(f) \tag{3.9}$$

or

$$|H(-f)| = |H(f)| \qquad \theta(-f) = -\theta(f)$$

Thus the amplitude response is an even function of frequency, while the phase shift is odd.

The interpretation of amplitude response and phase shift is enhanced by taking as an input the phasor signal

$$x(t) = A_x e^{j(\omega_c t + \phi_x)} = A_x e^{j\phi_x} e^{j\omega_c t}$$

whose spectrum is $X(f) = A_x e^{j\phi_x} \delta(f - f_c)$. Recalling that $g(u)\delta(u - u_0) = g(u_0)\delta(u - u_0)$, the output spectrum can be written as

$$Y(f) = H(f_c) A_x e^{j\phi_x} \delta(f - f_c),$$

and so

$$y(t) = H(f_c) A_x e^{j\phi_x} e^{j\omega_c t} = A_y e^{j(\omega_c t + \phi_y)}$$

where

$$A_y = |H(f_c)| A_x \qquad \phi_y = \theta(f_c) + \phi_x$$

The amplitude of the phasor is thus multiplied by the network's amplitude response *at that frequency*, and the phase is shifted by the network's phase shift.

Had we considered the conjugate phasor, corresponding to a negative-frequency input, the amplitude response would be the same, $|H(-f_c)| = |H(f_c)|$, while the phase shift would be reversed, $\theta(-f_c) = -\theta(f_c)$.

These conclusions can be summarized in the statement: the transfer function equals the output amplitude and phase when the input is $e^{j\omega t}$. This statement may have a familiar ring; it is the basis of ac circuit analysis using complex impedance or admittance.

Steady-state response to periodic signals

From the interpretation developed above, network response to periodic signals becomes very simple if the input is written as an exponential Fourier series. For when $x(t) = \sum\limits_{n=-\infty}^{\infty} c_n e^{jn\omega_0 t}$, then, by superposition,

$$y(t) = \sum_{n=-\infty}^{\infty} H(nf_0) c_n e^{jn\omega_0 t} \tag{3.10}$$

Using Parseval's theorem, the input power is $P_x = \sum\limits_{n=-\infty}^{\infty} |c_n|^2$, and the output power is

$$P_y = \sum_{n=-\infty}^{\infty} |H(nf_0)|^2 |c_n|^2 \tag{3.11}$$

As an elementary but important example, consider the sinusoidal input $x(t) = A_x \cos(\omega_c t + \phi_x) = (A_x/2) e^{j\phi_x} e^{j\omega_c t} + (A_x/2) e^{-j\phi_x} e^{-j\omega_c t}$.

Applying (3.10),

$$y(t) = H(f_c) \frac{A_x}{2} e^{j\phi_x} e^{j\omega_c t} + H(-f_c) \frac{A_x}{2} e^{-j\phi_x} e^{-j\omega_c t}$$

$$= |H(f_c)| A_x \cos\left[\omega_c t + \theta(f_c) + \phi_x\right] \tag{3.12}$$

Similarly,

$$P_y = |H(f_c)|^2 \frac{A_x{}^2}{2}$$

Having gone through this type of calculation once, the reader henceforth should be able to write down similar results by inspection.

Impulse response

Signal analysis with nonperiodic inputs requires one more step than that for periodic inputs, namely, finding the output time function from the inverse transformation $y(t) = \int_{-\infty}^{\infty} H(f)X(f)e^{j\omega t}\,df$. An alternate method, involving only the time domain, is based on the *impulse response* of a two-port.

The impulse response $h(t)$ is simply the output time function when the input is a unit impulse applied at $t = 0$, that is, when $x(t) = \delta(t)$. Since $\delta(t) \leftrightarrow 1$, the output is

$$h(t) = \int_{-\infty}^{\infty} H(f)e^{j\omega t}\,df = \mathcal{F}^{-1}[H(f)] \tag{3.13a}$$

which follows from (3.5) upon setting $X(f) = 1$. Conversely, given $h(t)$, the transfer function is found from

$$H(f) = \mathcal{F}[h(t)] = \int_{-\infty}^{\infty} h(t)e^{-j\omega t}\,dt \tag{3.13b}$$

In short, the impulse response and transfer function form a Fourier transform pair: $h(t) \leftrightarrow H(f)$.

Note that if $x(t) = \delta(t)$, then $y(t) = h(t)$, and the *output spectrum* is just $H(f)$. In this case, and this alone, the transfer function is itself the spectrum of a signal; specifically, $H(f)$ is the spectrum of the impulse response.

Now consider an arbitrary input $x(t)$ having spectrum $X(f)$. Applying the convolution theorem of (2.40) to $Y(f) = H(f)X(f)$ yields for the output signal

$$y(t) = h(t) * x(t) = x(t) * h(t)$$

$$= \int_{-\infty}^{\infty} h(t')x(t - t')\,dt' = \int_{-\infty}^{\infty} x(t')h(t - t')\,dt' \tag{3.14}$$

Thus, the output signal can be obtained by *convolving* the input with the impulse response.

An interesting physical interpretation of (3.14) is found by writing the input as $x(t) = x(t) * \delta(t) = \int_{-\infty}^{\infty} x(t')\delta(t - t')\,dt'$ and comparing with $y(t) = \int_{-\infty}^{\infty} x(t')h(t - t')\,dt'$. As a rough approximation, we can think of the input as a *sum* of impulses, spaced in time by Δt_k, whose weights are proportional to sample values of $x(t)$, that is,

$$x(t) \sim \sum_k x(t_k)\,\Delta t_k\,\delta(t - t_k)$$

as shown in Fig. 3.2a. By definition of impulse response, the input $A\,\delta(t)$ produces the output $Ah(t)$. Then if the network is both linear and

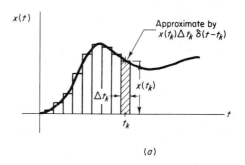

(a)

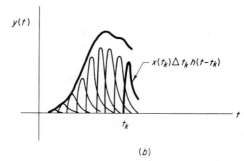

Fig. 3.2 *Graphical interpretation of input and output signals.* (a) *Input; (b) output.*

(b)

time-invariant

$$y(t) \sim \sum_k x(t_k)\,\Delta t_k\,h(t - t_k)$$

as shown in Fig. 3.2b. Letting the sum go to an integral gives $y(t) = \int_{-\infty}^{\infty} x(t')h(t - t')\,dt'$, as expected. From this interpretation, (3.14) has acquired the name *superposition integral*.

Determining transfer functions

Those who are familiar with analysis of ac circuits or linear systems should encounter little difficulty in determining transfer functions from network diagrams. There are, however, differences in notation which may have obscured this fact. To clear things up, four methods for finding $H(f)$ are given below.

From the integrodifferential equation: Going back to fundamentals, if the system integrodifferential equation is written in the form of (3.1), substitution of $j\omega = p$ and rearrangement will yield $H(f)$ as in (3.3).

From ac circuit analysis: We have shown that $H(f)$ is precisely the amplitude response and phase shift for the phasor input $e^{j\omega t}$. Thus $H(f)$ can be found using Kirchhoff's laws and complex impedance or admittance. As an example, for the two-port of Fig. 3.3, where $v_1(t)$ and $v_2(t)$ are

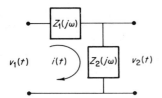

Fig. 3.3

voltages and $V_1(j\omega)$ and $V_2(j\omega)$ are phasors,

$$V_2(j\omega) = I(j\omega)Z_2(j\omega) = \frac{Z_2(j\omega)}{Z_1(j\omega) + Z_2(j\omega)} V_1(j\omega)$$

and hence

$$H(f) = \frac{V_2(j\omega)}{V_1(j\omega)} = \frac{Z_2(j\omega)}{Z_1(j\omega) + Z_2(j\omega)}$$

From time-domain analysis: In some cases, the impulse response of the network can be found with a minimum of effort; $H(f)$ is then the direct transform, $H(f) = \mathcal{F}[h(t)]$. This method has particular merit if the properties of the network are expressed in *functional* form such as $y(t) = g[x(t)]$, where g is a *linear operator*. One then sets $x(t) = \delta(t)$ and obtains

$$h(t) = g[\delta(t)] \tag{3.15}$$

The resulting transfer function $H(f) = \mathcal{F}\{g[\delta(t)]\}$ will then give the desired response to $x(t)$.

From linear-systems analysis: All other methods considered, the reader is probably most familiar with the linear-systems approach using Laplace

transform techniques.   In linear-system theory, the transfer function can
be defined by

$$H_L(s) = \mathcal{L}[h(t)] = \int_0^\infty h(t)e^{-st}\,dt = \int_0^\infty h(t)e^{-\sigma t}e^{-j\omega t}\,dt \qquad (3.16)$$

where $\mathcal{L}$ is the one-sided, or unilateral, Laplace transform and $s = \sigma + j\omega$.
In communication-system theory, the transfer function can be written as

$$H(f) = \mathcal{F}[h(t)] = \int_{-\infty}^\infty h(t)e^{-j\omega t}\,dt$$

For both cases $h(t)$ is the same physical entity, namely, the output signal
when the input is a unit impulse at $t = 0$.   For any real physical network,
the output cannot appear before the input is applied, so $h(t) = 0$ for
$t < 0$.   Comparing integrals, then, the only difference is the convergence
factor $e^{-\sigma t}$ in (3.16).   Hence, if $H_L(s)$ converges for $\sigma = 0$, then

$$H(f) = H_L(s = j2\pi f) \qquad (3.17)$$

While (3.17) is correct, we must emphasize that it applies only to
physically realizable networks and only if $H_L(s = j2\pi f)$ converges.
Furthermore, a similar relationship between the unilateral Laplace transform and the Fourier transform of a *signal* is not true in general.

Example 3.1 RC lowpass filter

Consider the network of Fig. 3.4, where $x(t)$ and $y(t)$ are voltages.   Since
this is virtually a classic example in communications, let us calculate the
transfer function from each of the above methods.

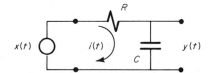

Fig. 3.4 *An RC lowpass filter.*

1. In terms of the loop current $i(t)$, $x(t) = Ri(t) + y(t)$.   But $i(t) = C\,dy(t)/dt$, so $x(t) = RC\,dy(t)/dt + y(t)$, and $X(f) = RC(j\omega)Y(f) + Y(f) = Y(f)(1 + j\omega RC)$.   Hence

$$H(f) = \frac{Y(f)}{X(f)} = (1 + j2\pi fRC)^{-1}$$

2. Using phasor analysis and the voltage-divider relation for complex impedance gives immediately

$$H(f) = \frac{V_y(j\omega)}{V_x(j\omega)} = \frac{1/j\omega C}{R + 1/j\omega C}$$

3. For the impulse response we note that an impulsive applied voltage produces a current of $\delta(t)/R$ through the capacitor. At $t = 0+$, the voltage across C is then $\dfrac{1}{C} \int_{-\infty}^{0+} \delta(t)/R \, dt = \dfrac{1}{RC}$, which thereafter decays exponentially with the time constant RC. Hence

$$h(t) = \frac{1}{RC} e^{-t/RC} \qquad \text{for } t > 0$$

4. From linear-systems analysis $H_L(s) = 1/(1 + sRC)$, and this converges at $s = j\omega$. Thus

$$H(f) = \frac{1}{1 + j2\pi fRC}$$

Taking the direct transform of $h(t)$ or the inverse transform of $H(f)$, all four methods yield the same results, namely,

$$h(t) = \frac{1}{RC} e^{-t/RC} \qquad t > 0 \tag{3.18a}$$

$$H(f) = \frac{1}{1 + j2\pi fRC} = \frac{1}{1 + j(f/f_3)}$$

$$= \frac{1}{\sqrt{1 + (f/f_3)^2}} e^{-j \arctan (f/f_3)} \tag{3.18b}$$

where $f_3 = 1/2\pi RC$. The amplitude response $|H(f)|$ and phase shift $\theta(f)$ are sketched in Fig. 3.5.

This network is called an RC lowpass filter since it passes without appreciable effect only "low" frequencies, say $|f| < f_3$, as shown by Fig.

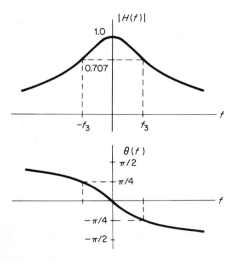

Fig. 3.5 *Amplitude response and phase shift of an RC lowpass filter.*

3.5.   Although the amplitude response has no pronounced demarcation points, f_3 is the conventional measure of the filter's frequency response, and is called the 3-db *bandwidth*.   The name stems from the fact that $|H(f_3)| = 1/\sqrt{2}$, so the power of a sinusoid at frequency f_3 would be reduced by the factor $(1/\sqrt{2})^2 = \frac{1}{2}$.   In decibels, $10 \log \frac{1}{2} = -3$ db, and the *attenuation* at f_3 is said to be 3 db.   (Note that the decibel value was calculated from a *power ratio*, a policy that will be followed throughout this text.   Henceforth then $u_{db} = 10 \log u$, where u is a power ratio.)

To illustrate how far one can go just in terms of the frequency domain, let the input signal be a rectangular pulse of duration τ.   The input spectrum is $X(f) = A\tau \operatorname{sinc} f\tau$, which we know has most of its spectral content in $|f| < 1/\tau$.   Clearly the shape of the output spectrum $Y(f) = H(f)X(f)$ will depend on the relative values of f_3 and $1/\tau$.   If $f_3 \gg 1/\tau$, the filter will "pass" all the significant frequency components of the input, and hence $Y(f) \approx X(f)$; in this case the output time function should look very much like the input.   On the other hand, if $f_3 \ll 1/\tau$, the filter "rejects" most of the input frequency components, and $y(t)$ will differ greatly from $x(t)$; the output will be *distorted* in the sense that it does not resemble the input.   More precise statements than these require going back to the time domain, but such qualitative conclusions are often all that is necessary.

Actually this network and input are so simple that the output signal can be found by inspection of the circuit diagram.   For convenience, let the pulse start at $t = 0$ rather than $t = -\tau/2$.   Then

$$y(t) = \begin{cases} 0 & t < 0 \\ A(1 - e^{-t/RC}) & 0 < t < \tau \\ A(1 - e^{-\tau/RC})e^{-(t-\tau)/RC} & t > \tau \end{cases}$$

which is sketched in Fig. 3.6 for $f_3 = 1/2\pi RC \gg 1/\tau$ and $f_3 \ll 1/\tau$.   The figure supports our previous conclusions.

Before leaving this example there is one other item to note.   When

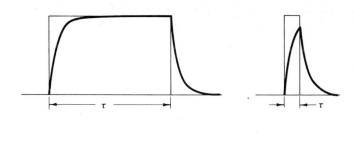

(a)   (b)

Fig. 3.6   *Pulse response of an RC lowpass filter.   (a) $\tau \gg RC$;*
(b) $\tau \ll RC$.

the pulse duration is very small compared to RC, then

$$Y(f) \approx \frac{A\tau}{1 + j(f/f_3)} = A\tau H(f) \qquad \tau \ll \frac{1}{f_3}$$

and $y(t) \approx A\tau h(t)$, which would be the response to an *impulse* of weight $A\tau$. Under such conditions, the output is controlled primarily by the two-port rather than the input, and $y(t)$ approximates the impulse response (see Fig. 3.6b).

Extrapolating to other networks and other pulse shapes, we can state the following rule of thumb: if the input spectrum is essentially constant over the bandwidth of a two-port network, the output signal is essentially the impulse response.

Example 3.2 An integrating filter

To keep things in balance, it is now appropriate to do a problem using time-domain analysis. This particular example not only illustrates an application of impulse-response techniques but also is important in its own right.

It is desired to have a network whose output is the time average of the input over the past T sec, that is,

$$y(t) = \frac{1}{T} \int_{t-T}^{t} x(t') \, dt'$$

The design problem is to find the appropriate transfer function $H(f)$.

We have $y(t)$ in functional form, so, from (3.15), the impulse response is found by setting $x(t') = \delta(t')$. By inspection, when the range of integration $[t - T, t]$ includes the impulse, $h(t) = 1/T$, otherwise $h(t) = 0$. Since $\delta(t')$ is located at $t' = 0$,

$$h(t) = \begin{cases} \dfrac{1}{T} & 0 < t < T \\ 0 & \text{otherwise} \end{cases}$$

as shown in Fig. 3.7. The desired transfer function is therefore

$$H(f) = \mathfrak{F}[h(t)] = \operatorname{sinc} fT \, e^{-j\omega T/2}$$

and the network is a *sinc-function* filter.

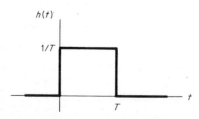

Fig. **3.7** *Impulse response of an integrating filter.*

Although the network cannot be exactly synthesized from lumped elements, it can be approximated. In fact, filters of this type are regularly employed to smooth random signals for the reduction and analysis of experimental data.

Example 3.3 A matched filter ★

Under certain conditions a filter is said to be *matched* to the signal $x(t)$ if the filter impulse response is

$$h(t) = Kx(-t + t_0)$$

where K and t_0 are constants. Since $x(-t + t_0) = x[-(t - t_0)]$, the impulse response takes the shape of $x(t)$ delayed by t_0 and reversed in time. Such filters prove to be useful for pulsed signals, notably in radar and digital data systems.

By definition, the transfer function is

$$H(f) = \int_{-\infty}^{\infty} h(t)e^{-j\omega t}\, dt = K \int_{-\infty}^{\infty} x(-t + t_0)e^{-j\omega t}\, dt$$

so setting $t' = -t + t_0$ yields

$$H(f) = K(-1) \int_{+\infty}^{-\infty} x(t')e^{-j2\pi f(-t' + t_0)}\, dt'$$
$$= Ke^{-j\omega t_0} \int_{-\infty}^{\infty} x(t')e^{-j2\pi(-f)t'}\, dt'$$

But the integral can be recognized as the Fourier transform of $x(t)$ with f replaced by $-f$. Hence

$$H(f) = Ke^{-j\omega t_0}X(-f) = Ke^{-j\omega t_0}X^*(f)$$

where we have used the hermitian property $X(-f) = X^*(f)$.

Now suppose that $x(t)$ is applied at the input; the output spectrum is then

$$Y(f) = H(f)X(f) = Ke^{-j\omega t_0}X^*(f)X(f) = K|X(f)|^2 e^{-j\omega t_0}$$
$$= KS_x(f)e^{-j\omega t_0}$$

Note that the output *spectrum* takes the shape of $S_x(f) = |X(f)|^2$, the *energy spectral density* of the input. Taking the inverse transform and recalling that $S_x(f)$ is an even function, the output signal is

$$y(t) = \int_{-\infty}^{\infty} KS_x(f)e^{-j\omega t_0}e^{+j\omega t}\, df$$
$$= K \int_{-\infty}^{\infty} S_x(f) \cos \omega(t - t_0)\, df$$

Further recalling that $S_x(f)$ is never negative, it follows that the above integral is *maximum* when $t = t_0$. Therefore, if a nonperiodic signal is applied to a matched filter, the output signal has a peak at t_0, at which point its value is

$$y(t_0) = K \int_{-\infty}^{\infty} S_x(f) \, df = K \int_{-\infty}^{\infty} |X(f)|^2 \, df$$
$$= K \int_{-\infty}^{\infty} x^2(t) \, dt = KE_x$$

The latter relations follow from Rayleigh's energy theorem, Eqs. (2.21) and (2.22), E_x being the total input energy. This maximization of the output at a specific time is the desired function of the matched filter.

To determine $y(t)$ at times other than t_0 it is more convenient to use the convolution relationship $y(t) = x(t) * h(t)$ per Eq. (3.14). Hence

$$y(t) = Kx(t) * x[-(t - t_0)]$$
$$= K \int_{-\infty}^{\infty} x(t')x(t + t' - t_0) \, dt'$$

Illustrating these results, if $x(t)$ is the rectangular pulse $x(t) = A\Pi(t/\tau)$, then the impulse response of a matched filter is the delayed rectangular pulse $h(t) = KA\Pi[(t - t_0)/\tau]$, the transfer function is $H(f) = KA\tau$ sinc $f\tau \, e^{-j\omega t_0}$, and the output signal is a *triangular* pulse of duration 2τ centered at $t = t_0$, where $y(t_0) = KE_x = KA^2$.

Output power spectra ★

Looking at the expression for output *energy* spectra, $S_y(f) = |H(f)|^2 S_x(f)$, it seems almost obvious that input and output *power* spectra should be related by

$$G_y(f) = |H(f)|^2 G_x(f) \tag{3.19}$$

which is indeed correct. However, proof of (3.19) is not at all trivial, and is given below for the sake of completeness.

We begin with the correlation of the output $R_y(\tau) = \langle y(t)y(t + \tau)\rangle$, into which we insert $y(t)$ written in terms of $x(t)$ and $h(t)$. Letting u and v be dummy variables of integration,

$$y(t) = \int_{-\infty}^{\infty} h(u)x(t - u) \, du \qquad y(t + \tau) = \int_{-\infty}^{\infty} h(v)x(t + \tau - v) \, dv$$

Substituting and averaging over time gives

$$R_y(\tau) = \int_{-\infty}^{\infty} du \int_{-\infty}^{\infty} dv \, h(u)h(v)\langle x(t - u)x(t + \tau - v)\rangle$$

But the time-average term can be identified as $R_x[(t + \tau - v) - (t - u)] = R_x(\tau - v + u)$, and thus

$$R_y(\tau) = \int_{-\infty}^{\infty} du \int_{-\infty}^{\infty} dv \, h(u)h(v)R_x(\tau - v + u) \tag{3.20}$$

We note in passing that we have obtained a time-domain relation for the input and output correlation functions.

For power spectrum, we invoke the Wiener-Kinchine theorem and transform (3.20)

$$G_y(f) = \int_{-\infty}^{\infty} R_y(\tau)e^{-j\omega\tau}\,d\tau$$
$$= \int_{-\infty}^{\infty} d\tau \int_{-\infty}^{\infty} du \int_{-\infty}^{\infty} dv\, h(u)h(v)R_x(\tau - v + u)e^{-j\omega\tau}$$

Further simplification requires a rather ingenious trick: multiply inside the integral by $e^{j\omega u}e^{j\omega v}e^{-j\omega u}e^{-j\omega v}$, which equals 1 and does not change anything. But now the integrals can be rearranged as

$$G_y(f) = \int_{-\infty}^{\infty} h(u)e^{j\omega u}\,du \int_{-\infty}^{\infty} h(v)e^{-j\omega v}\,dv$$
$$\int_{-\infty}^{\infty} R_x(\tau - v + u)e^{-j\omega(\tau - v + u)}\,d\tau$$

the first two integrals being $H(-f)$ and $H(f)$, respectively [see (3.13b)]. Finally, setting $\tau' = \tau - v + u$, the last integral is seen to be the Fourier transform of R_x, which is $G_x(f)$. Therefore

$$G_y(f) = H(-f)H(f)G_x(f) = |H(f)|^2 G_x(f)$$

which completes the proof.

3.2 SIGNAL DISTORTION IN TRANSMISSION

Whenever a signal is sent through a two-port network, the output differs from the input, at least to some degree. For purposes of signal *processing*, the differences may be intended, as in Example 3.2; but in signal *transmission* systems the goal is usually an undistorted output. But *distortionless transmission* does not necessarily imply that the output is identical to the input. Certain differences can be tolerated and not classified as distortion.

Our purpose here is to formalize the meaning of distortionless transmission and the requirements for it. With this background, the various types of distortion can be defined and their effects investigated. The emphasis will be on those aspects which are pertinent to communication systems.

Stated crudely, for distortionless transmission the output should "look like" the input. More precisely, given an input signal $x(t)$, we say that the output is undistorted if it differs from the input only by a multiplying constant and a finite time delay. Analytically if

$$y(t) = Kx(t - t_0) \tag{3.21}$$

we have distortionless transmission—of course t_0 must be positive or zero. (Why?)

The properties of a distortionless network are easily found by examining the output spectrum

$$Y(f) = \mathfrak{F}[y(t)] = Ke^{-j\omega t_0}X(f)$$

(This result follows directly from the time-delay theorem.) Now by definition of transfer function, $Y(f) = H(f)X(f)$, and so for distortionless transmission

$$H(f) = Ke^{-j\omega t_0} \qquad\qquad (3.22)$$

In words, a network giving distortionless transmission must have *constant amplitude response* and negative *linear phase shift*, that is,

$$|H(f)| = \text{const} \qquad \theta(f) = -\omega t_0 \pm m\pi$$

We have added the $\pm m\pi$ term to allow for the constant K's being positive or negative. Zero phase is of course allowable since it implies zero time delay. One more qualification can be added to (3.22): these conditions are required only over those frequencies for which the input signal has nonzero spectrum. Thus, if $x(t)$ is bandlimited in W, (3.22) need be satisfied only for $|f| < W$.

In practice (3.22) is a stringent condition which, at best, can be only approximately satisfied. Returning to the RC lowpass filter, for example, we see that the conditions are never fulfilled. Nonetheless, if the input is bandlimited in $W \ll f_3$, we can write $H(f) \approx 1e^{-j(f/f_3)}$ for $|f| \ll f_3$, implying distortionless transmission, or nearly so, with a time delay of $t_0 = (f/f_3)/\omega = 1/2\pi f_3 = RC$. Likewise, a pulse of sufficiently large duration will suffer minimal distortion since its spectrum is essentially zero for $|f| \gg 1/\tau$; this is the case shown in Fig. 3.6a.

An inescapable fact of signal transmission is that distortion will occur, though it can be minimized by proper design.[1] We should therefore be concerned with the *degree* of distortion, measured in some quantitative fashion. Unfortunately, quantitative measures prove to be rather unwieldy and impractical for engineering purposes. As an alternate approach, distortion has been classified as to type, and each type considered separately. But before discussing the various types, it must be emphasized that *distortion is distortion;* a severely distorted output will differ significantly from the input, regardless of the specific cause.

---

[1] A discussion of distortion correction by means of *equalizing networks* is beyond the scope of this book. See, for example, Bode (1945) or Guillemin (1935).

The three major classifications of distortion are:

1. Amplitude distortion: $|H(f)| \neq$ const
2. Phase (delay) distortion: $\theta(f) \neq -\omega t_0 \pm m\pi$
3. Nonlinear distortion

In the third case the network includes nonlinear elements, and its transfer function is not defined. We now examine these individually.

Amplitude distortion

Amplitude distortion is easily described in the frequency domain; it means simply that the output frequency components are not in correct proportion. Since this is caused by $|H(f)|$'s not being constant with frequency, amplitude distortion is sometimes called *frequency distortion*.

The most common forms of amplitude distortion are excess attenuation or enhancement of extreme high or low frequencies in the signal spectrum. Less common, but equally bothersome, is disproportionate response to a band of frequencies within the spectrum. While the frequency-domain description is easy, the effects in the time domain are far less obvious, save for very simple signals. For illustration, a suitably simple test signal is $x(t) = \cos \omega_0 t - \frac{1}{3} \cos 3\omega_0 t + \frac{1}{5} \cos 5\omega_0 t$ (Fig. 3.8), a rough approximation to a square wave. If the low-, middle-, or high-frequency component is attenuated by one-half, the resulting outputs are as shown in Fig. 3.9. As expected, loss of the high-frequency term reduces the "sharpness" of the waveform.

Beyond such qualitative observations, there is little more that can be said about amplitude distortion without experimental study of specific signal types. Results of such studies are usually couched in terms of required *frequency response*, i.e., the range of frequencies over which $|H(f)|$ must be constant to within a certain tolerance (say ± 1 db) so that the amplitude distortion is sufficiently small. Table 3.1 lists typical

Table 3.1 Typical frequency-response requirements

Signal type	Frequency range, Hz
Telegraph: 500 letters per minute	0–120
Voice and music:	
High fidelity	20–20,000
Average broadcast quality	100–5,000
Average telephone quality	200–3,200
Barely intelligible speech	500–2,000
Television video	60–4,200,000

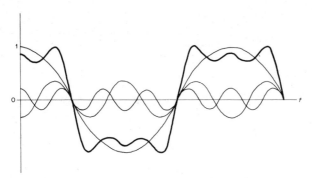

Fig. 3.8 *Test signal* $x(t) = \cos \omega_0 t - \frac{1}{3} \cos 3\omega_0 t + \frac{1}{5}$
$\cos 5\omega_0 t$.

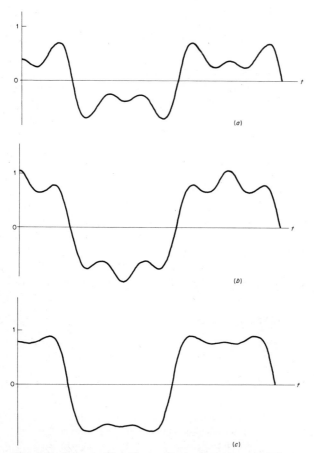

(a)

(b)

(c)

Fig. 3.9 *Test signal with amplitude distortion.* *(a) Low
frequency attenuated; (b) middle frequency attenuated;
and (c) high frequency attenuated.*

frequency-response requirements for signals commonly encountered in communication systems.

Amplitude distortion and paired echoes ★

One quantitative approach to amplitude distortion, called *paired-echo analysis*, is especially suited to those cases where $|H(f)|$ can be described in terms of *ripples* in frequency.

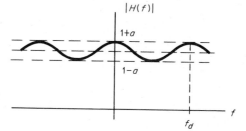

Fig. 3.10 *Amplitude response with ripples, $|H(f)| = 1 + a \cos (2\pi f/f_d)$.*

Consider the amplitude response shown in Fig. 3.10, where $|H(f)| = 1 + a \cos (2\pi f/f_d)$. Letting $t_d = 1/f_d$ and including a linear phase shift, we have

$$H(f) = (1 + a \cos \omega t_d)e^{-j\omega t_0}$$

$$= e^{-j\omega t_0} + \frac{a}{2}e^{-j\omega (t_0 - t_d)} + \frac{a}{2}e^{-j\omega (t_0 + t_d)}$$

Now for an arbitrary input having $X(f)$, the output spectrum $Y(f) = H(f)X(f)$ will be the sum of three terms. With the aid of the time-delay and linearity theorems, the output time function is

$$y(t) = x(t - t_0) + \frac{a}{2}x(t - t_0 + t_d) + \frac{a}{2}x(t - t_0 - t_d)$$

Since an undistorted output would be $x(t - t_0)$, the effect of ripple in the amplitude response is to produce a pair of echoes displaced $\pm t_d$ from the desired output (see Fig. 3.11). Note that when the ripples are widely

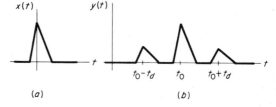

Fig. 3.11 *(a) Input signal; (b) output signal with paired echoes.*

spaced (large f_d, small t_d), the echoes are close together and may overlap the desired signal, obscuring the fact that the distortion is due to echoes.

The paired-echo technique can be extended to more complicated transfer functions by decomposing the amplitude response into a sum of ripple terms, via trigonometric Fourier series, for example, each term producing its own echo pair. The original paper by Wheeler (1939) covers the technique in detail.

Echoes can be particularly disturbing for pulsed data transmission and television video. In the former they cause errors, while in the latter they appear as ghosts. (The ghosts usually seen in television, however, are due to multiple radio-transmission paths which cause amplitude-response ripples.)

Phase shift and delay distortion

A linear phase shift yields a constant time delay for all frequency components in the signal. This, coupled with constant amplitude response, yields an undistorted output. If the phase shift is not linear, the various frequency components suffer different amounts of time delay, and the resulting distortion is termed *phase* or *delay distortion*.

For an arbitrary network phase shift $\theta(f)$, the time delay is a function of frequency, call it $t_0(f)$, and can be found by writing $\theta(f) = -\omega t_0(f) = -2\pi f t_0(f)$, so

$$t_0(f) = - \frac{\theta(f)}{2\pi f} \tag{3.23}$$

which is independent of frequency if $\theta(f)$ is linear with frequency. To illustrate (3.23), the RC lowpass filter has $\theta(f) = -\arctan(f/f_3)$ and $t_0(f) = (2\pi f)^{-1} \arctan(f/f_3) = RC(f_3/f)\arctan(f/f_3)$. Frequency components at $f = \frac{1}{2}f_3, f_3,$ and $2f_3$ would be delayed by $0.92RC, 0.79RC,$ and $0.55RC$, respectively.

A common area of confusion is *constant time delay* versus *constant phase shift*. The former is desirable and is required for distortionless transmission. The latter, in general, causes distortion. Suppose a network has the constant phase shift $\theta(f) = -2\pi\alpha$. Then each signal frequency component will be delayed by α *cycles* of its own frequency; this is the meaning of constant phase shift. But the time delays will be different, the frequency components will be scrambled in time, and distortion will result. However, the constant phase shifts $\theta(f) = 0$ and $\theta(f) = \pm m\pi$ are acceptable.

That constant phase shift does give distortion is simply illustrated by returning to the test signal of Fig. 3.8 and shifting each component by one-fourth cycle, that is, $\theta(f) = -\pi/2$. Whereas the input was roughly a square wave, the output will look like a triangular wave (Fig. 3.12). With an arbitrary nonlinear phase shift, the deterioration of waveshape can be even more severe.

One should also note from Fig. 3.12 that the *peak* excursions of the phase-shifted signal are substantially greater (by about 50 percent) than those of the input test signal.   This is not due to amplitude response, since the output amplitudes of the three frequency components are, in fact, unchanged; rather, it is because the components of the distorted signal all attain maximum or minimum values at the same time, which was not true of the input.   Conversely, had we started with Fig. 3.12 as the

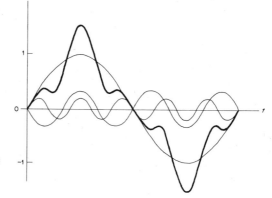

Fig. 3.12   *Test signal with phase (delay) distortion,* $\theta(f) = -\pi/2$.

test signal, a constant phase shift of $+\pi/2$ would yield Fig. 3.8 for the output waveform.   Thus we see that delay distortion *alone* can result in an increase or decrease of peak values as well as other waveshape alterations.

Clearly, delay distortion can be critical in pulse transmission, and much labor is spent *equalizing* transmission delay for digital data systems and the like.   On the other hand, the human ear is curiously insensitive to delay distortion; the waveforms of Figs. 3.8 and 3.12 would sound just about the same when driving a loudspeaker.   Thus, delay distortion is seldom of concern in voice and music transmission.

Envelope delay ★

The large majority of communication signals are transmitted in modulated form, such that all frequency components are contained in a relatively narrow band about the carrier frequency f_c.   If the modulation is of the linear type, it is convenient to distinguish between the *envelope delay* and the *carrier delay*.   (The specific meanings of *carrier frequency* and *linear modulation* will be discussed in Chap. 5; for the time being, the following will suffice.)

To define the concept of envelope, consider a modulated signal in phasor form $x_c(t) = x(t)e^{j\omega_c t}$ whose projection on the real axis is $x(t) \cos \omega_c t$, as sketched in Fig. 3.13. The dashed line connecting the peaks of the

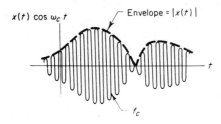

Fig. 3.13   *The envelope of $x(t) \cos \omega_c t$.*

wave is called the *envelope* and is identical to $|x(t)|$. The modulated spectrum, obtained from the translation theorem, is just $X_c(f) = X(f - f_c)$ and will look like Fig. 2.13*b* if $x(t)$ is bandlimited in W. We shall further assume that $f_c \gg W$, so the relative bandwidth of $X(f - f_c)$ is small, i.e., $2W/f_c \ll 1$.

Now suppose $x_c(t)$ is transmitted over a system having constant amplitude response but arbitrary phase shift, $H(f) = Ke^{j\theta(f)}$. Expanding the phase shift in a Taylor's series about f_c yields

$$\theta(f) = \theta_c + (f - f_c)\theta_c' + \frac{(f - f_c)^2}{2}\theta_c'' + \cdots$$

where $\theta_c = \theta(f_c)$, $\theta_c' = d\theta(f)/df \big|_{f=f_c}$, etc. For the system to be of any practical use, the phase cannot vary too rapidly over the frequency range of interest, namely, $f_c - W$ to $f_c + W$. Hence we approximate $\theta(f)$ in this region by the first two terms of the series. The output spectrum $Y_c(f) = H(f)X_c(f)$ is then

$$Y_c(f) = KX(f - f_c)e^{j\theta_c}e^{j(f-f_c)\theta_c'}$$
$$= KX(f - f_c)e^{-j2\pi(f-f_c)t_{0e}}e^{j\theta_c}$$

where $t_{0e} = -\theta_c'/2\pi$.

Combining the time-delay and frequency-translation theorems, it is easily shown that

$$x(t - t_0)e^{j\omega_c t} \leftrightarrow X(f - f_c)e^{-j2\pi(f-f_c)t_0}$$

so the output-signal phasor is

$$y_c(t) = Kx(t - t_{0e})e^{j(\omega_c t + \theta_c)}$$

corresponding to the real time function

$$Kx(t - t_{0e}) \cos (\omega_c t + \theta_c)$$

The carrier frequency has therefore been phase-shifted by $\theta_c = \theta(f_c)$, which was obvious from the start, and the carrier has been delayed in time by $t_{0c} = -\theta_c/2\pi f_c$. But the envelope of the output $|x(t - t_{0e})|$ is the input envelope delayed by $t_{0e} = -\theta'_c/2\pi$. Since this delay is independent of frequency, the envelope has *not* suffered delay distortion, at least to the extent of the approximations involved.

For simplicity, the above calculations were carried out with signals in phasor form. Had we used the real signal $x_c(t) = x(t) \cos \omega_c t = \frac{1}{2}x(t)(e^{j\omega_c t} + e^{-j\omega_c t})$, the results would have been identical, but the manipulations are more complicated.

In practice, many such modulated signals, spaced in frequency, may be transmitted over a system. It is then common practice to say that the system carrier delay is

$$t_{0c} = - \left. \frac{\theta(f)}{2\pi f} \right|_{f=f_c} \tag{3.24a}$$

while the envelope delay is

$$t_{0e} = - \left. \frac{1}{2\pi} \frac{d\theta(f)}{df} \right|_{f=f_c} \tag{3.24b}$$

Plots of $d\theta/df$ versus f are then used as an aid in evaluating the system envelope-delay characteristics. If this curve is not reasonably flat over a proposed frequency range, excessive envelope distortion will result. (The student of electromagnetic theory may recognize a parallel between carrier and envelope delay and phase and group velocity of guided waves. They are analogous concepts.)

Nonlinear distortion

A network having nonlinear elements cannot be described by a transfer function, since its integrodifferential equation is not linear. Instead, the instantaneous values of input and output are related by a curve or function, commonly called the *transfer characteristic*. Figure 3.14 is a representative transfer characteristic; the flattening out of the output for large input excursions is the familiar saturation-and-cutoff effect of vacuum-tube or transistor amplifiers. We shall consider only *memoryless* devices, for which the transfer characteristic is a complete description.

Under small-signal input conditions, it may be possible to linearize the transfer characteristic in a piecewise fashion, as shown by the thin lines in the figure. The more general approach is a polynomial approximation to the curve, of the form

$$y(t) = a_0 + a_1 x(t) + a_2 x^2(t) + a_3 x^3(t) + \cdots \tag{3.25a}$$

It is the higher powers of $x(t)$ in this equation which give rise to the non-linear distortion.

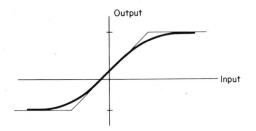

Fig. 3.14 *Transfer characteristic of a nonlinear device.*

Even though we have no transfer function, the output spectrum can be found, at least in a formal way, by transforming (3.25a). Specifically, invoking the convolution theorem,

$$Y(f) = a_0 \delta(f) + a_1 X(f) + a_2 X(f) * X(f)$$
$$+ a_3 X(f) * X(f) * X(f) + \cdots \tag{3.25b}$$

Now if $x(t)$ is bandlimited in W, the output of a linear network will contain no frequencies beyond $|f| < W$. But in the nonlinear case, we see from (3.25b) that the output includes $X(f) * X(f)$, which is band-limited in $2W$, $X(f) * X(f) * X(f)$, which is bandlimited in $3W$, etc. The nonlinearities have therefore created output frequency components which were not present in the input. Furthermore, since $X(f) * X(f)$ may contain components for $|f| < W$, this portion of the spectrum overlaps that of $X(f)$.

Using filtering techniques, the added components at $|f| > W$ can be removed, as shown in the next section. But there is no convenient way to get rid of the added components at $|f| < W$. These, in fact, constitute the nonlinear distortion.

A quantitative measure of nonlinear distortion is provided by taking a simple cosine wave, $x(t) = \cos \omega_0 t$, as the input. Inserting in (3.25a) and expanding yields

$$y(t) = \left(a_0 + \frac{a_2}{2} + \frac{3a_4}{8} + \cdots \right) + \left(a_1 + \frac{3a_3}{4} + \cdots \right) \cos \omega_0 t$$
$$+ \left(\frac{a_2}{2} + \frac{a_4}{4} + \cdots \right) \cos 2\omega_0 t + \cdots$$

Therefore, the nonlinear distortion appears as *harmonics* of the input wave. The amount of second-harmonic distortion is the ratio of the amplitude of this term to that of the fundamental, or in percent

$$\text{Second-harmonic distortion} = \frac{a_2/2 + a_4/4 + \cdots}{a_1 + 3a_3/4 + \cdots} \times 100\%$$

Higher-order harmonics are treated similarly. However, their effect is usually much less, and many can be removed entirely by filtering.

One final point here. If the input is a sum of two cosine waves, say $\cos \omega_1 t + \cos \omega_2 t$, the output will include all the harmonics of f_1 and f_2, plus cross-product terms which yield $f_2 - f_1$, $f_2 + f_1$, $f_2 - 2f_1$, etc. These sum and difference frequencies are designated as *intermodulation distortion*.

Generalizing the intermodulation effect, if $x(t) = x_1(t) + x_2(t)$, then $y(t)$ contains the *cross product* $x_1(t)x_2(t)$ (and higher-order products, which we ignore here). In the frequency domain $x_1(t)x_2(t)$ becomes $X_1(f) * X_2(f)$; and even though $X_1(f)$ and $X_2(f)$ may be separated in frequency, $X_1(f) * X_2(f)$ can overlap both of them, producing one form of *cross talk*. This aspect of nonlinear distortion is of particular concern in telephone transmission systems. On the other hand, the cross-product term is the desired result when nonlinear devices are used for modulation purposes.

3.3 FILTERS AND FILTERING

An *ideal bandpass filter* (BPF) is a fictitious network whose transfer function is shown in Fig. 3.15. The filter passes without distortion all frequencies between its *lower cutoff frequency* f_l and its *upper cutoff frequency* f_u. The frequency range $f_l < |f| < f_u$ is called the *passband;* the filter has zero response outside its passband. By convention, the filter *bandwidth* B is the width of the passband measured in positive frequency only, so $B = f_u - f_l$. Symbolically, an ideal BPF is described by the transfer function

$$H(f) = \begin{cases} Ke^{-j\omega t_0} & f_l < |f| < f_u \\ 0 & \text{otherwise} \end{cases} \tag{3.26}$$

where K is the *voltage gain* and t_0 is the time delay of the filter.

In like fashion, an *ideal lowpass filter* (LPF) has $f_l = 0$ and $B = f_u$ (Fig. 3.16a), whereas an *ideal highpass filter* (HPF) has $f_u = \infty$ and $B = \infty$ (Fig. 3.16b). Ideal lowpass and bandpass filters are sometimes called *zonal* filters, the zone being their finite passband.

Since we shall most often be involved with ideal lowpass filters, it is worthwhile to write the transfer function more succinctly as

$$H(f) = Ke^{-j\omega t_0}\Pi\left(\frac{f}{2B}\right) \tag{3.27}$$

keeping in mind that B has two meanings, bandwidth and upper cutoff frequency.

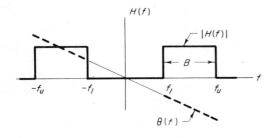

Fig. 3.15 *An ideal bandpass filter.*

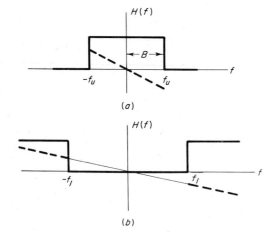

Fig. 3.16 *(a) Ideal lowpass filter; (b) ideal highpass filter.*

Example 3.4

To show an application of ideal filters in conjunction with nonlinear devices, suppose the signal $x(t) = x_m(t) + \cos \omega_c t$ is applied to a device having the transfer characteristic $y(t) = a_1 x(t) + a_2 x^2(t)$. Recalling that $\cos^2 \alpha = \frac{1}{2} + \frac{1}{2} \cos 2\alpha$, the output signal and spectrum are

$$y(t) = \frac{a_2}{2} + a_1 x_m(t) + a_2 x_m{}^2(t) + a_1 \cos \omega_c t$$

$$+ 2a_2 x_m(t) \cos \omega_c t + \frac{a_2}{2} \cos 2\omega_c t$$

$$Y(f) = \frac{a_2}{2} \delta(f) + a_1 X_m(f) + a_2 X_m(f) * X_m(f)$$

$$+ \frac{a_1}{2} [\delta(f - f_c) + \delta(f + f_c)] + a_2 [X_m(f - f_c) + X_m(f + f_c)]$$

$$+ \frac{a_2}{4} [\delta(f - 2f_c) + \delta(f + 2f_c)]$$

Though $Y(f)$ may seem rather formidable, it is easily sketched assuming $x_m(t)$ is bandlimited in $W \ll f_c$. Figure 3.17 is the positive-frequency portion of this spectrum and should be carefully studied until the source of each spectral component is grasped. Note, for example, that the impulse at $f = 0$ stems from squaring the cosine wave, an entirely different source than that of the dc impulse in (3.25b).

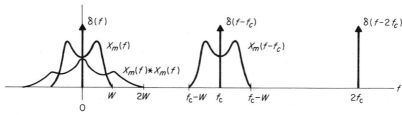

Fig. 3.17

Passing $y(t)$ through various types of ideal filters yields several possible final outputs, depending on the objectives, which, in turn, determine the filter type. Designating the filtered time function by $y_F(t)$, if the filter is an ideal LPF having bandwidth $B = W$, then by *inspection* of Fig. 3.17

$$y_F(t) = \frac{Ka_2}{2} + Ka_1 x(t - t_0)$$
$$+ \mathfrak{F}^{-1} \left\{ Ke^{-j\omega t_0} \Pi \left(\frac{f}{2W} \right) [a_2 X_m(f) * X_m(f)] \right\}$$

The last term corresponds to unfilterable second-harmonic distortion; but if $a_2 \ll a_1$, then $y_F(t) \approx Ka_1 x(t - t_0)$.

On the other hand, using an ideal BPF with $f_l = f_c - W$ and $f_u = f_c + W$ (so $B = 2W$),

$$y_F(t) = a_1 \cos \omega_c t + 2a_2 x_m(t) \cos \omega_c t$$

where we have taken unity gain ($K = 1$) and zero time delay ($t_0 = 0$) for simplicity. This result is very significant since, for as we shall see, it corresponds to amplitude modulation. The combination of the nonlinear device and ideal bandpass filter has thus produced a *modulator*.

We have by no means exhausted the possibilities here, and the reader may wish to examine the output time functions when two or more ideal filters are used in parallel, their individual outputs being recombined by summing networks.

Real and ideal filters

In advanced network theory it is shown that ideal filters cannot be physically realized (the approach is outlined in the next section). We skip the general proof here and give instead an argument based on impulse response. Consider, for example, the impulse response of an ideal LPF. By definition and using (3.27)

$$h(t) = \mathcal{F}^{-1}[H(f)] = \mathcal{F}^{-1}\left[Ke^{-j\omega t_0}\Pi\left(\frac{f}{2B}\right) \right]$$
$$= 2BK \operatorname{sinc}\left[2B(t - t_0)\right] \tag{3.28}$$

which is a sinc pulse in time and nonzero for $t < 0$. Since $h(t)$ is the response to $\delta(t)$, *the output appears before the input is applied*. Such a filter is said to be *anticipatory*, and the portion of the output appearing before the input is called a *precursor*. Without doubt, such behavior is

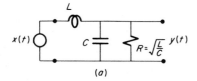

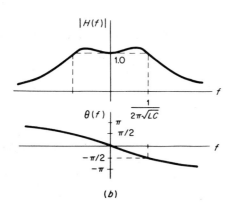

Fig. **3.18** *A practical lowpass filter.* (a) *Circuit;* (b) *transfer function.*

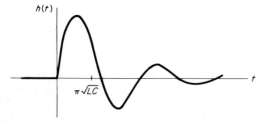

Fig. **3.19** *The impulse response of a practical lowpass filter.*

physically impossible, and hence the filter must be nonrealizable.   Similar results are found for the bandpass and highpass case.

Fictitious though they may be, ideal filters are still conceptually useful in the study of communication systems.   In practice, filters can be designed which come quite close to being ideal, at least for engineering purposes.   Figure 3.18 shows a practical lowpass filter which is both simple and has a reasonably sharp cutoff.   Like all real filters, the cutoff is not perfectly straight, so the bandwidth is conventionally specified in terms of the 3-db frequency points.   The impulse response (Fig. 3.19) is seen to be similar to a sinc pulse minus the precursors.

More complicated designs, such as Butterworth and Chebyshev filters, more closely approximate the ideal—see, for example, Guillemin (1957, chap. 14).   In fact, as the number of reactive elements increases without limit, the transfer function can be made arbitrarily close to that of an ideal filter.   But at the same time, the filter time delay increases without limit, making the filter useless.   As a side point we observe that the infinite time delay means the precursors will always appear *after* the input is applied, which must be true of a real filter.

In future work we shall often assume filters to be virtually ideal. It is thus natural to examine the effects of ideal filters on signal transmission.   In the frequency domain the effect is obvious: frequency components falling outside the filter passband are removed from the spectrum. But what this implies for the time domain may or may not be easily determined, as illustrated by the following examples.

Example 3.5

A sinc pulse $x(t) = A \operatorname{sinc} 2Wt$ is applied to an ideal LPF having bandwidth B, zero time delay, and unity gain.   The input spectrum is $X(f) = (A/2W)\Pi(f/2W)$; the output spectrum is

$$Y(f) = \frac{A}{2W} \Pi\left(\frac{f}{2B}\right) \Pi\left(\frac{f}{2W}\right) = \begin{cases} \dfrac{A}{2W} \Pi\left(\dfrac{f}{2W}\right) & B > W \\[3mm] \dfrac{A}{2W} \Pi\left(\dfrac{f}{2B}\right) & B < W \end{cases}$$

Therefore, the output time function is also a sinc pulse.   Specifically

$$y(t) = \begin{cases} A \operatorname{sinc} 2Wt & B > W \\[3mm] \dfrac{B}{W} A \operatorname{sinc} 2Bt & B < W \end{cases}$$

The case of $B > W$ is trivial; the input signal is bandlimited in W, and all its frequency components are passed by the filter.   The output is therefore undistorted.   But for $B < W$, the filter rejects high-frequency components of the input, and the output is affected in two ways. First, the maximum signal amplitude is reduced by the factor B/W. Second, and far more critical, the *duration* of the output pulse is *increased*. Since the output-pulse duration can be measured by $1/B$ (see Fig. 2.12), the smaller the filter bandwidth, the more the output pulse is stretched in time. (Again we have the reciprocal-spreading phenomenon.) Note particularly that in this case the output-pulse duration is determined by the filter bandwidth rather than by the input signal.

Example 3.6

While the previous example had the advantage of simplicity, the sinc pulse is not always a suitable pulse model. At the other extreme, let us consider a rectangular pulse as input to the same filter. Turning directly to the output time function, via (3.5),

$$y(t) = \int_{-\infty}^{\infty} H(f)X(f)e^{j\omega t}\, df = \int_{-B}^{B} A\tau \text{ sinc } f\tau\, e^{j\omega t}\, df$$

The integral can be simplified somewhat by invoking Euler's theorem and taking account of symmetry. This leads to

$$y(t) = 2A\tau \int_{0}^{B} \frac{\sin \pi f\tau}{\pi f\tau} \cos 2\pi ft\, df$$

$$= \frac{A}{\pi} \int_{0}^{B} \frac{1}{f}\left[\sin 2\pi f\left(t + \frac{\tau}{2}\right) - \sin 2\pi f\left(t - \frac{\tau}{2}\right)\right] df$$

which is still nonelementary and requires series evaluation.

However, the result can be expressed in terms of the tabulated *sine integral* Si (v) defined by

$$\text{Si }(v) = \int_{0}^{v} \frac{\sin u}{u}\, du \tag{3.29}$$

Changing integration variables, we finally obtain

$$y(t) = \frac{A}{\pi}\left\{\text{Si}\left[2\pi B\left(t + \frac{\tau}{2}\right)\right] - \text{Si}\left[2\pi B\left(t - \frac{\tau}{2}\right)\right]\right\} \tag{3.30}$$

which is shown in Fig. 3.20 for three values of the product $B\tau$. Notice the precursors caused by the ideal filter.

Despite the rather involved mathematics, the conclusions which can be drawn from Fig. 3.20 are quite similar to Example 3.5. We have said that the spectral width of a rectangular pulse is about $1/\tau$. For $B \gg 1/\tau$, the output signal is essentially undistorted; whereas for $B \ll 1/\tau$, the output pulse is stretched and has a duration which depends more on the filter bandwidth than on the input signal.

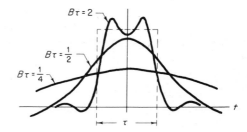

Fig. 3.20 *The pulse response of an ideal lowpass filter.*

Bandlimiting, timelimiting, and signal models ★

Ideal filters and bandlimited signals are concepts which go hand in hand. Indeed, passing an arbitrary signal through an ideal LPF produces a bandlimited signal at the output. It has been seen that the impulse response of an ideal LPF is a sinc pulse, which exists for all time. We now assert that any signal which emerges from an ideal LPF will exist for all time; stated another way, a bandlimited signal cannot be strictly time-limited. Conversely, a strictly timelimited signal cannot be bandlimited, as nicely demonstrated by the rectangular pulse, whose spectrum is a sinc function. In short, *bandlimiting and timelimiting are mutually incompatible.*

A general proof of the above assertion is difficult and will not be attempted here. Instead we offer the following plausibility argument.

Suppose $x(t)$ is strictly timelimited in the interval t_1 to t_2, as in Fig. 3.21a, so that

$$x(t) = 0 \qquad \text{for } t < t_1 \text{ and } t > t_2$$

Also shown in the figure is another time function, $z(t) = 1$ everywhere save for $|t| < \tau/2$, where $z(t) = 0$. Analytically

$$z(t) = 1 - \Pi\left(\frac{t}{\tau}\right)$$

and hence

$$Z(f) = \delta(f) - \tau \text{ sinc } f\tau$$

Now if $\tau/2 < t_1$, then multiplying $x(t)$ by $z(t)$ has no effect, since $z(t) = 1$ wherever $x(t) \neq 0$. Thus we can write

$$x(t) = x(t)z(t)$$

and applying the convolution theorem yields

$$X(f) = X(f) * Z(f) = X(f) * [\delta(f) - \tau \text{ sinc } f\tau]$$
$$= X(f) - X(f) * (\tau \text{ sinc } f\tau)$$

where we have invoked the replication property $g(u) * \delta(u) = g(u)$. Therefore, if $x(t)$ is identically zero over the finite interval $|t| < \tau/2$, its spectrum must satisfy

$$X(f) * \text{sinc} f\tau = 0$$

Contrary to what we wish to prove, assume that $x(t)$ is bandlimited in W as well as being timelimited. The frequency functions to be convolved are sketched in Fig. 3.21b, where we have taken $1/\tau > W$. Recalling the graphical interpretation of convolution, it is clear that $X(f) *$

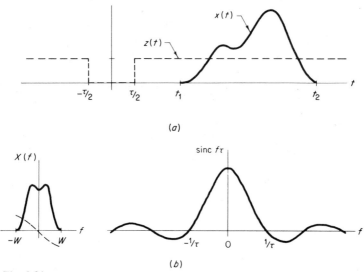

(a)

(b)

Fig. 3.21

$\text{sinc} f\tau$ cannot equal zero for all f. True, at some particular frequency f_1 the result may be zero, that is, $\int_{-\infty}^{\infty} X(f') \text{sinc} (f_1 - f')\tau \, df' = 0$; but a slight change to $f_1 + \epsilon$ will destroy the delicate balance that gave zero net area before. The assumption of $x(t)$'s being both bandlimited and timelimited leads to the contradictions

$$X(f) * \text{sinc} f\tau = 0 \qquad \text{and} \qquad X(f) * \text{sinc} f\tau \neq 0$$

and therefore the assumption is invalid. [In fact, the only signal satisfying both requirements is the null signal $x(t) = 0$ for all time.] Hence, we conclude that simultaneous bandlimiting and timelimiting is impossible, a notion which is discussed in much greater detail by Slepian, Landau, and Pollak (1961).

This observation has implications for the signal and network models used in the study of communication systems. Since a signal cannot be simultaneously bandlimited and timelimited, we should either abandon bandlimited signals (and ideal filters) or accept signal models which exist for all time. But a physically real signal *is* strictly timelimited; it starts and it stops, or is turned on and off. On the other hand, the concept of bandlimited spectra is too powerful and appealing for engineering purposes to dismiss entirely.

Fortunately, resolution of the dilemma is really not so difficult, requiring but a small compromise. Although a strictly timelimited signal is not strictly bandlimited, its spectrum can be *essentially* zero outside a certain frequency range, in the sense that the neglected frequency components contain an inconsequential portion of the total energy, e.g., consider $|f| \gg 1/\tau$ in the spectrum of a rectangular pulse. Similarly, a strictly bandlimited signal can be virtually zero outside a certain time interval, for example, sinc $2Wt \approx 0$ for $|t| \gg 1/2W$. Therefore, it is not inappropriate to speak of signals which are both bandlimited and time-limited for most practical purposes.

3.4 TRANSMISSION BANDWIDTH REQUIREMENTS

It is apparent from our earlier studies that insufficient system bandwidth can seriously impair signal transmission. We therefore wish to establish system bandwidth requirements for signal transmission, expressed in terms of the signal parameters.

The case of bandlimited signals and systems having ideal-filter characteristics is trivial; the system bandwidth should equal or exceed the signal bandwidth. This case is obvious because both the system and signal have clearly defined bandwidths.

More realistic—but more difficult to handle—is the case of non-bandlimited signals and systems having gradual roll-off characteristics, like an RC filter. Assuming that relatively distortionless transmission is desired, the system bandwidth (say, measured in terms of the 3-db frequency response) should certainly encompass the significant spectral width of the signal. Beyond this general guideline little more can be said unless the signal type and system objectives are specified.

If the goal is faithful signal reproduction, it must be recognized that some amount of distortion is inevitable. To determine whether or not the distortion is tolerable, the analysis techniques of Sec. 3.2 can be invoked. Experimental testing may also be necessary, because the definition of *tolerable* distortion is often arbitrary and subjective.

But there are many applications for which signal fidelity is not necessary. This is true, for example, in pulse transmission systems, where one is usually concerned with detecting the presence or absence of pulses, resolving two or more closely spaced pulses, or accurately determining the time location of a pulse, rather than reproducing the pulse shape per se. Relative to these objectives there are two detrimental influences, system frequency response and noise. We shall defer problems of noise to Chap. 7 and here concentrate on the bandwidth requirements for pulse transmission.

Pulse detection and resolution

Short pulses have large spectral widths, as we have seen time and again. Reversing this observation, it can be said that given a system of fixed bandwidth, there is a lower limit on the duration of pulses at the output, i.e., a *minimum output-pulse duration*. Consequently, the maximum number of distinct output pulses that can be resolved per unit time is limited by the system bandwidth.

To put the matter on a quantitative footing, consider the lowpass system transfer function of Fig. 3.22a. (We assume that any delay distortion has been sufficiently equalized over the passband, a necessity for

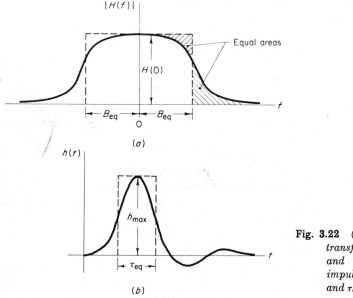

Fig. 3.22 (a) *A lowpass transfer function and B_{eq}; (b) the impulse response and τ_{eq}.*

pulse transmission.) Because the 3-db bandwidth is awkward analytically, let us define the *equivalent* bandwidth

$$B_{eq} = \frac{1}{2H(0)} \int_{-\infty}^{\infty} |H(f)| \, df$$

whose geometric interpretation is also shown in the figure. The factor of 2 reflects our convention that bandwidth is measured in terms of positive frequency only.

Now if an impulse is applied at the system input, the output will be the impulse response $h(t) = \mathfrak{F}^{-1}[H(f)]$. We are particularly interested in the *duration* of this response, since, clearly, no other input signal can produce a shorter output. [That $h(t)$ will have a pulsed shape is consistent with our assumption of negligible delay distortion.] Suppose then that $h(t)$ is as sketched in Fig. 3.22b, and we define its equivalent duration in the same manner as equivalent bandwidth, that is,

$$\tau_{eq} = \frac{1}{h_{max}} \int_{-\infty}^{\infty} h(t) \, dt$$

where h_{max} is assumed to be positive and $\int_{-\infty}^{\infty} h(t) \, dt \neq 0$. Geometrically, a rectangular pulse of amplitude h_{max} and duration τ_{eq} would have the same net *area* as $h(t)$.

Relating B_{eq} and τ_{eq}, we recall that $H(f) = \int_{-\infty}^{\infty} h(t) e^{-j\omega t} \, dt$, so $H(0) = \int_{-\infty}^{\infty} h(t) \, dt$. Furthermore, since $h(t) = \int_{-\infty}^{\infty} H(f) e^{j\omega t} \, df$, it follows that

$$|h(t)| \leq \int_{-\infty}^{\infty} |H(f)| \, df$$

This inequality stems from the fact that $|h(t)|$ cannot exceed the resultant of all the components in its spectrum, $H(f)$, added together *in phase*. Therefore, $h_{max} \leq \int_{-\infty}^{\infty} |H(f)| \, df$, and

$$\tau_{eq} = \frac{1}{h_{max}} \int_{-\infty}^{\infty} h(t) \, dt \geq \frac{H(0)}{\int_{-\infty}^{\infty} |H(f)| \, df} = \frac{1}{2B_{eq}}$$

so

$$B_{eq}\tau_{eq} \geq \frac{1}{2}$$

Of course there are other ways of defining equivalent bandwidth and duration, but all lead to similar inequalities—see, for example, Bracewell (1965, chap. 8).

As it stands, the above result seems rather academic unless B_{eq} is known. However, with reasonably selective transfer functions, the type usually employed in communication systems, the 3-db bandwidths and equivalent bandwidths are approximately equal. Therefore, as a rule of thumb, one can say that the transmission bandwidth B required for an output-pulse duration of τ_{min} or greater is

$$B \geq \frac{1}{2\tau_{min}}$$

(This does not contradict our earlier comment that the spectral width of a pulse is about $1/\tau$, since we are not concerned here with reproducing the pulse *shape*.) If the input-pulse duration is much less than $1/2B$, the input is essentially an impulse, as far as the system is concerned, and the output approximates the impulse response, but with reduced amplitude. If the input duration is much greater that $1/2B$, the output approximates the input in both duration and shape.

Going somewhat further, we can also say that the maximum number of resolved output pulses per unit time is about $1/\tau_{min} = 2B$. This is achieved using input pulses of duration less than $1/2B$ and spaced in time by $1/2B$, a topic further pursued in Chaps. 7 and 8.

We have spoken so far of lowpass or dc pulses, in the sense that their spectra are concentrated about zero frequency. Such signals are often called *baseband* signals, a term given further meaning in Chap. 5. But pulses are usually modulated before transmission, which translates the spectrum so as to be concentrated about some carrier frequency f_c. For *modulated* pulses, like the RF pulse of Example 2.6, the resolution requirements are slightly modified, the modifications being twofold. First, the transmission system must be of the *bandpass* variety centered at f_c. Second, we saw from the frequency-translation theorem that modulation doubles the spectral width of a signal; the bandwidth requirement must likewise be doubled, giving $B \geq 1/\tau_{min}$, where τ_{min} has the same interpretation as before.

Summarizing pulse-resolution requirements in terms of the minimum output-pulse duration,

$$B \geq \begin{cases} \dfrac{1}{2\tau_{min}} & \text{baseband pulses} \\[2mm] \dfrac{1}{\tau_{min}} & \text{modulated pulses} \end{cases} \tag{3.31}$$

When (3.31) is satisfied, one can recognize pulses of duration τ_{min} or longer and resolve pulses spaced by τ_{min} or more seconds.

Pulse-location requirements—rise time

Locating the time position of pulses is somewhat more subtle than detecting their presence or absence. Given a permanent record of the output waveform, an oscilloscope photograph or a strip chart, one can determine the *center* position of each pulse to almost any desired accuracy, neglecting noise. But in those communication systems where pulse position is important, it must be determined as swiftly as possible, usually in real time. Position measurements therefore commonly are based on the *leading edge* of the pulse.

For leading-edge measurement, rectangular pulses are desired, since the edge has a unique position. But realizable pulse shapes rise more gradually toward their peak value, causing the position of the leading edge to be ambiguous and its measurement less certain. The conventional rule about uncertainty is stated in terms of the *rise time*, which is defined as the interval required for the pulse to go from zero to full amplitude or from 10 to 90 percent of full amplitude. We then say that the uncertainty of the pulse-position measurement approximately equals the rise time t_r.

Unfortunately, it is difficult to formulate a general expression relating bandwidth and rise time. But analysis of typical lowpass transfer functions shows that t_r is inversely proportional to B, the proportionality constant being approximately $\frac{1}{2}$ but varying somewhat with the definition of t_r and the specific transfer function. As an approximate rule of thumb we can take for the pulse-location requirement

$$B \geq \begin{cases} \dfrac{1}{2t_r} & \text{baseband pulses} \\[2ex] \dfrac{1}{t_r} & \text{modulated pulses} \end{cases} \tag{3.32}$$

Interpreting this requirement, when (3.32) is satisfied the output pulses will have rise times no less than t_r, and the minimum location uncertainty is about t_r. Alternately, if the input rise time is greater than t_r, the output rise time will be approximately the same as the input.

Example 3.7 The Heisenberg uncertainty principle

Fourier transform theory has applications to many areas of physical science. In quantum mechanics, for example, it is used to derive the famous *Heisenberg uncertainty principle* (also called the *indeterminacy principle*). The derivation can be based on the bandwidth–rise time relationship just obtained and is given below for illustration and entertainment.

Suppose a wave packet (photon or electron) is emitted with energy such that the wave falls in the radio-frequency portion of the electromagnetic spectrum. The wave packet is then essentially a modulated pulse. A radio engineer sets out to make measurements on the wave packet to determine its frequency and time of emission. For this purpose he has a bandpass filter connected to an RF voltmeter; the filter has bandwidth B and variable center frequency. He plans to vary the center frequency until a response is seen on the meter.

Assuming the meter is sufficiently sensitive to respond to a single photon (which is somewhat questionable), what can be learned from the experiment? First, the engineer knows that the wave packet got through the filter, so its frequency must be in the range $f_c \pm B/2$. The uncertainty in the frequency measurement is thus $\Delta f = B$. Second, he records the time at which the meter showed response, which indicates emission time. From (3.32), the uncertainty in this measurement is $\Delta t = t_r \geq 1/B$. Therefore $\Delta f \, \Delta t = B t_r \geq 1$, and multiplying both sides by Planck's constant h yields $h \, \Delta f \, \Delta t \geq h$. But quantum theory says that the wavepacket energy is $E = hf$, so $h \, \Delta f = \Delta E$ is the energy uncertainty. Hence

$$\Delta E \, \Delta t \geq h$$

which is one form of Heisenberg's principle.

3.5 THE HILBERT TRANSFORM AND ITS APPLICATIONS ★

Aside from Fourier and Laplace techniques, there are many other linear transforms which have their own special areas of application in electrical engineering. And though Fourier transforms serve most of the needs of communications, the *Hilbert transform* gives a much more convenient approach to two particular topics, namely, quadrature phase shifting and single-sideband modulation. We first examine phase shifting, which leads to the Hilbert transform, and then develop the analytic-signal concept in preparation for our later study of single sideband. Finally, with the aid of Hilbert transforms, we demonstrate the nonrealizability of ideal filters.

Quadrature phase shifting

To begin with, consider a two-port network which does nothing more than shift the phase of all input frequency components by $-\pi/2$, an operation which is called *quadrature phase shifting*. Since a $-\pi/2$ phase shift is equivalent to multiplying by $e^{-j\pi/2} = -j$, and recalling that phase shift must be an odd function of frequency, the network transfer function is as sketched in Fig. 3.23 and can be written in the form

$$H(f) = -j \, \text{sgn} f \tag{3.33}$$

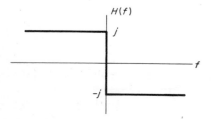

Fig. **3.23** *The transfer function of a quadrature phase shifter.*

where sgn f, read "signum f," is the *sign function*

$$\text{sgn } u = \begin{cases} 1 & u > 0 \\ -1 & u < 0 \end{cases} \tag{3.34}$$

If $X(f)$ is the input spectrum to the network, the output spectrum is $-j \text{ sgn } f X(f)$; the corresponding time function will be determined shortly. For the moment we note that if the process is repeated, the total phase shift is $-\pi$, a complete *phase reversal* of all frequency components. The output spectrum will be $(-j \text{ sgn } f)^2 X(f) = -X(f)$. The output time function is then $-x(t)$, which agrees with the notion of phase reversal.

But quadrature phase shift is not so easily interpreted in the time domain, and, for this purpose, we turn to the impulse response of the quadrature phase shifter.

Taking the inverse transform of (3.33) requires either contour integration or a limiting operation. A simpler approach to the impulse response is through the use of transform theorems and a little ingenuity. Recall the differentiation theorem, $dx(t)/dt \leftrightarrow j2\pi f X(f)$. Its dual is $-j2\pi t x(t) \leftrightarrow dX(f)/df$, and so

$$x(t) = \frac{1}{-j2\pi t} \mathcal{F}^{-1}\left[\frac{dX(f)}{df}\right]$$

With the aid of Fig. 3.23, $H(f) = -j \text{ sgn } f$ can be differentiated, yielding $dH(f)/df = -j2\delta(f)$. Therefore the corresponding impulse response is

$$h(t) = \frac{1}{-j2\pi t} \mathcal{F}^{-1}[-j2\delta(f)] = \frac{1}{\pi t}$$

Generalizing and applying duality, we have derived the transform pairs

$$\frac{1}{\pi t} \leftrightarrow -j \text{ sgn } f \tag{3.35a}$$

$$\text{sgn } t \leftrightarrow \frac{-j}{\pi f} \tag{3.35b}$$

We can now say that if $x(t)$ is the input to a quadrature phase shifter, the output time function is

$$\hat{x}(t) = x(t) * \frac{1}{\pi t} = \frac{1}{\pi} \int_{-\infty}^{\infty} \frac{x(t')}{t - t'} dt' \tag{3.36a}$$

We shall define $\hat{x}(t)$ as the *Hilbert transform* of $x(t)$, though some authors use the negative of (3.36a) corresponding to a $+\pi/2$ phase shift. $\hat{x}(t)$ is also called the *harmonic conjugate* of $x(t)$.

In like fashion, $\hat{\hat{x}}(t)$ is the Hilbert transform of the Hilbert transform of $x(t)$ and corresponds to the output of two cascaded phase shifters. But this output is known to be $-x(t)$, so $-x(t) = \hat{\hat{x}}(t)$, and we arrive at the *inverse* Hilbert transform relation

$$x(t) = -\hat{\hat{x}}(t) = -\hat{x}(t) * \frac{1}{\pi t} = -\frac{1}{\pi} \int_{-\infty}^{\infty} \frac{\hat{x}(t')}{t - t'}\, dt' \qquad (3.36b)$$

Because of the singularity at $t' = t$, the integrals in (3.36a) and (3.36b) may not be single-valued. One then takes the *Cauchy principal value* by approaching the singularity point from both sides, that is,

$$\int_{-\infty}^{\infty} = \lim_{\epsilon \to 0} \left(\int_{-\infty}^{t-\epsilon} + \int_{t+\epsilon}^{\infty} \right)$$

We also should note that Hilbert transformation is basically convolution and does not produce a change of *domain;* if x is a function of time, then $\hat{x}$ is also a function of time.

Bearing in mind the phase-shift interpretation, Hilbert transforms of sinusoidal functions are trivial. Some examples are

$$\cos \omega_0 t = \cos \left(\omega_0 t - \frac{\pi}{2} \right) = \sin \omega_0 t \qquad \widehat{\sin \omega_0 t} = - \cos \omega_0 t$$

$$\widehat{e^{j\omega_0 t}} = \sin \omega_0 t - j \cos \omega_0 t = -je^{j\omega_0 t}$$

The implications of these relations to periodic signals are left for the reader.

Summarizing the results so far, the impulse response of a quadrature phase shifter is $h(t) = 1/\pi t$. When all the frequency components of a signal $x(t)$ are phase-shifted by $-\pi/2$, the resulting time function is its Hilbert transform $\hat{x}(t)$. The spectrum of $\hat{x}(t)$ is $-j \operatorname{sgn} f X(f)$. Repeating the process produces $\hat{\hat{x}}(t) = -x(t)$.

The analytic signal

Let $x(t)$ be a real function of time and $X(f)$ its spectrum. In discussing single-sideband modulation we shall encounter the spectrum

$$X_+(f) = \begin{cases} X(f) & f > 0 \\ 0 & f < 0 \end{cases}$$

or, more compactly,

$$X_+(f) = \tfrac{1}{2}(1 + \operatorname{sgn} f) X(f) \qquad (3.37)$$

$X_+(f)$ is therefore the positive-frequency portion of the spectrum of $x(t)$. Since $X_+(f)$ is not hermitian, its inverse transform will be a complex function of time, but a very useful one.

We shall call $\varphi_x(t) = \mathcal{F}^{-1}[X_+(f)]$ the *analytic signal* derived from the real signal $x(t)$, and taking the inverse transform of (3.37) provides the time-domain relationship of $\varphi_x(t)$ to $x(t)$

$$\varphi_x(t) = \tfrac{1}{2}\{\mathcal{F}^{-1}[X(f)] + \mathcal{F}^{-1}[\operatorname{sgn} f\, X(f)]\}$$

From $(3.35a)$, $j/\pi t \leftrightarrow \operatorname{sgn} f$, so

$$\varphi_x(t) = \frac{1}{2}\left[x(t) + x(t) * \frac{j}{\pi t}\right] = \frac{1}{2}[x(t) + j\hat{x}(t)] \qquad (3.38)$$

The real and imaginary parts of the $\varphi_x(t)$ are thus a Hilbert transform pair. Interesting though this may be, the real significance of the analytic signal lies in its frequency-domain property. Reiterating for emphasis, the spectrum of $\varphi_x(t)$ is the positive-frequency portion of $X(f)$.

Taking the complex conjugate of $\varphi_x(t)$ yields $\varphi_x^*(t) = \tfrac{1}{2}[x(t) - j\hat{x}(t)]$, whose spectrum is

$$\mathcal{F}[\varphi_x^*(t)] = \tfrac{1}{2}(1 - \operatorname{sgn} f)X(f) = \begin{cases} 0 & f > 0 \\ X(f) & f < 0 \end{cases}$$

which will be designated by $X_-(f)$ for obvious reasons. Clearly, $\varphi_x(t) + \varphi_x^*(t) = x(t)$, and $X_+(f) + X_-(f) = X(f)$.

Illustrating use of the analytic signal, suppose $x(t)$ has the band-limited spectrum shown in Fig. 3.24a. Then what signal has the spectrum of Fig. 3.24b? The positive-frequency portion can be identified as $X_+(f - f_c)$, while the negative-frequency part is $X_-(f + f_c)$. Figure 3.24b is thus the sum $X_+(f - f_c) + X_-(f + f_c)$. From the frequency translation theorem, the inverse transform of $X_+(f - f_c)$ is $\varphi_x(t)e^{j\omega_c t}$,

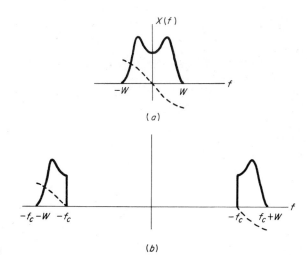

Fig. 3.24

(a)

(b)

etc., so the time function in question is

$$\varphi_x(t)e^{j\omega_c t} + \varphi_x^*(t)e^{-j\omega_c t} = \frac{1}{2}x(t)(e^{j\omega_c t} + e^{-j\omega_c t}) + \frac{j}{2}\hat{x}(t)(e^{j\omega_c t} - e^{-j\omega_c t})$$

$$= x(t)\cos\omega_c t - \hat{x}(t)\sin\omega_c t \qquad (3.39)$$

The second term can also be written as $-\hat{x}(t)\overbrace{\cos\omega_c t}$. Note that (3.39) is a real time function, as it should be since its spectrum is hermitian.

Causality and realizable networks

As a final application of the Hilbert transform, we investigate the conditions necessary for a transfer function to be physically realizable. We emphasize at the start that these will be *necessary* but not necessarily *sufficient* conditions for realizability.

What can be said in general about a real two-port network? Speaking in terms of its impulse response, the following two statements must certainly be true:

1. $h(t)$ is a real function of time, i.e., no imaginary component
2. $h(t) = 0$ for $t < 0$

As demonstrated earlier, the first condition is satisfied if the transfer function is hermitian, $H^*(f) = H(-f)$. The second condition expresses the notion of *causality;* i.e., the effect $h(t)$ cannot precede its cause $\delta(t)$, which occurs at $t = 0$. Responses satisfying this condition are said to be *causal*, and it is causality which will shortly lead us again to Hilbert transforms.

Before getting too involved we can get an indication of what is ahead from some elementary calculations. First, the transfer function can be written as

$$H(f) = |H(f)|\cos\theta(f) + j|H(f)|\sin\theta(f) = H_r(f) + jH_i(f)$$

From the hermitian condition, the real and imaginary components $H_r(f)$ and $H_i(f)$ will have even and odd symmetry, respectively. Using this symmetry and the definition of impulse response,

$$h(t) = \int_{-\infty}^{\infty} H(f)e^{j\omega t}\,df = 2\int_0^{\infty} H_r(f)\cos\omega t\,df$$

$$- 2\int_0^{\infty} H_i(f)\sin\omega t\,df$$

But from the causality condition, $h(t) = 0$ for $t < 0$, so

$$\int_0^{\infty} H_r(f)\cos\omega t\,df = \int_0^{\infty} H_i(f)\sin\omega t\,df \qquad t < 0 \qquad (3.40)$$

Since (3.40) must be satisfied if the two-port is to be realizable, one cannot arbitrarily specify both $H_r(f)$ and $H_i(f)$. The same can be said for $|H(f)|$ and $\theta(f)$.

Unfortunately, (3.40) is not very useful, the relationship being implicit rather than explicit. An explicit relation can be obtained in the following manner.

Define the *even* time function $h_e(t)$ by

$$h_e(t) = \begin{cases} \frac{1}{2}h(t) & t > 0 \\ \frac{1}{2}h(-t) & t < 0 \end{cases}$$

In this fashion an even function can be constructed from any time function, causal or not. But because $h(t)$ is *causal*, we can also write

$$h(t) = (1 + \operatorname{sgn} t)h_e(t)$$

since $h(t) = 0$ for $t < 0$ and $h(t) = 2h_e(t)$ for $t > 0$.

With the aid of $\operatorname{sgn} t \leftrightarrow -j/\pi f$ the transform of $h(t)$ is

$$H(f) = \mathfrak{F}[(1 + \operatorname{sgn} t)h_e(t)] = H_e(f) - jH_e(f) * \frac{1}{\pi f}$$

where $H_e(f) = \mathfrak{F}[h_e(t)]$. But we have previously written $H(f) = H_r(f) + jH_i(f)$. Noting that $H_e(f)$ is real, because $h_e(t)$ is even, we can equate real and imaginary parts of $H(f)$ to yield

$$H_r(f) = H_e(f) = \mathfrak{F}[h_e(t)]$$

and

$$H_i(f) = -H_e(f) * \frac{1}{\pi f} = -H_r(f) * \frac{1}{\pi f} = -\hat{H}_r(f)$$

Therefore the real and imaginary components of a realizable transfer function are a Hilbert transform pair

$$H_i(f) = -\hat{H}_r(f) = -\frac{1}{\pi}\int_{-\infty}^{\infty} \frac{H_r(f')}{f - f'}\, df'$$

$$H_r(f) = \hat{H}_i(f) = \frac{1}{\pi}\int_{-\infty}^{\infty} \frac{H_i(f')}{f - f'}\, df' \tag{3.41}$$

Since the real and imaginary parts of $H(f)$ are explicitly related by (3.41), there must also be an explicit relationship between the amplitude response $|H(f)|$ and phase shift $\theta(f)$. And although the result does not just drop out of (3.41), starting from a slightly different point, it can be shown that

$$\theta(f) = -\widehat{\ln |H(f)|} = -\frac{1}{\pi}\int_{-\infty}^{\infty} \frac{\ln |H(f')|}{f - f'}\, df' \tag{3.42a}$$

$$\ln |H(f)| = \widehat{\theta(f)} = \frac{1}{\pi}\int_{-\infty}^{\infty} \frac{\theta(f')}{f - f'}\, df' \tag{3.42b}$$

Like (3.41), (3.42) is a necessary but not always sufficient condition for realizability. A derivation of (3.42) and other realizability conditions is given by Tuttle (1958, chap. 8).

Finally we can see by inspection that all forms of ideal filters are nonrealizable. For an ideal filter, regardless of type, $|H(f)| = 0$ over a nonzero frequency range (or ranges, for a BPF). But if $|H(f)| = 0$, then $\ln |H(f)| = -\infty$, which, when integrated over a finite range, will cause (3.42a) to diverge for *all* f. Under these conditions, a realizable phase shift cannot be found, so the network itself cannot be realized physically.

PROBLEMS

3.1 Starting with Eq. (3.3), show that $H(f)$ is hermitian.

3.2 Suppose a transfer function is written in terms of its real and imaginary parts as $H(f) = H_r(f) + jH_i(f)$. Find H_r and H_i in terms of the impulse response $h(t)$. Since $h(t)$ is real, what are the symmetry properties of H_r and H_i?

3.3 Figure P 3.1 shows a number of RC lowpass filters connected in cascade. Explain why the overall transfer function is *not* $[1 + j(f/f_3)]^{-N}$. Under what conditions

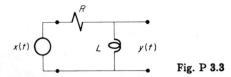

Fig. P 3.1

is the overall transfer function of N cascaded sections given by

$$H(f) = H_1(f)H_2(f) \cdots H_N(f)?$$

3.4 Find $H(f)$, $h(t)$, and f_3 for the LR filter of Fig. P 3.2.

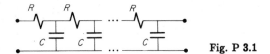

Fig. P 3.2

3.5 Find and sketch the amplitude response and phase shift for the RL filter of Fig. P 3.3.

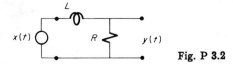

Fig. P 3.3

3.6   Calculate the transfer function of the filter shown in Fig. 3.18.

3.7   Given the transfer function $H_l(f)$ and the corresponding impulse response $h_l(t)$, find $h_b(t)$ for $H_b(f) = H_l(f - f_0) + H_l(f + f_0)$. (This is essentially the lowpass-to-bandpass transformation used in tuned-amplifier design.) If $H_l(f)$ represents an RC lowpass filter, sketch $H_b(f)$, $\theta_b(f)$, and $h_b(t)$ assuming $f_0 \gg 1/2\pi RC$.

3.8   A pentode amplifier with a parallel-tuned load has the equivalent circuit of Fig. P 3.4. Show that

$$H(f) = \frac{g_m R}{1 + jQ(f/f_0 - f_0/f)}$$

where

$$f_0 = \frac{1}{2\pi \sqrt{LC}} \quad \text{and} \quad Q = \frac{R}{2\pi f_0 L} = R\sqrt{\frac{C}{L}}$$

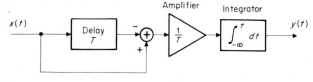

Fig. P 3.4

3.9   Assuming that the pentode amplifier of the above problem has $Q \gg 1$, justify the approximation

$$H(f) \approx \frac{g_m R}{1 + j(2Q/f_0)(f - f_0)} + \frac{g_m R}{1 + j(2Q/f_0)(f + f_0)}$$

and show that the 3-db bandwidth is $B \approx f_0/Q$. Compare these results with Prob. 3.7.

3.10   The input to an RC lowpass filter is a rectangular pulse. Find the output signal $y(t)$ by convolving $x(t)$ and $h(t)$.

3.11   The input to an RC lowpass filter is $x(t) = \text{sinc } 2Wt$. Plot the energy ratio E_y/E_x as a function of f_3/W.

3.12 RC lowpass filters are sometimes used as integrators. With the aid of the integration theorem, Eq. (2.36), determine the conditions for which this is possible.

3.13   Find the transfer function of the system in Fig. P 3.5 and compare with Example 3.2. Sketch the output for a rectangular pulse input of duration $\tau = T$, $\tau \ll T$, and $\tau \gg T$ by carrying out the system operations graphically.

Fig. P 3.5

3.14   Suppose a filter is matched in the sense of Example 3.3 to $x(t) = \text{sinc } 2Wt$. Find $H(f)$ and sketch $y(t)$. *Hint:* Find $y(t)$ by transforming $H(f)X(f)$, not by convolution.

3.15 A measure of signal distortion, from any cause, is the mean-square error $\epsilon = \langle[x(t) - Cy(t - \lambda)]^2\rangle$, where C and λ are constants to be appropriately chosen. Investigate this measure by finding C and λ when transmission is distortionless. Consider the difficulties of this measure for generalized analysis by taking $x(t) = \delta(t)$ and $y(t) = h(t)$.

3.16 The input to an RC lowpass filter is the test signal $x(t) = \cos \omega_0 t - \frac{1}{3} \cos 3\omega_0 t + \frac{1}{5} \cos 5\omega_0 t$. Taking $f_0 = \omega_0/2\pi = \frac{1}{3}f_3$, plot the output waveform and compare with Fig. 3.8.

3.17 A Butterworth lowpass filter has $|H(f)| = [1 + (f/f_3)^{2n}]^{-\frac{1}{2}}$, where n is the number of reactive components. Find n so that $|H(f)|^2$ is constant to within ± 1 db over the range $|f| \leq 0.8f_3$ and $|f| \leq 0.9f_3$.

3.18 A network having ripples in its phase shift can also produce echoes. Demonstrate this by considering $|H(f)| = 1$, $\theta(f) = -\omega t_0 + b \sin \omega t_d$, where $b \ll 1$. *Hint:* Use a series expansion of $\exp(jb \sin \omega t_d)$.

3.19 A nonlinear device has $y(t) = a_0 + a_1x(t) + a_2x^2(t) + a_3x^3(t)$. If $x(t) = \cos \omega_1 t + \cos \omega_2 t$, list all the frequency components present in $y(t)$.

3.20 Show that, as $n \to \infty$, the amplitude response of the Butterworth filter in Prob. 3.17 approaches that of an ideal LPF.

3.21 A gaussian filter has $H(f) = \exp(-af^2)$. Calculate the 3-db bandwidth and

$$B_{eq} = \frac{1}{2H(0)} \int_{-\infty}^{\infty} |H(f)|\, df$$

3.22 The finite-duration cosine wave of Prob. 2.12 is to be amplified by the pentode amplifier of Prob. 3.9, where $f_0 = f_c$. Find the necessary amplifier bandwidth B, in terms of N and Q, for reasonable recognition of the output waveform. State your assumptions.

3.23 Calculate the rise time t_r of the impulse response of a gaussian filter (Prob. 3.21) and of an ideal LPF. Express your answers in terms of the 3-db bandwidths.

3.24 Using the integral expression for the Hilbert transform, Eq. (3.36a), show that $\widehat{\cos \omega_c t} = \sin \omega_c t$.

3.25 Sketch the spectrum of $y(t) = x(t) \cos \omega_c t + \hat{x}(t) \sin \omega_c t$, taking $X(f)$ from Fig. 3.24a. Compare with Fig. 3.24b.

3.26 Using Eq. (3.40), show that when a two-port is realizable, its impulse response must satisfy

$$h(t) = 4 \int_0^{\infty} H_r(f) \cos \omega t\, df = -4 \int_0^{\infty} H_i(f) \sin \omega t\, df \qquad \text{for } t > 0$$

3.27 If $H_r(f) = \Pi(f/2W)$, find $H_i(f)$ from (3.41) aided by the graphical interpretation of convolution. Sketch the resulting $|H(f)|$.

SELECTED SUPPLEMENTARY READING

Further discussions of transfer function and impulse response in the context of signal analysis can be found in Javid and Brenner (1963), Lathi (1965), and the somewhat more mathematical Papoulis (1962). Most introductory circuit-analysis texts, e.g., Close (1966), give a treatment of transfer functions related to circuit elements and complex impedance, while the Laplace transform approach is emphasized in the study of automatic control systems, D'Azzo and Houpis (1965), for example.

For an appreciation of the difficulties encountered when the network is nonlinear or has time-varying parameters, the reader is referred to Gibson (1963, chaps. 5 and 6). The recently developed state-variable method of system analysis is an attractive alternative to transfer functions for certain problems, since it is equally applicable to nonlinear, time-varying, or discrete systems—see DeRusso, Roy, and Close (1965) or Cooper and McGillem (1967, chap. 8).

For network analysis and design for the specific tasks of electrical communication, Guillemin (1935), Albert (1940), Bode (1945), and Everitt and Anner (1956) are among the standard works. Goldman (1948, chap. 4) gives an especially well illustrated coverage of pulse responses and filter requirements.

Some of the modern techniques for network and filter synthesis are described by Guillemin (1957), Tuttle (1958), etc., while a convenient tabulation of Butterworth and Chebyshev filter designs has been compiled by Christian and Eisenmann (1966).

4

Random Signals and Noise

The signals dealt with in previous chapters were *deterministic signals;* i.e., writing an explicit time function $x(t)$ presumes that the behavior of a signal is known or determined for all time. On the other hand, one characteristic of a *random signal* is that its behavior cannot be predicted in advance, the signal is a priori unknown. In this sense, and particularly from the receiving-end viewpoint, all meaningful communication signals are random; for if the signal were known beforehand, there would be no point in transmitting it. Stated another way, a known signal conveys no information, an observation which is the keystone of information theory and which will be returned to in Chap. 8. In this chapter, however, the primary emphasis is on *unwanted* random signals, or *noise.* Because noise is present in every communication system, it must be included in any realistic description of the system performance.

Our purpose here is to develop mathematical representations, or models, of random signals for later use. And because such signals cannot be expressed as explicit functions of time, additional analytic techniques are required. In particular, when examined over a long period, a random signal may exhibit certain regularities which are best described in terms of probabilities and statistical averages. Thus, lacking an exact description, we speak instead of average values and the probability that a random signal will be in a given range at some specific time.

This chapter therefore begins with the elementary concepts of probability theory. Random variables and their statistical description are then introduced, with emphasis on the random-signal interpretation. Having obtained a statistical description, we relate it to the frequency domain by means of autocorrelation functions and the Wiener-Kinchine theorem. Finally, the properties of white noise and filtered noise are examined in some detail.

Keeping within the objectives of this text, the coverage of probability and statistics is directed specifically toward those aspects which will be used in later chapters. Furthermore, we shall rely heavily on intuitive reasoning rather than mathematical rigor. Those wishing to pursue the subject in more detail will find a wealth of material in the literature.

4.1 INTRODUCTION TO PROBABILITIES

Probability theory is a mathematical approach to random phenomena. Because random means unpredictable, learning something about such a phenomenon requires experimental observation of its manifestations. Indeed, the language of probability theory is couched in terms of experiments and their outcomes; and this is the approach we shall take to define the concept of probability.

Consider an experiment whose outcome varies from trial to trial and is a priori unknown, called a *random experiment*. Let the event A be one of the possible outcomes. Tossing a coin is such an experiment, the possible outcomes being heads and tails. The experiment is repeated n times and n_A, the number of times A occurred, is recorded. The ratio n_A/n is the *relative frequency of occurrence* of the event A for that particular set of experiments, or trials. Now if the experiment has statistical regularity, then n_A/n may approach some definite value as n becomes very large; i.e., further increase in n causes no significant change in n_A/n. Assuming that a unique limiting value exists, we define $P(A)$, the probability of the event A, as the relative frequency in the limit as $n \to \infty$,

$$P(A) = \lim_{n \to \infty} \frac{n_A}{n} \tag{4.1}$$

Thus the assumption of statistical regularity leads intuitively to our definition of probability.

From the definition, it is clear that a probability P is a positive number bounded by

$$0 \leq P \leq 1 \tag{4.2}$$

If an event is *certain* to occur on any trial, its probability is $P = 1$; if the event is *impossible*, its probability is $P = 0$. However, if an event has $P = 1$, it is not necessarily certain to happen; likewise, an event having $P = 0$ is not necessarily impossible. The reader should examine (4.1) with $n_A = n - 1$ and $n_A = 1$ for justification of these comments.

The experimentally confirmed fact that n_A/n may approach a constant as the number of trials increases is called the *empirical law of large numbers*. The crucial phrase here is "large numbers," for it is only in the context of a large number of repetitions that probability becomes meaningful. For example, if a coin is tossed 10 times, heads may come up 7 times, $n_H/n = 0.7$. But we feel certain that as n is increased, $n_H/n \to 0.5$ if the coin is honest. This feeling is expressed in the familiar statement that "the chance of heads is 50 percent."

In many cases, such as tossing an honest coin, it is possible to calculate probabilities on the basis of symmetry conditions, etc., without actually performing an experiment. In a sense, the experiments are done mentally. But if the coin is biased, by virtue of uneven wear or intentional weighting, the probabilities would require actual empirical determination. A third case arises when the coin's honesty is unknown and one is forced to estimate a probability, say of heads, without the aid of experimentation. Lacking further information, the only logical guess would be $P(H) = \frac{1}{2}$ since there is no evidence that heads is more or less likely than tails. (This rather simple-minded approach is sometimes useful, and has been given the refined name *the principle of insufficient reason*.)

Mutually exclusive events

Now consider an experiment which has m *mutually exclusive* outcomes, $A_1, A_2, \ldots, A_m$, where by mutually exclusive we mean that the occurrence of any one outcome A_j prohibits the simultaneous occurrence of any other outcome. We then ask: What is the probability of A_1 *or* A_2 on a given trial? The probability is written as $P(A_1 + A_2)$. If $n_1 + n_2$ is the number of occurrences of A_1 or A_2 in n trials, then

$$P(A_1 + A_2) = \lim_{n \to \infty} \frac{n_1 + n_2}{n} = P(A_1) + P(A_2)$$

where $P(A_1) = \lim_{n \to \infty} (n_1/n)$, etc. By direct extension to two or more outcomes

$$P(A_i + A_j + A_k + \cdots) = P(A_i) + P(A_j) + P(A_k) + \cdots \tag{4.3}$$

and if all the possible outcomes are included,

$$P(A_1 + A_2 + \cdots + A_m) = \sum_{j=1}^{m} P(A_j) = 1 \tag{4.4}$$

Equation (4.4) merely reflects the fact that, on a given trial, one of the possible outcomes must occur. If all the outcomes have equal probabilities, in which case they are said to be *equally likely*, then

$$\sum_{j=1}^{m} P(A_j) = mP(A_j) = 1$$

hence

$$P(A_j) = \frac{1}{m} \tag{4.5}$$

It must be emphasized that (4.3) to (4.5) apply only to mutually exclusive events.

On occasion we shall be interested not in the occurrence of the event A but in its *nonoccurrence*. For convenience we designate A^c, read "A complement," as the event *not A*, similar to the corresponding terms of Boolean algebra. Obviously A and A^c are mutually exclusive, and, on a given trial, one or the other must occur. Applying (4.4), $P(A) + P(A^c) = 1$, or

$$P(A^c) = 1 - P(A) \tag{4.6}$$

which is true in general.

Example 4.1

To illustrate some of these ideas, suppose two honest coins are tossed together. There are four mutually exclusive and equally likely outcomes, HH, HT, TH, TT (note that TH and HT are considered to be different outcomes). Since the number of equally likely outcomes is $m = 4$, the probability of any one is 1/4. Thus $P(HH) = 1/4$, and, using complements, $P(\text{not } HH) = 1 - 1/4 = 3/4$. Two of the outcomes give a *match*, so $P(\text{match}) = P(HH) + P(TT) = 1/2$.

Joint and conditional probabilities

In the above example, we spoke of some events which were not identical to the basic outcomes of the experiment, e.g., "match" and "not HH."

This is often a useful practice but does have a danger in that the derived events may not be mutually exclusive.   For instance, the derived events "not HH" and "match" can occur simultaneously if the outcome is TT. To deal with events which are not mutually exclusive, we must speak in terms of *joint probability*, the probability of the joint occurrence of two (or more) events.

Consider the events A and B, which may or may not occur together, and let their joint probability be $P(AB)$.   We repeat the experiment n times, where n is very large, and record n_{AB}, the number of times A and B occur together.   The joint probability is thus

$$P(AB) = \frac{n_{AB}}{n}$$

In this process A has occurred n_A times, with or without B, so n_A includes n_{AB} and $n_A \geq n_{AB}$.

It is quite possible that the occurrence of B depends in some way on the occurrence of A.   (The probability of a match in the coin-tossing experiment is obviously reduced if it is known that the outcome is not HH.)   Given that A has occurred, we should take this information into account in determining the probability that B also has occurred, for that probability may be increased or decreased.   To include such additional information we define the *conditional probability* $P(B|A)$, the probability of B given that A has occurred.   Again applying the relative-frequency definition to the n_A trials in which A occurs,

$$P(B|A) = \frac{n_{AB}}{n_A} = \frac{n_{AB}/n}{n_A/n} = \frac{P(AB)}{P(A)} \tag{4.7a}$$

Since n_{BA} is the same as n_{AB}, reversing the argument yields

$$P(A|B) = \frac{P(AB)}{P(B)} \tag{4.7b}$$

Combining (4.7a) and (4.7b) gives two relations for the joint probability

$$P(AB) = P(B|A)P(A) = P(A|B)P(B) \tag{4.8}$$

When (4.8) is written in the form

$$P(B|A) = \frac{P(B)P(A|B)}{P(A)}$$

it is known as *Bayes' theorem*, and plays an important role in statistical decision theory.

Statistically independent events

Suppose the event B is independent of A, so the occurrence of A does not influence the occurrence of B. Symbolically, $P(B|A) = P(B)$. Inserting this in (4.8) gives $P(AB) = P(B)P(A)$. But $P(A|B) = P(B|A)P(A)/P(B) = P(A)$, so if B is independent of A, then A is independent of B. In short, two events are said to be *statistically independent* if

$$P(AB) = P(A)P(B) \tag{4.9}$$

Note carefully that statistical independence is quite different from mutual exclusiveness. In fact, if A and B are mutually exclusive, then $P(AB) = 0$ by definition.

It is a simple matter to extend (4.9) to more than two events. Thus, if $A, B, C, \ldots$ are all statistically independent, the probability of their joint occurrence is

$$P(ABC \cdots) = P(A)P(B)P(C) \cdots \tag{4.10}$$

Note carefully the difference between (4.10) and (4.3).

Example 4.2

Continuing Example 4.1, let A designate the event "not HH" and B designate "match"; $P(A) = \frac{3}{4}$ and $P(B) = \frac{1}{2}$. The joint event AB is "match and not HH," which must be TT. Therefore $P(AB) = \frac{1}{4}$ and

$$P(B|A) = \frac{\frac{1}{4}}{\frac{3}{4}} = \frac{1}{3} \qquad P(A|B) = \frac{\frac{1}{4}}{\frac{1}{2}} = \frac{1}{2}$$

(Do these results agree with intuitive calculations?) The reader should also note that in this example $P(B|A) < P(B)$ while $P(B|A^c) = 1 > P(B)$.

Obviously A and B are not statistically independent, since

$$P(A)P(B) = \frac{3}{8} \neq P(AB)$$

But the coins themselves are statistically independent, if tossed in a fair manner, and (4.9) can be applied to the calculations of Example 4.1. Identifying the two coins by subscripts for clarity, the probability of the joint event H_1H_2 is

$$P(H_1H_2) = P(H_1)P(H_2) = (\tfrac{1}{2})(\tfrac{1}{2}) = \tfrac{1}{4}$$

since we have taken the coins to be honest. Similarly, when N honest coins are tossed, the probability of all heads is $\frac{1}{2}^N$.

4.2 RANDOM VARIABLES AND PROBABILITY FUNCTIONS

The various games of chance—matching coins, playing cards, shooting craps, etc.—are natural and fascinating subjects for probability calculations. But in the study of communications, we are more concerned with random experiments or processes which have *numerical* outcomes, e.g., the value of a noise voltage at some instant of time or the number of errors in a digital data message. The description and analysis of such *numerical-valued random phenomena* are greatly facilitated by speaking of random variables and probability functions, as developed below.

Consider a random process or experiment where the result of any one observation is expressed as a single numerical quantity. The totality of *all* possible outcomes forms a set of real numbers $x_1, x_2, \ldots, x_j, \ldots$, and the x_j's can be thought of as points on the real line $-\infty < x < +\infty$. The possible results of this process are thus encompassed by a one-dimensional space, i.e., a line, called the *sample space*. The various points in the sample space represent mutually exclusive events since $x_i \neq x_j$.

As illustration, suppose four coins are tossed and the outcome is taken to be the number of heads which turn up. We then have a numerical-valued random process whose possible outcomes are the integers $x_j = 0, 1, \ldots, 4$, and the sample space is the set of these integers.

A variable is then assigned to the process representing the outcome of a given observation and is called a *random variable* or *variate*. The random variable X therefore may assume any of the possible values of the process with which it is associated. In other words, X is the general symbol for *observed* values, whereas x_j represents the *possible* values.

In this sense, the random-variable concept leads to a useful shorthand notation, for if the process has statistical regularity, we can describe it by *probability functions* succinctly defined in terms of X. To illustrate, $P(X = x_k)$ is the probability that an observed value will be exactly equal to the number x_k. Likewise, the probability that an observed value is greater than x_k but less than or equal to x_l is written $P(x_k < X \leq x_l)$. We note that in the first case the probability is a function of x_k; in the second it is a function of x_k and x_l. Curiously, probability functions which are defined using the random-variable approach are *not* functions of the random variable.

Depending on the nature of the process, a random variable may assume discrete or continuous values. If, in any finite interval of the real line, X can assume only a finite number of distinct values, it is called a *discrete* random variable. The coin-tossing experiment mentioned above is an example. (This definition does not preclude an infinite number of discrete values, but in a *finite* range the number must be *finite*.) On the other hand, if X can take on *any* value in a given range of the real line, it is said to be a *continuous* random variable. A random process is often classified according to the type of variate associated with it. We shall give separate attention to both types, emphasizing the continuous, after a brief discussion of probability distribution functions.

Distribution functions

The cumulative probability distribution function, or *distribution function* for short, of a random process is defined as

$$F(x) = P(X \leq x) \tag{4.11}$$

$F(x)$ is simply the probability that an observed value will be less than or equal to the quantity x and applies to either continuous or discrete processes. Since it is a probability, the distribution function is bounded by

$$0 \leq F(x) \leq 1 \tag{4.12}$$

Other properties of $F(x)$ are obtained from the following consideration. For $x_1 < x_2$, the *events* $X \leq x_1$ and $x_1 < X \leq x_2$ are mutually exclusive, and the event $X \leq x_2$ is the combination of the two, that is, $X \leq x_2$ implies $X \leq x_1$ or $x_1 < X \leq x_2$. (Note the importance of *open* and *closed* inequalities for specifying events.) From (4.3) we have

$$P(X \leq x_2) = P(X \leq x_1) + P(x_1 < X \leq x_2)$$

or

$$P(x_1 < X \leq x_2) = F(x_2) - F(x_1) \tag{4.13}$$

Thus, if the distribution function is known for all values of x, the various probabilities associated with the random variable are completely known.

Since probabilities are nonnegative and (4.13) is a probability for any $x_1 < x_2$, we conclude that the distribution function is always a *nondecreasing function of x*. Analytically, $dF(x)/dx \geq 0$. Combining this observation with (4.12) gives

$$F(-\infty) = 0 \qquad F(+\infty) = 1$$

If we were concerned just with probabilities, the distribution function of a random process would be a complete and convenient description. However, when it comes to statistical averages, $F(x)$ is less useful than other probability functions which take account of whether the process is of the discrete or continuous type. In particular, we shall now define the frequency function for a discrete variate and the density function for a continuous variate and then show how statistical averages are obtained therefrom.

Discrete random variables—frequency functions

A discrete random variable can be described by the *frequency function*

$$P(x_j) = P(X = x_j)$$

where x_j is one of the discrete values which X may assume and $P(X = x_j)$ is its probability.   The frequency function is therefore a function of the possible values x_j; the name stems from the relative-frequency interpretation of probability, Eq. (4.1).

Relating frequency function and distribution function, we note that the x_j's are mutually exclusive, so, combining (4.3) and (4.11),

$$F(x) = \sum_{x_j \leq x} P(x_j) \tag{4.14}$$

If there are m possible outcomes all told,

$$\sum_{j=1}^{m} P(x_j) = 1$$

i.e., one of the possible outcomes must occur on a given observation.

Further inspection of (4.14) reveals several interesting points with respect to distribution functions.   Let x_{k-1} and x_k be two *adjacent* possible outcomes.   Then

$$F(x_k) - F(x_{k-1}) = \sum_{j=1}^{k} P(x_j) - \sum_{j=1}^{k-1} P(x_j) = P(x_k)$$

which should be obvious from (4.13).   Now consider $x = x_k - \epsilon$, where $x_{k-1} < x_k - \epsilon < x_k$.   Clearly

$$F(x_k - \epsilon) = \sum_{j=1}^{k-1} P(x_j)$$

so

$$\lim_{\epsilon \to 0} F(x_k - \epsilon) = \sum_{j=1}^{k-1} P(x_j) = F(x_{k-1})$$

Conversely

$$\lim_{\epsilon \to 0} F(x_k + \epsilon) = \sum_{j=1}^{k} P(x_j) = F(x_k)$$

Therefore, the distribution function of a discrete random variable has *stepwise discontinuities* at every $x = x_j$, the "height" of the step being $P(x_j)$.   Between steps $F(x)$ is constant.   At a discontinuity point the value of $F(x = x_j)$ is taken as the value to the right of $x = x_j$ (the larger value).   Briefly, $F(x)$ is *continuous to the right* and *discontinuous to the left*.

Finally, for any point *between* possible values, such as $x = a \neq x_j$, we can write (4.13) in the form

$$P(X = a) = \lim_{\epsilon \to 0} [F(a + \epsilon) - F(a - \epsilon)] = 0$$

which shows that at any point for which $F(x = a)$ is continuous, the probability of $X = a$ is zero.

Example 4.3

Suppose four honest coins are tossed and X is designated as the number of heads which turn up.   The possible outcomes and their probabilities are summarized in tabular form as

x_j	$P(x_j)$
0	$\frac{1}{16}$
1	$\frac{4}{16}$
2	$\frac{6}{16}$
3	$\frac{4}{16}$
4	$\frac{1}{16}$

A technique for obtaining these probabilities is given in the next section; for the moment we are interested in $F(x)$ as shown in Fig. 4.1.

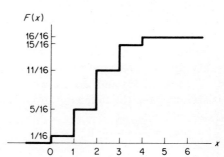

Fig. 4.1   *Distribution function for a discrete random variable.*

The figure nicely demonstrates the properties of distribution functions, both general properties and the discontinuous characteristics when the variate is discrete.   A careful examination of the figure is recommended.

Continuous random variables—density functions

To have something definite to talk about, suppose coins are tossed at a line drawn on the floor, a popular pastime among schoolboys known as penny pitching. The distance from coin to line is a random number which may be any value, positive or negative, within reasonable limits, and we can assign to the process a *continuous* random variable whose range is essentially $[-\infty, +\infty]$.

Whether or not the variate's range is bounded, in the continuum of the range there is an uncountably infinite number of possible values, and distinction between "adjacent" values clearly is impossible. (In contrast, adjacent values of a *discrete* variate are uniquely identifiable, a property implied in the definition of frequency function.) Therefore, the description of continuous variates in terms of frequency function is unsatisfactory. Indeed, even if x_1 is a possible outcome of a continuous random process, $P(X = x_1)$ is usually zero except in certain special cases. The reader who questions this assertion should consider the probability of a tossed coin's landing *exactly* 1 in. from the line.

Although $P(X = x_1) = 0$ for any x_1, the coin must land somewhere, and we can speak of such events as $X \leq x_1$ or $x_1 < X \leq x_2$ if $x_1 < x_2$. In short, the distribution-function description is still valid for continuous variates. And so is its derivative, which turns out to be a better description. Specifically, we define the probability *density function* by

$$p(x) = \frac{dF(x)}{dx} \tag{4.15a}$$

Then, recalling that $F(-\infty) = 0$,

$$F(x) = \int_{-\infty}^{x} p(x)\, dx = P(X \leq x) \tag{4.15b}$$

In words, the *area* under the density function from $-\infty$ to x is the probability that an observed value will be less than or equal to x.

From (4.15) and the known properties of $F(x)$, the density function is seen to have the following properties:

$$p(x) = \frac{dF(x)}{dx} \geq 0$$

$$\int_{-\infty}^{\infty} p(x)\, dx = 1 \tag{4.16}$$

$$P(x_1 < X \leq x_2) = F(x_2) - F(x_1) = \int_{x_1}^{x_2} p(x)\, dx \tag{4.17}$$

Thus, $p(x)$ is a *nonnegative* function whose area from x_1 to x_2 is simply the probability of X's being observed in this range. Since X must be found somewhere in $[-\infty, \infty]$, the total area under $p(x)$ is unity. Note that since $p(x)$ is never negative, if $P(x_1 < X \leq x_2) = 0$, then $p(x) = 0$ over this range.

As a special case of (4.17), let $x_1 = x - dx$ and $x_2 = x$; the integral then reduces to $p(x)\,dx$, so

$$p(x)\,dx = P(x - dx < X \le x) \tag{4.18}$$

Equation (4.18) provides another and very useful interpretation of $p(x)$ and further emphasizes its nature as a probability *density*.

The following examples illustrate the use and interpretation of density functions and underscore the implications of Eq. (4.18). Other examples will be given in Sec. 4.4.

Example 4.4

Suppose the density function for a game of penny pitching is known to be $p(x) = Ke^{-|x|}$. To evaluate the constant K we invoke the unit-area property, Eq. (4.16): $\int_{-\infty}^{\infty} p(x)\,dx = \int_0^{\infty} Ke^{-x}\,dx + \int_{-\infty}^0 Ke^x\,dx = 2K = 1$, and therefore $K = \frac{1}{2}$. The density function is sketched in Fig. 4.2.

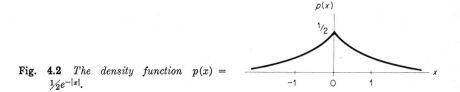

Fig. 4.2 *The density function* $p(x) = \frac{1}{2}e^{-|x|}$.

[The reader may wish to determine the corresponding distribution function $F(x)$.]

From the even symmetry and unit area of $p(x)$ it is immediately seen that $P(X \le 0) = P(X > 0) = \frac{1}{2}$, in agreement with the fact that $P(X \le 0) + P(X > 0) = 1$. The probability of a coin's landing within ± 1 unit of the line is found using (4.17), that is,

$$P(-1 < X \le 1) = \int_{-1}^{1} p(x)\,dx = 2\int_0^1 \frac{1}{2}e^{-x}\,dx$$
$$= 1 - e^{-1} = 0.632$$

But the probability of the coin's landing *exactly* one unit from the line is zero.

Now let us define a new random variable $Y = aX + b$, where a and b are constants. (Physically this is equivalent to measuring the coin's position in a new set of units and relative to a different line.) What then is the density function for Y?

Noting that each value of Y corresponds to a *unique* value of X, it follows that the probability of finding Y in some differential range dy equals the probability that X is in the corresponding range dx. Applying

Eq. (4.18) and designating the density function for Y as $q(y)$ yields

$$q(y)\,dy = p(x)\,dx \qquad \text{so} \qquad q(y) = p(x)\left|\frac{dx}{dy}\right|$$

where the absolute value is used to ensure that $q(y)$ will be nonnegative. Since $y = ax + b$, we have $x = (y - b)/a$, $dx/dy = 1/a$, and $p(x) = \tfrac{1}{2}\exp\left[-|(y - b)/a|\right]$. Therefore

$$q(y) = \frac{1}{2|a|}\,e^{|(y-b)/a|}$$

The reader can verify that $q(y)$ has unit area, etc.

Example 4.5

The previous example shows that different random variables can be assigned to the same experiment, according to the needs at hand. Pursuing this idea further, let X be as above and let

$$Z = \begin{cases} 0 & X \le 0 \\ X & X > 0 \end{cases}$$

Clearly, the density function for Z is the same as that of X when Z is positive, $q(z) = \tfrac{1}{2}e^{-z}$, $z > 0$. Furthermore, $q(z) = 0$ for $z < 0$, since $P(Z < 0) = 0$

But we must give careful attention to $z = 0$ because $P(Z = 0) = P(X \le 0) = \tfrac{1}{2}$, and a finite amount of *area* must be concentrated in the density function at $z = 0$. Noting that $P(Z = 0) = \lim_{\epsilon \to 0} \int_{-\epsilon}^{\epsilon} q(z)\,dz = \tfrac{1}{2}$ suggests that $q(z)$ includes an *impulse* of weight $\tfrac{1}{2}$ at the origin. Thus

$$q(z) = \begin{cases} 0 & z < 0 \\ \tfrac{1}{2}[\delta(z) + e^{-z}] & z \ge 0 \end{cases}$$

as shown in Fig. 4.3.

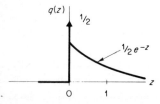

Fig. **4.3**

In the past we have been little concerned with the difference between open and closed inequalities that specify events for continuous variates. However, when impulses are present in the density function, particular care must be taken to see whether impulses are included in the range of integration. For instance, $P(Z < 0) = \int_{-\infty}^{0-} q(z)\, dz = 0$, but $P(Z \leq 0) = \int_{-\infty}^{0+} q(z)\, dz = \frac{1}{2}$.

This and the previous example have touched lightly upon the matter of *functional transformations* of random variables: given $p(x)$ and $Y = g(X)$, what is the density function for y? Such transformations may be relatively simple or quite difficult. A general treatment of the problem can be found in Davenport and Root (1958, chap. 3) and others.

Density functions for discrete random variables

On occasion it simplifies life to have a density-function representation for *discrete* variates. Recognizing that $P(x_j) = \lim_{\epsilon \to 0} [F(x_j + \epsilon) - F(x_j - \epsilon)]$,

$$\lim_{\epsilon \to 0} \int_{x_j - \epsilon}^{x_j + \epsilon} p(x)\, dx = P(x_j)$$

so

$$p(x) = \sum_j P(x_j) \delta(x - x_j) \tag{4.19}$$

a relation which can also be obtained by applying (4.15a) to (4.14). The impulse symbol is required to indicate that a finite probability, or area under $p(x)$, is concentrated at a discrete point, as in Example 4.5. It may be noted that (4.19) has a mathematical form similar to the continuous spectra of periodic signals, Eq. (2.50), but it is dangerous to push the analogy any farther.

As an exercise, the reader should sketch $p(x)$ for Example 4.3.

Joint and conditional density functions

Some random processes are characterized by two or more random variables, which may or may not be independent. (A process with n random variables has an n-dimensional sample space.) The density-function description is readily extended in terms of joint and conditional density functions. For simplicity, we consider only the two-dimensional case.

Let X and Y be continuous variates which can be observed simultaneously. Their joint density function $p(x,y)$ is developed along the lines of (4.18), that is,

$$p(x,y)\, dx\, dy = P(x - dx < X \leq x,\, y - dy < Y \leq y) \tag{4.20}$$

(Note that a comma in the argument is interpreted as "and.") Hence, the probability of finding X in the range $x_1 < X \leq x_2$ while at the same time $y_1 < Y \leq y_2$ is

$$P(x_1 < X \leq x_2, y_1 < Y \leq y_2) = \int_{y_1}^{y_2} \int_{x_1}^{x_2} p(x,y) \, dx \, dy$$

which is the two-dimensional equivalent of (4.17).

If the two variates are *statistically independent*, then, analogous to (4.9), it can be shown that

$$p(x,y) = p(x)p(y) \tag{4.21}$$

If they are not independent, the conditional density functions are needed in the form

$$p(x,y) = p(x)p(y|x) = p(y)p(x|y) \tag{4.22}$$

In (4.20) to (4.22) the functions $p(x)$, $p(y)$, etc., may be *different functions* as well as having different arguments. When it is necessary to clearly distinguish functional differences, we shall write $p_x(x)$, and so forth. For the present it is sufficient to keep in mind that different arguments generally imply different functions.

Finally, given a joint density function $p(x,y)$, the density function for X alone is obtained by the following procedure. If we are interested only in X, the value of Y does not matter, so $F(x) = P(X \leq x)$ is equivalent to $P(X \leq x, -\infty < Y \leq +\infty)$. Thus

$$F(x) = \int_{-\infty}^{\infty} \int_{-\infty}^{x} p(x,y) \, dx \, dy$$

and differentiating with respect to x yields

$$p(x) = \int_{-\infty}^{\infty} p(x,y) \, dy \tag{4.23}$$

4.3 STATISTICAL AVERAGES

For many of our purposes the probability-function representation of a random variable provides more information than is really necessary. In fact, such a complete description can prove to be an embarrassment of riches, more confusing than illuminating. Thus, we shall often find it simpler and more convenient to describe a random variable by a few characteristic *numbers* that briefly summarize the probability function. These numbers are the various statistical averages, or mean values.

Means and moments

The statistical average, or mean value, $\bar{x}$ of a random variable X is the numerical average of the values X can assume weighted by their probabilities. Essentially, $\bar{x}$ is the conventional arithmetic average of the observations in the sense that for a very large number of trials N, the *sum* of the observed values is expected to be approximately $N\bar{x}$. (Again, the large-number condition is crucial.) For this reason, mean values are also called *expected values* or *expectations*, sometimes written as $\bar{x} = E(x)$.

Consider a *discrete* random process upon which n observations have been made. If $X = x_1$ is observed n_1 times, etc., the sum of the observed values is just

$$n_1 x_1 + n_2 x_2 + \cdots + n_m x_m = \sum_j n_j x_j$$

Dividing by the number of trials and letting $n \to \infty$ yields

$$\bar{x} = \lim_{n \to \infty} \sum_j \frac{n_j}{n} x_j = \sum_j x_j P(x_j) \tag{4.24}$$

which we shall take as the definition of $\bar{x}$ for discrete variates.

As to continuous variates, we can apply (4.24) by partitioning the sample space into intervals of width Δx centered on x_j such that $P(x_j) \approx p(x_j)\,\Delta x$ and $\bar{x} \approx \sum_j x_j p(x_j)\,\Delta x$. Letting the intervals become vanishingly small, $\Delta x \to dx$, the summation approaches integration over all x, and

$$\bar{x} = \int_{-\infty}^{\infty} x p(x)\, dx \tag{4.25}$$

which *defines* the mean of a continuous random variable.

Hereafter, all statistical averages will be written in terms of density functions, recognizing that discrete variates are included through the use of impulses. To illustrate, (4.24) can be obtained from (4.25) by inserting $p(x) = \sum_j P(x_j)\delta(x - x_j)$, per Eq. (4.19).

If X is a random variable and $g(\ )$ is any analytic function, then $g(X)$ is a random variable whose density function can be found by transformation. But if we are concerned only with the mean value $\overline{g(x)}$, it is given directly by

$$\overline{g(x)} = \int_{-\infty}^{\infty} g(x) p(x)\, dx \tag{4.26}$$

When $g(X) = X^n$, its mean value $\overline{x^n}$ is designated as the nth *moment* of X. The name comes by analogy to the concept of moment in mechanics, if we think of $p(x)$ as the mass density of a one-dimensional rod having unit total mass. The first moment $\bar{x}$ is then the center of gravity, while the second moment $\overline{x^2}$ is the moment of inertia relative to the origin, etc. In statistics $\overline{x^2}$ is called the *mean-square* value and must be clearly distinguished from $\bar{x}^2$, the mean value squared. The order of operations, averaging and squaring, is not interchangeable.

Time and statistical averaging have certain similarities; in particular, both are *linear* operations. Thus, if a and b are constants, then

$$\overline{ax + b} = a\bar{x} + b$$

as is easily proved using (4.26) with $g(x) = ax + b$. While this result seems rather trivial, it leads to the not so obvious relation

$$\overline{\bar{x}x} = \bar{x}^2$$

since $\bar{x}$ is a constant.

Multiple random variables

In the case of two or more random variables, mean values are found using multiple integration and joint density functions in the general form

$$\overline{g(x,y)} = \iint\limits_{-\infty}^{\infty} g(x,y)p(x,y) \, dx \, dy$$

Thus, for example,

$$\overline{x + y} = \iint\limits_{-\infty}^{\infty} (x + y)p(x,y) \, dx \, dy$$

$$= \int_{-\infty}^{\infty} x \left[\int_{-\infty}^{\infty} p(x,y) \, dy \right] dx + \int_{-\infty}^{\infty} y \left[\int_{-\infty}^{\infty} p(x,y) \, dx \right] dy$$

But from (4.23), $\int_{-\infty}^{\infty} p(x,y) \, dy = p(x)$, etc., so

$$\overline{x + y} = \int_{-\infty}^{\infty} xp(x) \, dx + \int_{-\infty}^{\infty} yp(y) \, dy = \bar{x} + \bar{y} \qquad (4.27)$$

a most simple and satisfying result. Generalizing, it can be said that the *mean of a sum is the sum of the means*, regardless of statistical dependence.

A parallel conclusion applies to the *product* of two or more variates, providing they are *statistically independent*. Thus if X and Y are independent random variables, then

$$\overline{xy} = \bar{x}\bar{y} \qquad (4.28)$$

The proof is left as an exercise for the reader.

Variance, standard deviation, and Chebyshev's inequality

A slightly modified form of the second moment is called the *variance*, defined as

$$\sigma_x{}^2 = \overline{(x - \bar{x})^2} \qquad (4.29a)$$

where the order of operations must be carefully noted.   In the mechanics analogy σ_x^2 is the second central moment, or the moment of inertia about the center of gravity.   In statistics, the square root of the variance is called the *standard deviation* σ_x and is a measure of the *spread* of observed values.

Before demonstrating this assertion, let us obtain a more convenient expression for σ_x^2 by invoking the linearity of statistical averaging

$$\sigma_x^2 = \overline{x^2 - 2x\bar{x} + \bar{x}^2} = \overline{x^2} - 2\bar{x}^2 + \bar{x}^2$$

or

$$\sigma_x^2 = \overline{x^2} - \bar{x}^2 \tag{4.29b}$$

Hence the variance is simply the mean square minus the mean squared. Note that this result was obtained without resorting to integral forms; the bar notation, combined with (4.27), (4.28), and linearity, often simplifies such calculations.

For the interpretation of variance and standard deviation, we shall use *Chebyshev's inequality*

$$P(|X - \bar{x}| \leq k\sigma_x) \geq 1 - \frac{1}{k^2} \qquad k > 0 \tag{4.30}$$

which states that the probability of finding X within $\pm k$ standard deviations of the mean is at least $1 - 1/k^2$, *regardless of $p(x)$*.   For example, with $k = 2$, $P \geq 0.75$; with $k = 10$, $P \geq 0.99$.   Thus, in terms of σ one can specify an interval, say $\bar{x} \pm 2\sigma_x$, within which observed values are expected with high probability.   The probability of an observed value's being outside $\bar{x} \pm 2\sigma_x$ is correspondingly small.   Moreover, a *large standard deviation* implies a *large spread* of likely values for any given observation, and vice versa.   When we get to random electric signals, a more physical interpretation of σ is possible.

The proof of (4.30) is not very difficult and begins with the mean square of a random variable

$$\overline{y^2} = \int_{-\infty}^{\infty} y^2 p(y) \, dy$$

For any positive constant a it is obvious that

$$\overline{y^2} \geq \int_{-\infty}^{-a} y^2 p(y) \, dy + \int_{a}^{\infty} y^2 p(y) \, dy$$

But in the above we see that $y^2 \geq a^2$ over the range of integration, so $\int_{a}^{\infty} y^2 p(y) \, dy \geq a^2 \int_{a}^{\infty} p(y) \, dy$, etc., hence

$$\overline{y^2} \geq a^2 \left[\int_{-\infty}^{-a} p(y) \, dy + \int_{a}^{\infty} p(y) \, dy \right]$$

The term in brackets is simply the probability $P(|Y| \geq a)$, so $\overline{y^2} \geq a^2 P(|Y| \geq a)$, or

$$P(|Y| \geq a) \leq \frac{\overline{y^2}}{a^2}$$

Setting $y = x - \bar{x}$, so $\overline{y^2} = \sigma_x^2$, and letting $a = k\sigma_x$ yields

$$P(|X - \bar{x}| \geq k\sigma_x) \leq \frac{1}{k^2}$$

from which (4.30) follows directly.

Example 4.6

Illustrating some of the above techniques, let $Z = X + Y$, where X and Y are independent variates. Clearly, $\bar{z} = \bar{x} + \bar{y}$, but what is the variance σ_z^2? To answer this we first find $\overline{z^2}$ from

$$\overline{z^2} = \overline{(x + y)^2} = \overline{x^2 + 2xy + y^2} = \overline{x^2} + 2\overline{xy} + \overline{y^2}$$

which takes advantage of the independence in the form $\overline{xy} = \bar{x}\bar{y}$. Now $\overline{z^2} = \overline{x^2} + 2\bar{x}\bar{y} + \overline{y^2}$, so

$$\sigma_z^2 = \overline{z^2} - \bar{z}^2 = \overline{x^2} - \bar{x}^2 + \overline{y^2} - \bar{y}^2$$
$$= \sigma_x^2 + \sigma_y^2$$

and

$$\sigma_z = \sqrt{\sigma_x^2 + \sigma_y^2}$$

In general, the variance of the sum is the sum of the variances if the components are statistically independent.

4.4 USEFUL PROBABILITY DISTRIBUTIONS

There are numerous probability laws, or distributions, both discrete and continuous, just a few of which are given below. Fortunately, the few we shall discuss cover most of the cases encountered in our later work.

Binomial distribution

Consider an experiment having only two possible outcomes A and $B = A^c$, tossing a single coin for instance. Let the probabilities be $P(A) = p$ and $P(B) = q = 1 - p$. If the experiment is repeated m times and we total the number of times A occurred, we have generated a numerical-valued discrete random process. Assigning the random variable N as the number of occurrences of A, then N may assume the $m + 1$ discrete values $0, 1, \ldots, m$.

The corresponding frequency function $P(N = n)$ is called the *binomial distribution*, written

$$P_m(n) = \binom{m}{n} p^n q^{m-n} \qquad (4.31)$$

where $\binom{m}{n}$ is the *binomial coefficient*

$$\binom{m}{n} = \frac{m!}{n!(m-n)!}$$

Derivation of (4.31) can be found in any standard probability text; it is omitted here for the sake of brevity.

The symbol $\binom{m}{n}$, read "m choose n," is the coefficient of the $(n+1)$st term in the expansion of $(a+b)^m$. It is readily seen that $\binom{m}{m-n} = \binom{m}{n}$ and $\binom{m}{1} = m$. Some trouble arises with $\binom{m}{0}$, but it is resolved by defining $0! = 1$, so $\binom{m}{0} = \binom{m}{m} = 1$ in agreement with the expansion coefficients. When m and n are not too large, $\binom{m}{n}$ can be found from *Pascal's triangle* or from tables. For brute-force evaluation, the following forms are convenient:

$$\binom{m}{n} = \begin{cases} \dfrac{m(m-1)(m-2) \cdots (m-n+1)}{n!} & m > 2n \\[2ex] \dfrac{m(m-1)(m-2) \cdots (n+1)}{(m-n)!} & m < 2n \end{cases}$$

As an exercise, the reader may wish to check the values of $P(x_j)$ given in Example 4.3, where $m = 4$ and $p = q = \frac{1}{2}$.

Inserting (4.31) into (4.24), etc., one obtains after some labor

$$\bar{n} = mp \qquad \sigma_n{}^2 = mpq = \bar{n}(1 - p)$$

for the mean and variance of the binomial distribution. Since $\sigma_n/\bar{n}$ is proportional to $1/\sqrt{m}$, the *relative spread* decreases with increasing m.

While tossing coins is not of particular interest to us, error occurrence in digital data transmission definitely is. And if the error probability per digit is the same for all digits, say p, then $P_m(n)$ gives the probability of n errors in a message of m digits, etc. Likewise

$$F_m(k) = \sum_{n=0}^{k} P_m(n)$$

is the probability of k or fewer errors in m digits.

But equally important for quick calculations, the expected number of errors is in the range of about $\bar{n} \pm 2\sigma_n = mp \pm 2\sqrt{mp(1-p)}$. For example, if 10,000 digits are sent with $p = 0.01$, then $\bar{n} = 100$ and $\sigma_n{}^2 = 99$. We would therefore expect roughly 80 to 120 errors in the message. Incidentally, this is not a very good error rate.

Poisson distribution

When m is very large and p very small, the binomial distribution becomes awkward to handle. But if the product mp remains finite, (4.31) can be approximated by the *Poisson distribution*

$$P(n) = e^{-\bar{n}}\frac{\bar{n}^n}{n!} \tag{4.32}$$

where, as before,

$$\bar{n} = mp \qquad \sigma_n{}^2 = \bar{n}q \approx \bar{n}$$

Neither m nor p appears in (4.32), for they are absorbed in $\bar{n}$.

The Poisson distribution is obviously pertinent to the transmission of many data digits with very small error probabilities. Other applications include the description of shot noise in electron tubes and radioactive decay.

For the latter two applications, the Poisson distribution is exact rather than approximate. In particular, consider an event whose occurrence probability in the very small time interval ΔT is proportional to ΔT, say $P = \mu\,\Delta T$. Then, if multiple occurrences are statistically independent, the probability of n events in time T is

$$P(n) = e^{-\mu T}\frac{(\mu T)^n}{n!}$$

which is Poisson with $\bar{n} = \mu T$.

Uniform distribution

When a continuous variate is equally likely to take on any value in the range $[a,b]$, it is said to be *uniformly distributed*. The density function, shown in Fig. 4.4, is

$$p(x) = \begin{cases} \dfrac{1}{b-a} & a \le x \le b \\ 0 & \text{otherwise} \end{cases} \tag{4.33}$$

Since this function is easily integrated, we shall use it to illustrate some of the earlier work. First, checking to see whether $P(-\infty <$

$X \leq +\infty) = 1$, in consequence of (4.16),

$$\int_{-\infty}^{\infty} p(x)\, dx = \int_a^b \frac{1}{b-a}\, dx = \frac{b-a}{b-a} = 1$$

The mean, mean square, and variance are

$$\bar{x} = \int_{-\infty}^{\infty} \frac{x}{b-a}\, dx = \frac{b^2 - a^2}{2(b-a)} = \frac{1}{2}(b+a)$$

$$\overline{x^2} = \int_{-\infty}^{\infty} \frac{x^2}{b-a}\, dx = \frac{b^3 - a^3}{3(b-a)} = \frac{1}{3}(b^2 + ab + a^2)$$

$$\sigma_x{}^2 = \overline{x^2} - \bar{x}^2 = \frac{1}{12}(b-a)^2$$

The mean falls just where one would expect it, right in the middle of the range.   On the other hand, the variance depends only on the *size* of the range, $b - a$, and not on its absolute position.

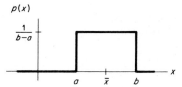

Fig. 4.4   *Uniform density function.*

Applying Chebyshev's inequality to the uniform distribution is possible but not very meaningful.   By definition, we can expect *any* value in the range $[a,b]$ on a given observation; all values are equally likely.   This conclusion is confirmed by the fact that $\bar{x} \pm 2\sigma_x$ encompasses the entire range, so $P(|X - \bar{x}| \leq 2\sigma_x) = 1$.

Example 4.7

As we shall develop later, certain noise waveforms can be treated as sinusoids with random phase angles, the phase angle θ having a uniform distribution over $[-\pi,\pi]$.   Hence $p(\theta) = 1/2\pi$ for $-\pi \leq \theta \leq \pi$.

Now suppose that θ is a random angle, uniformly distributed, and we observe $X = \cos \Theta$.   The density function $q(x)$ is found as follows. First, $-1 \leq \cos \theta \leq 1$, so $q(x) = 0$ for $|x| > 1$.   Second, the transformation is not one to one; $\cos \theta = \cos(-\theta)$, and there are two values of Θ for each value of X.   Noting that positive and negative angles have equal probabilities, it follows that

$$q(x) = 2p(\theta) \left| \frac{d\theta}{dx} \right| \qquad |x| \leq 1$$

Inverting the transformation, $\theta = \arccos x$, and

$$|d\theta/dx| = (1 - x^2)^{-\frac{1}{2}}$$

which yields

$$q(x) = \begin{cases} \dfrac{1}{\pi \sqrt{1 - x^2}} & |x| \leq 1 \\ 0 & |x| > 1 \end{cases}$$

The statistical averages of X are then

$$\bar{x} = 0 \qquad \sigma_x{}^2 = \overline{x^2} = \tfrac{1}{2}$$

reminiscent of the *time* averages of $x(t) = \cos \omega t$, for which $\langle x(t) \rangle = 0$ and $\langle x^2(t) \rangle = \tfrac{1}{2}$.

Gaussian distribution

The most widely known probability function is the *gaussian*, or *normal*, *distribution*. This function occurs in so many applications of statistical analysis because of a remarkable phenomenon called the *central-limit theorem*, which, for our purposes, can be stated as follows:

> **Theorem:** If $X_1, X_2, \ldots, X_M$ are independent random variables and the random variable Y is assigned as their sum, $Y = \sum\limits_{k=1}^{M} X_k$, then as M becomes very large, the distribution of Y approaches a gaussian distribution. The result is independent of the distributions of the individual components as long as the contribution of each is small compared to the sum.

We shall not attempt to prove the theorem, nor shall we use it explicitly in later work, for great care must be exercised in its application. (It has been said that the central-limit theorem is a dangerous tool in the hands of amateurs.) However, we can observe that electrical noise is often the summation of effects of a large number of randomly moving electrons and therefore will have a gaussian distribution. Similarly, random errors in experimental measurements are due to many irregular and fluctuating causes; such errors will cause measured values to be random with a gaussian distribution about the true value.

The density function of a gaussian-distributed variate is the familiar bell-shaped curve

$$p(x) = \frac{1}{\sqrt{2\pi}\,\sigma}\, e^{-(x-m)^2/2\sigma^2} \tag{4.34}$$

As seen from Fig. 4.5, $p(x)$ describes a continuous random variable that may take on any value in $[-\infty, +\infty]$ but is most likely to be found near $x = m$.   In fact, as the reader can verify, m is the mean value $\bar{x}$.   And, as the notation implies, the standard deviation is σ.

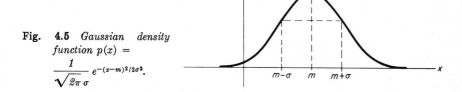

Fig.   4.5 *Gaussian   density function $p(x) =$*
$$\frac{1}{\sqrt{2\pi}\,\sigma}\,e^{-(x-m)^2/2\sigma^2}.$$

One should also note the even symmetry of $p(x)$ about $x = m$.   An immediate consequence is

$$P(X \le m) = P(X > m) = \tfrac{1}{2} \tag{4.35}$$

Suppose then that a random-noise voltage has a gaussian distribution and we wish to find the probability that an observed value will fall in the range $m \pm k\sigma$.   (Remember that Chebyshev's inequality gives only a lower bound for this probability.)   To simplify a bit, the symmetry of $p(x)$ implies $P(|X - m| \le k\sigma) = 2P(m < X \le m + k\sigma)$.   Setting up the latter as an integral,

$$P(m < X \le m + k\sigma) = \frac{1}{\sqrt{2\pi}\,\sigma} \int_{m}^{m+k\sigma} e^{-(x-m)^2/2\sigma^2}\, dx$$

which, unfortunately, cannot be solved in closed form and requires numerical evaluation.   For this purpose, the function $\Phi(k)$ is defined by

$$\Phi(k) = \frac{1}{\sqrt{2\pi}} \int_{0}^{k} e^{-u^2/2}\, du \tag{4.36}$$

and is commonly tabulated as the area under the *normal curve of error* (see Table D for a condensed listing).   Setting $u = (x - m)/\sigma$, it is seen that

$$P(m < X \le m + k\sigma) = P(m - k\sigma < X \le m) = \Phi(k) \tag{4.37a}$$

and

$$P(|X - m| \le k\sigma) = 2\Phi(k) \tag{4.37b}$$

We can now compare some exact values of $P(|X - m| \leq k\sigma)$ with Chebyshev's inequality, $P(|X - \bar{x}| \leq k\sigma) \geq 1 - 1/k^2$. Clearly, the

k	$2\Phi(k)$	$1 - \dfrac{1}{k^2}$
1.0	0.68	0.00
1.5	0.76	0.56
2.0	0.96	0.75
2.5	0.99	0.84

lower bound $1 - 1/k^2$ is a conservative estimate of $2\Phi(k)$. Moreover, the expected range of observed values is somewhat less than $m \pm 2\sigma$; a more realistic and frequently used range is $m \pm \sigma$, for which the probability is 0.68.

At the other extreme, we consider the probability $P(X > m + k\sigma)$. Again symmetry and the area interpretation of the density function are helpful, for we can divide $p(x)$ into three areas, as shown in Fig. 4.6a. The probability in question is the area under the tail of the curve, and, recalling that the total area is unity,

$$P(X > m + k\sigma) = 1 - \frac{1}{2} - \Phi(k)$$

so

$$P(X > m + k\sigma) = P(X \leq m - k\sigma) = \frac{1}{2} - \Phi(k) \qquad (4.38)$$

When $k \gg 1$, the above probability is very small, so small as to make most tables of $\Phi(k)$ useless for the calculation. In this case, there is an analytic approximation to (4.38) obtained from (4.36) by integration by parts. The approximation is

$$\frac{1}{2} - \Phi(k) \approx \frac{1}{\sqrt{2\pi}\,k}\, e^{-k^2/2} \qquad k \gg 1 \qquad (4.39)$$

Figure 4.6b is a plot of $\frac{1}{2} - \Phi(k)$ showing the exact value and the above approximation. The latter is seen to be quite accurate for $k \geq 3$.

Rayleigh distribution

Suppose two *independent* variates X and Y are gaussian-distributed, both with zero mean and the same variance σ^2. Their joint density function is

$$p(x,y) = p(x)p(y) = \frac{1}{2\pi\sigma^2}\, e^{-(x^2+y^2)/2\sigma^2} \qquad (4.40)$$

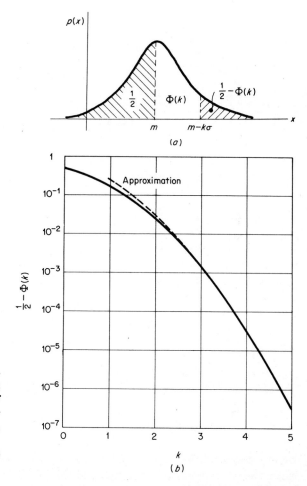

Fig. 4.6 *Gaussian probabilities. (a) Area interpretation of* $\Phi(k)$*; (b) area under the gaussian tail,* $P(X > m + k\sigma) = \frac{1}{2} - \Phi(k)$*, as a function of k.*

It is helpful to think of (X,Y) as the *cartesian coordinates* of a *random point* in the xy plane. Similarly, $p(x,y)$ is a bell-shaped solid on the xy plane, while the probability of finding the point in a specified area on the plane equals the *volume* of the solid over that area.

The point (X,Y) is just as well specified by random *polar coordinates* (R,Θ), where $R^2 = X^2 + Y^2$ and $\Theta = \arctan Y/X$ (Fig. 4.7). Given $p(x,y)$, as above, we wish to find the corresponding joint density function $q(r,\theta)$.

Since (X,Y) and (R,Θ) define the same point, then $P(x_1 < X \le x_2, y_1 < Y \le y_2) = P(r_1 < R \le r_2, \theta_1 < \Theta \le \theta_2)$, where $r_1^2 = x_1^2 + y_1^2$, etc. In differential form according to (4.20) we have

$$p(x,y)\, dx\, dy = q(r,\theta)\, dr\, d\theta \tag{4.41}$$

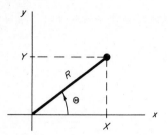

Fig. 4.7

Inserting (4.40) in (4.41) with $r^2 = x^2 + y^2$ and transforming differential areas $dx\,dy = r\,dr\,d\theta$ yields

$$q(r,\theta) = \frac{r}{2\pi\sigma^2}\, e^{-r^2/2\sigma^2} \tag{4.42}$$

This result requires two qualifications: (1) by definition, r cannot be negative, and (2) ambiguities in θ must be avoided. We accomplish this by limiting θ to the range $[-\pi, +\pi]$. Thus

$$q(r,\theta) = 0 \qquad r < 0 \text{ or } |\theta| > \pi$$

Equation (4.42) is still a trifle disturbing because it does not contain θ explicitly. However, with the aid of (4.23)

$$q(\theta) = \int_0^\infty \frac{r}{2\pi\sigma^2}\, e^{-r^2/2\sigma^2}\, dr = \frac{1}{2\pi} \int_0^\infty e^{-u}\, du$$

$$= \frac{1}{2\pi} \qquad |\theta| \leq \pi \tag{4.43}$$

Hence, Θ is *uniformly distributed* over $[-\pi, +\pi]$, a most logical conclusion when one reexamines the statement of the problem.

Proceeding in like fashion for the distribution of R alone, we integrate (4.42) over θ from $-\pi$ to $+\pi$ and get

$$q(r) = \frac{r}{\sigma^2}\, e^{-r^2/2\sigma^2} \qquad r \geq 0 \tag{4.44}$$

which is the *Rayleigh distribution* (Fig. 4.8). Note carefully that σ^2 is

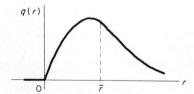

Fig. 4.8 *Rayleigh density function.*

the variance of X and Y but not R. The statistical averages of R are

$$\bar{r} = \sqrt{\frac{\pi}{2}}\,\sigma \qquad \overline{r^2} = 2\sigma^2 \qquad \sigma_r = \sqrt{2 - \frac{\pi}{2}}\,\sigma$$

A happy feature of the Rayleigh distribution is its ease of integration for determining probabilities. By the change of variable $r\,dr/\sigma^2 = d(r^2/2\sigma^2)$

$$\begin{aligned}
F(r) = P(R \le r) &= \int_0^{r^2/2\sigma^2} e^{-u}\,du \\
&= 1 - e^{-r^2/2\sigma^2}
\end{aligned}$$

4.5 RANDOM SIGNALS

Up to this point we have been speaking in rather general terms about random variables and their statistical averages. Now we concentrate on *random signals*, the manifestations of *random electrical processes* which take place in *time*. Thus, if $x(t)$ is a random signal, the random variable X may be assigned to represent values of $x(t)$ at the observation times.

Under certain conditions, termed *ergodicity*, an intuitively meaningful relationship between statistical averages and time averages will be obtained. This in turn leads to the spectral description of random signals via autocorrelation functions and the Wiener-Kinchine theorem.

We begin by considering ergodic processes and the implications of ergodicity.

Ergodic processes—ensemble averages

Let $x(t)$ be the voltage waveform produced by some noise generator, so that $x(t)$ is a random signal of finite average power. If the waveform is examined for a long period of time, its various *time* averages can be measured or calculated following the definitions of Chap. 2. Furthermore, the meaning of these averages is apparent: $\langle x(t) \rangle$ is the dc component of the noise, $\langle x^2(t) \rangle$ is its average power, and so forth.

Alternately, suppose there were available a very large number of identical generators, called an *ensemble*. The members of the ensemble are identical in the sense of having the same statistical description; nonetheless their individual signals, called *sample functions*, will be different. A particular signal, now designated $x_i(t)$ for clarity, is one of the infinitely many sample functions which a given noise generator can produce. The entire family of sample functions will be symbolized by $x(t)$.

A good analogy here is the coin-tossing experiment. A single coin can be tossed a large number of times, producing some pattern

$HTTHTHHHT \cdot \cdot \cdot$, which is one sample function of the random process. Or a large number of identical coins, an ensemble, can be tossed once. In either case, the empirically determined probabilities and statistical averages would be the same.

Returning to the noise waveform, instead of taking successive measurements in time on one sample function $x_i(t)$, conceptually we could examine all the sample functions at one specific time, say t_1. Averaging these measurements gives the *ensemble averages* $\overline{x(t_1)}$, $\overline{x^2(t_1)}$, etc. Clearly, the ensemble averages are the same as the statistical averages at $t = t_1$. But how do they relate to $\langle x_i(t) \rangle$, $\langle x_i^2(t) \rangle$, etc., the time averages of one sample function?

In general, ensemble averages may differ from such time averages. They surely would be different if the statistical properties of the generators were changing with time, for the time variation would be averaged out in the time averages but not in the ensemble averages. Moreover, ensemble averages at different times also would be different, for example, $\overline{x(t_2)} \neq \overline{x(t_1)}$, in which case the process is said to be *nonstationary*.

However, many of the random signals found in communication systems come from *ergodic processes*, meaning that *time and ensemble averages are identical*, $\overline{x(t)} = \langle x(t) \rangle$, $\overline{x^2(t)} = \langle x^2(t) \rangle$, etc. An ergodic process is also *stationary*, since ensemble averages are independent of the time of observation; but stationarity does not necessarily imply ergodicity.

In the remainder of this text we shall deal almost exclusively with random signals which are sample functions of ergodic processes. Therefore, the following statements can be made:

1. The *mean value* $\bar{x}$ is the *dc component* of the signal.
2. The *mean-squared* $\bar{x}^2$ is the power in the dc component, i.e., the *dc power*.
3. The *mean square* $\overline{x^2}$ is the total *average power*.
4. The *variance* $\sigma_x^2 = \overline{x^2} - \bar{x}^2$ is the power in the time-varying component, i.e., the *ac power*.
5. The *standard deviation* σ_x is the root mean square of the time-varying component, i.e., the *rms value*.

Although the above relations serve to make an electrical engineer feel more at home in the world of statistics, it must be emphasized that they apply only to the very special case of ergodic sources.

Autocorrelation

We have two reasons for considering the autocorrelation function of a random signal. First, $R(\tau)$ in its own right provides useful information about $x(t)$. Second, by the Wiener-Kinchine theorem, the frequency-domain description of a random signal is its power spectral density $G(f) = \mathfrak{F}[R(\tau)]$. Before tackling the latter, let us see how $R(\tau)$ can be interpreted when $x(t)$ is random. (At this point the reader may wish to review portions of Sec. 2.6 in preparation for the following discussion.)

For *deterministic* signals, the autocorrelation function was defined in Chap. 2 as the *time* average $R(\tau) = \langle x(t)x(t + \tau) \rangle$. Now if $x(t)$ is a *random* signal (or sample function) from an *ergodic* source, the time and statistical averages are identical, and its autocorrelation function is

$$R(\tau) = \langle x(t)x(t + \tau) \rangle = \overline{x(t)x(t + \tau)} \qquad (4.45)$$

[When $R(\tau)$ is written as an ensemble average, it is also called the *covariance function*, providing $\bar{x} = 0$.]

Clearly, all the mathematical properties previously ascribed to $R(\tau)$ also apply to the autocorrelation of random signals. In particular, we note that

$$R(0) = \overline{x^2} = \sigma^2 + \bar{x}^2 \geq |R(\tau)| \qquad (4.46a)$$
$$R(-\tau) = R(\tau) \qquad (4.46b)$$

In addition, for nonperiodic random processes, it is true that

$$R(\pm \infty) = \bar{x}^2 \qquad (4.46c)$$

In a sense, autocorrelation is a measure of both *time variation* and *statistical dependence*. Suppose, for example, we take a value of τ which is very small compared to the time intervals in which $x(t)$ has significant change; then $x(t + \tau) \approx x(t)$, and $R(\tau) \approx \overline{x^2(t)} = \sigma^2 + \bar{x}^2$. As $|\tau|$ increases, $R(\tau)$ will generally decrease at first (it cannot increase), and the range of τ around $\tau = 0$ for which $R(\tau) - \bar{x}^2 \approx \sigma^2$ is a measure of the time variation of $x(t)$. The smaller this range, the more rapid the time variations.

On the other hand, if $|\tau|$ is very large, we might find that $x(t)$ and $x(t + \tau)$ have so little in common that they are *statistically independent;* then $R(\tau) = \overline{x(t)}\ \overline{x(t + \tau)} = \bar{x}^2$. In general, $|R(\tau)| - \bar{x}^2 \approx \sigma^2$ indicates a strong linear dependence of $x(t + \tau)$ on $x(t)$, whereas $|R(\tau)| - \bar{x}^2 \ll \sigma^2$ indicates that the signal has undergone substantial changes in that interval.

When $|R(\tau)| - \bar{x}^2 = 0$, $x(t + \tau)$ and $x(t)$ are *uncorrelated* but not necessarily statistically independent. However, if the process is *gaussian*, uncorrelation also signifies statistical independence.

As written in (4.45), the autocorrelation function of a random signal appears deceptively simple. But unless $x(t)$ and $x(t + \tau)$ are statistically independent, calculation of $R(\tau)$ may be nontrivial. A systematic approach to the problem is facilitated by writing $x_1(t) = x(t)$ and $x_2(t) = x(t + \tau)$. Then

$$R_x(\tau) = \overline{x_1(t)x_2(t)} = \int\limits_{-\infty}^{\infty}\!\!\int x_1 x_2 p(x_1,x_2)\, dx_1\, dx_2 \tag{4.47}$$

where the joint density function $p(x_1,x_2)$ can be found from (4.22) as $p(x_1,x_2) = p(x_1)p(x_2|x_1)$, etc.

Now consider the crosscorrelation of two random signals, $R_{xy}(\tau) = \overline{x(t)y(t + \tau)}$. If they are statistically independent and one or the other has zero mean, then $R_{xy}(\tau) = \overline{x(t)}\,\overline{y(t + \tau)} = 0$. Therefore, *statistically independent random signals* are analogous to *incoherent deterministic signals*.

Finally, suppose we have a *desired signal* $x(t)$ contaminated by *unwanted noise* $n(t)$. If the signal and noise are *additive*, the resultant is the sum

$$y(t) = x(t) + n(t)$$

Treating $x(t)$ and $n(t)$ as sample functions from *independent* ergodic processes, then $R_{xn}(\tau) = R_{nx}(\tau) = 0$, assuming that $\bar{n} = 0$, which is usually true. The autocorrelation of the sum $y(t)$ is

$$
\begin{aligned}
R_y(\tau) &= \overline{[x(t) + n(t)][x(t + \tau) + n(t + \tau)]} \\
&= \overline{x(t)x(t + \tau)} + \overline{x(t)n(t + \tau)} + \overline{n(t)x(t + \tau)} + \overline{n(t)n(t + \tau)} \\
&= R_x(\tau) + R_n(\tau)
\end{aligned}
$$

since the crosscorrelation terms are zero. Setting $\tau = 0$ gives

$$\overline{y^2} = \overline{x^2} + \overline{n^2} = S + N$$

where $S = \overline{x^2}$ is the *average signal power* and $N = \overline{n^2}$ is the *average noise power*.

Therefore, under the assumption of independent additive noise with zero mean, *superposition* of autocorrelation functions and average powers applies. In such cases, it is meaningful to speak of a *signal-to-noise ratio* S/N, a power ratio. Had the signal and noise been *dependent*, we would have obtained

$$\overline{y^2} = S + R_{xn}(0) + R_{nx}(0) + N$$

The crosscorrelations involve both signal and noise, and a signal-to-noise ratio would be difficult to define.

Power spectra

Turning to the frequency domain, the sole reasonable description of a random signal is its *power spectral density* $G(f)$. This is because the only a priori knowledge we have about a random signal is its probability function and ensemble averages, the latter being directly related to average power if the source is ergodic.

Invoking the Wiener-Kinchine theorem, the power spectral density of the random signal $x(t)$ is

$$G(f) = \mathfrak{F}[R(\tau)] = \int_{-\infty}^{\infty} \overline{x(t)x(t+\tau)}e^{-j\omega\tau}\, d\tau \qquad (4.48)$$

Implicit in (4.48) is the condition that $\overline{x(t)x(t+\tau)}$ depends *only* on τ; that is, the source must be ergodic. For nonstationary sources, the *power spectrum* obtained from (4.48) would be a function of both frequency and time, in total contradiction to the concept of frequency-domain descriptions.

Assuming an ergodic source, the theorems stated in Sec. 2.6, namely, (2.67) to (2.71), apply to power spectra of random signals as well as deterministic ones. For example, multiplying the random signal $x(t)$ by $\cos \omega_c t$ produces a new random signal $y(t) = x(t) \cos \omega_c t$, whose power spectrum is $G_y(f) = \frac{1}{4}[G_x(f - f_c) + G_x(f - f_c)]$; frequency translation has therefore taken place.

While the Wiener-Kinchine theorem provides the *definition* of power spectra for random signals, an alternate expression not involving $R(\tau)$ can be derived as follows.

Suppose the output of a noise generator is recorded for a time interval $|t| < T/2$. Using graphical techniques, one could calculate a conventional Fourier transform for this "piece" of the signal, namely,

$$X_T(f) = \int_{-T/2}^{T/2} x_i(t)e^{-j\omega t}\, dt$$

We have written $x_i(t)$ to emphasize that we are dealing with but one of the many possible sample functions the generator can produce. Then, paralleling (2.57), $G(f)$ might be taken as $\lim_{T\to\infty} (1/T)|X_T(f)|^2$. However, to include *all* the possible sample functions, we write $G(f)$ as an *ensemble average*

$$G(f) = \lim_{T\to\infty} \frac{1}{T} \overline{|X_T(f)|^2} \qquad (4.49)$$

where the order of averaging and limiting cannot be interchanged. See Aseltine (1958) for further discussion of this matter.

The following two examples illustrate autocorrelation and power-spectra calculation techniques. One deals with an information-bearing random signal, the other discusses a type of noise found in vacuum tubes. Ergodicity is assumed for both examples.

Example 4.8 The random telegraph wave

Figure 4.9 shows a sample function of the random telegraph wave, a discrete random signal having two possible values, $+A$ and $-A$. We shall assume that $x(t)$ makes *independent* random shifts from one value to the other and that the two values occur with equal probability. Thus $\bar{x} = 0$ is the dc component, and $\overline{x^2} = A^2$ is the average power.

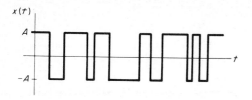

Fig. 4.9 *A sample function of the random telegraph wave.*

It is further assumed that the probability of n shifts in an interval of duration T is given by the Poisson frequency function

$$P(n) = e^{-\mu T} \frac{(\mu T)^n}{n!}$$

from which $\bar{n} = \mu T$. Hence the average time between shifts is $T/\bar{n} = 1/\mu$, and the *average shift rate* is μ shifts per second.

Because $x(t) = \pm A$, $R(\tau) = \overline{x(t)x(t+\tau)}$ can be determined as the mean of a *discrete* variate whose possible values are $x(t)x(t+\tau) = \pm A^2$. In particular, if $x(t)$ makes an even number of shifts in the interval $T = |\tau|$, then $x(t)x(t+\tau) = A^2$; for an odd number of shifts, $x(t)x(t+\tau) = -A^2$. Thus

$$R(\tau) = A^2 P(n \text{ even}) + (-A^2)P(n \text{ odd})$$
$$= A^2 \left[\sum_{n \text{ even}} P(n) - \sum_{n \text{ odd}} P(n) \right]$$

Taking $P(n)$ as above, with $T = |\tau|$, we have

$$R(\tau) = A^2 e^{-\mu|\tau|} \left[1 + \frac{(\mu|\tau|)^2}{2!} + \frac{(\mu|\tau|)^4}{4!} \right.$$
$$\left. + \cdots - \frac{\mu|\tau|}{1!} - \frac{(\mu|\tau|)^3}{3!} - \cdots \right]$$
$$= A^2 e^{-\mu|\tau|} e^{-\mu|\tau|} = A^2 e^{-2\mu|\tau|}$$

As expected $R(0) = \overline{x^2} = A^2$, and $R(\pm\infty) = \bar{x} = 0$. Moreover, $R(\tau) \approx 0$ for $|\tau| \gg 1/\mu$, so $x(t)$ and $x(t+\tau)$ are essentially uncorrelated under this condition.

Taking the Fourier transform of $R(\tau)$, the power spectrum is found to be

$$G(f) = 2 \int_0^\infty A^2 e^{-2\mu|\tau|} \cos 2\pi f\tau \; d\tau = \frac{A^2}{\mu} \frac{1}{1 + (\pi f/\mu)^2}$$

as illustrated in Fig. 4.10 along with the correlation function. It may be noted that though the average shift rate is μ shifts per second, $G(f)$ has

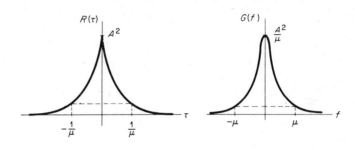

Fig. 4.10 *Random telegraph wave.* (a) *Correlation function;* (b) *power spectrum.*

appreciable high-frequency content: approximately 20 percent of the total power is in $|f| > \mu$. This is due to the occurrence, albeit rare, of adjacent shift times spaced by much less than $1/\mu$.

Example 4.9 Shot noise ★

The current in a vacuum tube is due to the transfer of electrons from cathode to plate. As it traverses the interelectrode path, each electron induces a *current pulse* $i_e(t)$ at the plate, like that shown in Fig. 4.11a

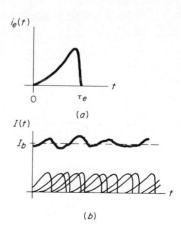

Fig. 4.11 *Shot noise.* (a) *Current pulse due to a single electron;* (b) *total current.*

where the emission time is taken as $t = 0$ and τ_e is the transit time. For our purposes the pulse *shape* is relatively unimportant; however, $\int i_e(t) \, dt = e$, the electronic charge, since that is the charge collected at the plate per electron.

Because the electrons are emitted at random times, the total current is a summation of many randomly spaced induced-current pulses and therefore exhibits small fluctuations about the dc value (Fig. 4.11b). The effect is known as *shot noise* and is a major cause of noise in vacuum tubes.

For a first-order study of the effect, we shall neglect control grids and space charge, implying that all emitted electrons are collected τ_e sec later and that emission times are statistically independent. Under these conditions, the probability of K electrons' being emitted in an interval T is Poisson, $P(K) = e^{-\bar{n}T}(\bar{n}T)^K/K!$, where $\bar{n}$ is the average number of electrons collected per unit time. (These assumptions are consistent with the operation of a *temperature-limited diode*.)

Proceeding to the analysis, we begin by considering the finite period of time, $|t| < T/2$. If T is much greater than the transit time τ_e, the total current can be written as a sum of pulses

$$I(t) = \sum_{k=1}^{K} i_e(t - t_k)$$

where K and t_k are random variables, K being the total number of electrons emitted in T, and $t_1, t_2, \ldots, t_k, \ldots, t_K$ being the emission times. We further assume that t_k is uniformly distributed in T,

$$p(t_k) = \frac{1}{T} \qquad |t| < \frac{T}{2}$$

so that any one electron may be emitted at any time in the interval, with equal probability.

Taking either a time average or statistical average, the dc plate current is

$$I_b = \langle I(t) \rangle = \bar{I} = e\bar{n}$$

consistent with $\bar{n} = \bar{K}/T = \lim_{T \to \infty} K/T$. In words, the dc current is just the charge per electron times the average number of electrons per second.

Turning to the autocorrelation of $I(t)$, we must recognize that there are $K + 1$ random variables to be dealt with, i.e., the K emission times and K itself. Hence

$$R_I(\tau) = \int_{-\infty}^{\infty} \int_{-\infty}^{\infty} \cdots \int_{-\infty}^{\infty} \left[\sum_k i_e(t - t_k) \right] \left[\sum_j i_e(t + \tau - t_j) \right]$$

$$p(t_1, t_2, \ldots, t_K, K) \, dt_1 \, dt_2 \cdots dt_K \, dK$$

However, all variates are statistically independent, so the joint density function is the product of the individual functions. Setting $p(t_1) = p(t_2) = \cdots = 1/T$ for $|t| < T/2$ and regrouping yields

$$R_I(\tau) = \int_{-\infty}^{\infty} \left[\sum_k \sum_j \int_T \frac{dt_1}{T} \cdots \right.$$

$$\left. \int_T \frac{dt_K}{T} i_e(t - t_k)i_e(t + \tau - t_j) \right] p(K)\, dK$$

The double summation contains K^2 terms, of which K have $k = j$, while the remaining $K^2 - K$ have $k \neq j$. Treating these two cases individually, for $k = j$

$$\int_T \frac{dt_1}{T} \cdots \int_T \frac{dt_K}{T} i_e(t - t_k)i_e(t + \tau - t_k)$$

$$= \frac{1}{T} \int_T i_e(t - t_k)i_e(t + \tau - t_k)\, dt$$

$$= \frac{1}{T} \int_{-\infty}^{\infty} i_e(t)i_e(t + \tau)\, dt$$

and for $k \neq j$

$$\int_T \frac{dt_1}{T} \cdots \int_T \frac{dt_K}{T} i_e(t - t_k)i_e(t + \tau - t_j)$$

$$= \left[\frac{1}{T} \int_T i_e(t - t_k)\, dt_k \right]\left[\frac{1}{T} \int_T i_e(t + \tau - t_j)\, dt_j \right]$$

$$= \left(\frac{e}{T} \right)^2$$

since each pulse is zero outside $|t| < T/2$ and its area is e.

Substituting these values, multiplied by K^2 and $K^2 - K$, respectively, we obtain

$$R_I(\tau) = \int_{-\infty}^{\infty} \left[\frac{K}{T} \int_{-\infty}^{\infty} i_e(t)i_e(t + \tau)\, dt + (K^2 - K)\left(\frac{e}{T} \right)^2 \right] p(K)\, dK$$

$$= \frac{\bar{K}}{T} \int_{-\infty}^{\infty} i_e(t)i_e(t + \tau)\, dt + \frac{\overline{K^2} - \bar{K}}{T^2} e^2$$

But K is Poisson with $\bar{K} = \sigma_K{}^2 = \bar{n}T$, so $\overline{K^2} - \bar{K} = (\bar{n}T)^2$, and we arrive at

$$R_I(\tau) = R_i(\tau) + I_b{}^2$$

where $R_i(\tau) = \bar{n} \int_{-\infty}^{\infty} i_e(t)i_e(t + \tau)\, dt$ represents the shot-noise fluctuations and $I_b{}^2 = (\bar{n}e)^2$ is the square of the direct, or average, current.

Of course the autocorrelation $R_i(\tau) = \bar{n} \int_{-\infty}^{\infty} i_e(t)i_e(t + \tau)\, dt$ is no trivial matter. It depends on the shape of $i_e(t)$, which, in turn, depends on the electrode configuration and applied potentials. However, referring to Fig. 4.11a, it is safe to say that $R_i(\tau) \approx 0$ for $|\tau| \gg \tau_e$, so $R_I(\tau)$ may take the typical form shown in Fig. 4.12a.

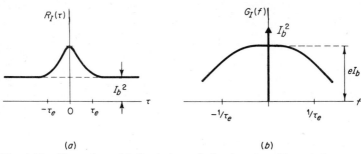

$$(a) \qquad\qquad\qquad (b)$$

Fig. 4.12 *Shot noise.* (a) *Typical correlation function;* (b) *typical power spectrum.*

As to the power spectral density of shot-noise current, $G_I(f)$ can be written in the form

$$G_I(f) = \bar{n}|I_e(f)|^2 + I_b{}^2\delta(f)$$

where $I_e(f) = \int_{-\infty}^{\infty} i_e(t)e^{-j\omega t}\,dt$ is the Fourier transform of the induced current pulse. The impulse represents the dc power. A typical $G_I(f)$ is sketched in Fig. 4.12b. Ignoring the dc component, the spectrum of the time-varying current is approximately

$$G_i(f) = \bar{n}|I_e(0)|^2 = \bar{n}e^2 = eI_b \qquad |f| \ll \frac{1}{\tau_e}$$

This elegantly simple relation, called the *Schottky formula*, emphasizes the fact that the noise variations are proportional to I_b.

In the interests of brevity, the above discussion has just touched the surface of this subject. For complete details, see Davenport and Root (1958, chap. 7) or van der Ziel (1954). An interesting alternate approach is given by Goldman (1948).

4.6 WHITE NOISE AND NOISE FILTERING

Unwanted electric signals come from a variety of sources, generally classified as man-made interference or naturally occurring noise. Man-made interference comes from other communication systems, ignition and commutator sparking, 60-cycle hum, and so forth; natural noise-producing phenomena include atmospheric disturbances, extraterrestrial radiation, and circuit noise.   By careful engineering, the effects of many unwanted signals can be reduced or eliminated completely.   But there always remain certain inescapable random signals, which present a fundamental limit to systems performance.

One unavoidable cause of electrical noise is the thermal motion of electrons in conducting media—wires, resistors, etc.   As long as communication systems are constructed from such material, this *thermal noise* will be with us.   A general derivation of the properties of thermal noise is omitted here, but, thanks to extensive theoretical and experimental studies, there is a sizable body of knowledge pertaining to the subject, from which we shall freely draw.

For instance, thermodynamic considerations have shown that the power spectrum of thermal noise is *constant* over a very wide frequency range.   Noise having such a spectrum contains all frequency components in equal proportion and is aptly designated *white noise*, by analogy to white light.

Other studies indicate that thermal noise has a *gaussian* probability distribution, in agreement with predictions from the central-limit theorem.   Hence, thermal noise is *gaussian white noise*.   Vacuum-tube and semiconductor noise also satisfy this description, to a good approximation, as illustrated by the shot-noise example, where $G(f)$ is essentially constant for $|f| < \tau_e$.   (The gaussian nature of shot noise likewise follows from the central-limit theorem.)

It can also be shown that linear transformation of a gaussian variate results in another gaussian variate.   Hence, assuming a *linear* two-port, filtered gaussian noise remains gaussian.   Therefore, we shall hereafter focus attention on gaussian white noise or filtered gaussian white noise, unless stated to the contrary.

White noise

We write the power spectrum of white noise, having zero mean, as

$$G(f) = \frac{\eta}{2} \tag{4.50a}$$

where the seemingly extraneous factor of $\frac{1}{2}$ is included to indicate that half the power is associated with positive frequency and half with negative frequency (Fig. 4.13a).   Alternately, η is the *positive-frequency* power density in watts per hertz.   Since $G(f)$ is known, the autocorrelation function follows immediately by Fourier transformation

$$R(\tau) = \int_{-\infty}^{\infty} \frac{\eta}{2} e^{j\omega\tau} \, df = \frac{\eta}{2} \delta(\tau) \tag{4.50b}$$

as in Fig. 4.13b.

From (4.50b) or Fig. 4.13b, we see that $R\ (\tau \neq 0) = 0$, so any two different samples of a gaussian white-noise signal are uncorrelated and statistically independent. This observation, coupled with the constant power spectrum, leads to an interesting conclusion: if white noise is displayed on an oscilloscope, successive sweeps are always different from each other; yet the waveform always "looks" the same, no matter what sweep speed is used, since all rates of time variation (frequency components) are contained in equal proportion.

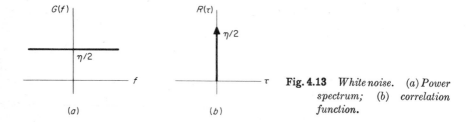

Fig. 4.13 *White noise.* *(a) Power spectrum; (b) correlation function.*

Similarly, if white noise drives a loudspeaker, it always sounds the same, somewhat like waterfall. As Pierce[1] puts it:

Mathematically, white gaussian noise . . . is the epitome of the various and unexpected. It is the least predictable, the most original of sounds. To a human being, however, all gaussian white noise sounds alike. Its subtleties are hidden from him, and he says that it is dull and monotonous.

Turning to power considerations, $\displaystyle\int_{-\infty}^{\infty} G(f)\,df = \int_{-\infty}^{\infty} \frac{\eta}{2}\,df = \infty$, implying that white noise has *infinite power* (or infinite variance) and must be *physically impossible*. The same conclusion is reached using (4.50b); $\overline{x^2} = R(0) = \infty$. This disturbing result is essentially the *ultraviolet catastrophe* of nineteenth-century physics, and reflects the fact that $G(f)$ cannot be strictly flat over all frequency.

Actually, the spectrum of thermal noise begins to decrease exponentially in the neighborhood of

$$f_0 = \frac{kT}{h}$$

where k = Boltzmann constant
h = Planck's constant
T = temperature, °K

[1] J. R. Pierce, "Symbols, Signals, and Noise," Harper & Row, Publishers, Incorporated, New York, 1961.

At room temperature, f_0 is of order 10^{13} Hz, which falls in the infrared. Inasmuch as conventional electrical systems have ceased to respond well below this frequency, the assumption of constant power spectra is justified as far as the system is concerned. Of course, one should bear in mind that the power is finite.

In any case, subsequent filtering of white noise will limit the power, among other effects, a subject we now examine.

Filtered white noise

For the study of filtered white noise, or filtering of random signals in general, a basic tool is the input-output power-spectra relationship

$$G_y(f) = |H(f)|^2 G_x(f) \qquad (4.51a)$$

$H(f)$ being the filter's transfer function. Thus, if the input to a linear two-port is the random signal $x(t)$, the output will be a random signal $y(t)$ with $G_y(f)$ as above, and

$$R_y(\tau) = \int_{-\infty}^{\infty} |H(f)|^2 G_x(f) e^{j\omega\tau} \, df \qquad (4.51b)$$

$$\overline{y^2} = R_y(0) = \int_{-\infty}^{\infty} |H(f)|^2 G_x(f) \, df \qquad (4.51c)$$

Proof of (4.51) follows that for the deterministic case with time averages replaced by ensemble averages. (Need it be mentioned that ergodicity is assumed?)

If white noise is the input to an ideal lowpass filter of unit gain and bandwidth B, then

$$G_y(f) = \frac{\eta}{2} \Pi\left(\frac{f}{2B}\right) \qquad (4.52a)$$

The output power spectrum is thus a rectangular function (Fig. 4.14a).

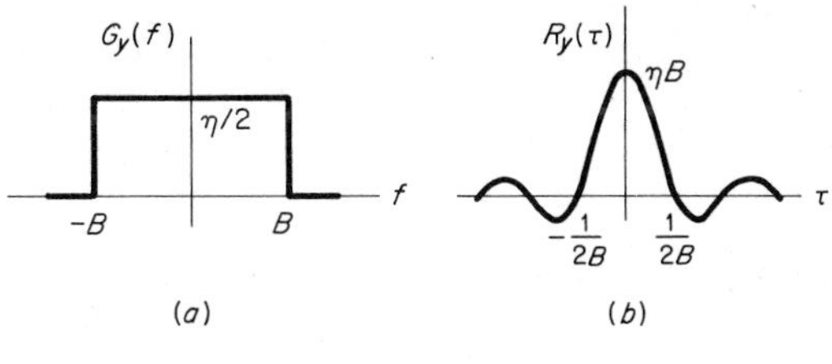

Fig. 4.14 *White noise passed by an ideal lowpass filter. (a) Power spectrum; (b) correlation function.*

Likewise, the output autocorrelation is a sinc function (Fig. 4.14b); specifically

$$R_y(\tau) = \eta B \text{ sinc } 2B\tau \qquad (4.52b)$$

As evident in the figures, the filtering process has done three things:

1. The power spectrum is no longer white, though it is constant over a finite frequency range.
2. The output power is finite.   In fact, $\overline{y^2} = \eta B$.
3. The output signal is correlated over time intervals of about $1/2B$.

Though the above conclusions were based on ideal lowpass filtering, similar results are obtained with any real filter.   One should particularly note that the spectrum of filtered white noise takes on the *shape* of $|H(f)|^2$.   Since the resulting spectrum is no longer white, filtered white noise is often called *colored noise*.

Example 4.10

Suppose white noise is filtered by an RC lowpass filter.   From Example 3.1, $|H(f)| = [1 + (f/f_3)^2]^{-\frac{1}{2}}$, where $f_3 = 1/2\pi RC$, hence

$$G_y(f) = \frac{\eta/2}{1 + (f/f_3)^2}$$

Comparing this result with Example 4.8 yields a rather surprising conclusion: RC-filtered white noise and the random telegraph wave have power spectra of the same shape!   Of course the actual sample functions (waveforms) are quite different.   This further emphasizes the comment made earlier with respect to deterministic signals, namely, that power spectral density does not uniquely characterize a signal.

As to the output autocorrelation function $R_y(\tau)$, Fourier transformation of $G_y(f)$ is not particularly easy; however, drawing upon the results of Example 4.8, it follows that

$$R_y(\tau) = \tfrac{1}{2}\eta\pi f_3 e^{-2\pi f_3|\tau|} = \frac{\eta}{4RC}\, e^{-|\tau|/RC}$$

The interval over which the filtered noise has appreciable correlation is therefore approximately equal to the circuit time constant RC, as might have been suspected.

Noise equivalent bandwidth

Filtered white noise usually has finite power.   To emphasize this, we designate *average noise power* by $N = \overline{y^2}$ and write (4.51c) in the form

$$N = \int_{-\infty}^{\infty} |H(f)|^2 \frac{\eta}{2}\, df = \eta \int_0^{\infty} |H(f)|^2\, df$$

Noting that the integral depends only on the filter transfer function, we can simplify discussion of noise power by defining a *noise equivalent bandwidth* B_N (not to be confused with B_{eq} of Sec. 3.4) as

$$B_N = \frac{1}{H_0{}^2} \int_0^\infty |H(f)|^2 \, df \qquad (4.53)$$

where $H_0 = |H(f)|_{\max}$ is the center-frequency amplitude response, i.e., the voltage gain. (This definition assumes the filter has a meaningful center frequency.) Hence the filtered noise power is

$$N = H_0{}^2 \eta B_N \qquad (4.54)$$

Examining (4.54) shows that the effect of the filter has been separated into two parts: (1) the *relative frequency selectivity*, as described by B_N, and (2) the *gain* (or attenuation), represented by H_0. Thus, as illustrated by Fig. 4.15 for a bandpass filter, B_N equals the bandwidth of an ideal rectangular filter that would pass as much white-noise power as the two-port network in question, their maximum gains being equal.

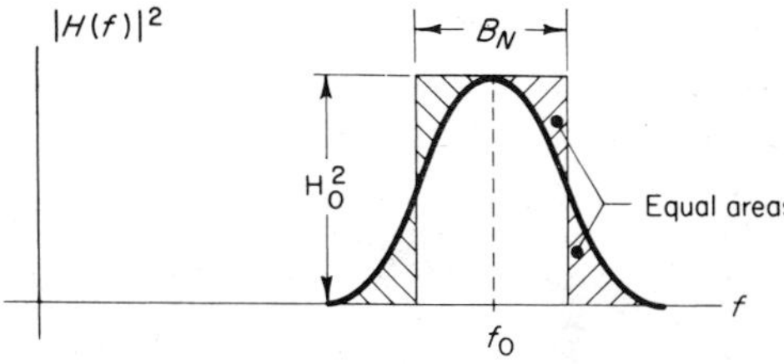

Fig. 4.15 *Noise equivalent bandwidth B_N of a bandpass filter.*

By definition, the noise equivalent bandwidth of an ideal filter is its actual bandwidth. For practical filters, B_N is somewhat greater than the 3-db bandwidth; e.g., an RC lowpass filter has $B_N = \pi f_3/2$, so B_N is about 50 percent greater than f_3. However, as the filter becomes more selective (sharper cutoff characteristics), its noise bandwidth approaches the 3-db bandwidth, and for most applications one is not too far off in taking them to be equal. Thus we now drop the subscript, letting B stand for noise bandwidth or 3-db bandwidth as the case may be. Furthermore, we shall always assume unit gain, $H_0 = 1$.

Summarizing, if $y(t)$ is filtered white noise of zero mean, then

$$\bar{y} = 0$$
$$\overline{y^2} = N = \eta B = \sigma_y{}^2 \qquad (4.55)$$
$$\sigma_y = \sqrt{N} = \sqrt{\eta B}$$

This means that given a source of white noise, an average-power meter (or mean-square voltage meter) will read $\overline{y^2} = N = \eta B$, where B is the noise equivalent bandwidth of the meter itself.   Working backward, the source power density can be inferred via $\eta = N/B$, providing one is sure that the noise is indeed white over the frequency-response range of the meter.

White noise and filter measurements   ★

Because white noise contains all frequencies in equal proportion, it is a convenient signal for filter measurements and experimental design work. Consequently, white-noise sources with calibrated power density have become standard laboratory instruments.   A few of the measurements that can be made with these sources are discussed below.

Noise equivalent bandwidth : Suppose the gain of an amplifier is known and we wish to find its noise equivalent bandwidth.   To do so we can apply white noise to the input and measure the average output power with a meter whose frequency response is essentially constant over the amplifier's passband.   The noise bandwidth in question is then, from Eq. (4.54), $B = N/H_0^2\eta$.

Amplitude response : To find the amplitude response of a given filter, we apply white noise to the input so the output power spectrum is proportional to $|H(f)|^2$.   Then we scan the output spectrum with a tunable bandpass filter whose bandwidth is constant and small compared to the variations of $|H(f)|^2$.   Thus, if the scanning filter is centered at f_c, the rms noise voltage at its output is proportional to $|H(f_c)|$.   By varying f_c, a point-by-point plot of $|H(f)|$ is obtained.

Impulse response : Figure 4.16 shows a method for measuring the impulse response $h(t)$ of a given system.   The instrumentation required is a white-noise source, a variable time delay, a multiplier, and an averaging device.

    Denoting the input noise as $x(t)$, the system output is $h(t) * x(t)$,

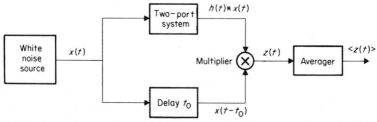

Fig. 4.16   *Instrumentation for measuring impulse response using white noise.*

and the delayed signal is $x(t - t_0)$.   Thus, the output of the multiplier is

$$z(t) = x(t - t_0)[h(t) * x(t)]$$
$$= x(t - t_0) \int_{-\infty}^{\infty} h(t')x(t - t') \, dt'$$

Now suppose that $z(t)$ is averaged over all time[1] to obtain $\langle z(t) \rangle$.   If the noise source is ergodic and the system is linear and time-invariant, the averaged output is the ensemble average

$$\overline{z(t)} = \int_{-\infty}^{\infty} h(t')\overline{x(t - t_0)x(t - t')} \, dt'$$

where we have interchanged the order of integration and averaging.
    Inspecting the integral, it can be seen that

$$\overline{x(t - t_0)x(t - t')} = \overline{x(t)x[t - (t' - t_0)]} = R_x(t' - t_0)$$

But, with $x(t)$ being white noise, Eq. (4.50b) says that $R_x(t' - t_0) = (\eta/2)\delta(t' - t_0)$.   Hence

$$\overline{z(t)} = \frac{\eta}{2} \int_{-\infty}^{\infty} h(t')\delta(t' - t_0) \, dt' = \frac{\eta}{2} h(t_0)$$

Therefore, $h(t)$ is measured by varying the time delay t_0.
    The reader may question the practicality of this method, particularly since $h(t)$ can be obtained immediately by applying an impulse (or a brief and intensive pulse) to the system.   While this is a valid conclusion for most *filter* measurements, there are many systems, such as industrial processing and control systems, for which an impulsive input cannot be readily achieved or, if achieved, might damage or destroy the system.

4.7 NARROWBAND NOISE

Narrowband noise is defined as the output of a bandpass filter whose bandwidth B is very small compared to the center frequency f_c, the input being white noise.   Clearly, as $B/f_c \to 0$, the only possible output is a *sinusoidal wave* of frequency f_c, albeit with infinitesimal amplitude.   For small but finite B/f_c, narrowband noise looks very much like such a wave but with slow and random amplitude and phase variations.
    To describe narrowband noise quantitatively, let the bandpass filter have characteristics as shown in Fig. 4.17a.   (The power spectrum of narrowband noise will have the same shape as Fig. 4.17a.)   Introducing an equivalent *lowpass* response function $H_l(f)$ (Fig. 4.17b), we can write

$$|H(f)|^2 = |H_l(f - f_c)|^2 + |H_l(f + f_c)|^2$$

---

[1] Needless to say, one cannot average over all time; however, a finite averaging period is usually sufficient.

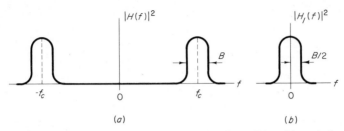

Fig. 4.17   (a) *Power response of a narrowband amplifier;* (b) *equivalent lowpass response.*

and the autocorrelation of narrowband noise is

$$R_y(\tau) = \frac{\eta}{2}\left[\int_{-\infty}^{\infty}|H_l(f-f_c)|^2 e^{j2\pi f\tau}\,df + \int_{-\infty}^{\infty}|H_l(f+f_c)|^2 e^{j2\pi f\tau}\,df\right]$$

$$= \eta B\rho(\tau)\cos 2\pi f_c\tau$$

where

$$\rho(\tau) = \frac{1}{B}\int_{-\infty}^{\infty}|H_l(u)|^2 e^{j2\pi u\tau}\,du$$

In essence $\rho(\tau)$ is the output autocorrelation function of the equivalent lowpass filter with a white-noise input.

For most filters of interest, it is true that $\rho(\tau)\approx 1$ when $|\tau|\ll 1/B$, witness Fig. 4.14b. Thus

$$R_y(\tau) \approx \eta B\cos 2\pi f_c\tau \qquad |\tau|\ll\frac{1}{B} \tag{4.56}$$

which is the same as the autocorrelation function of a sinusoid having amplitude $A = \sqrt{2\eta B}$ and frequency f_c (see Example 2.9).   This suggests that for time intervals of duration small compared to $1/B$, narrowband noise looks very much like a simple sinusoidal wave.   Moreover, with $f_c \gg B$, such intervals contain many cycles at frequency f_c.

But, as the observation time is extended, random variations in amplitude and phase will be seen, these variations being slow compared to f_c.   To conclude this chapter we examine the effect in more detail, concentrating on the amplitude and phase variations.

The statistical description of narrowband noise

Practically all communication systems are narrowband.   Consequently, in our later work we shall frequently encounter a desired signal contaminated by narrowband noise.   Meaningful treatment of this situation requires appropriate probability density functions and statistical averages. Here we consider the noise alone.

Although there is no direct route from power spectra to probability density functions, which is what we are after, advanced techniques do yield a solution for this particular case. We shall give a qualitative description of the approach and discuss the results.

Let the power spectrum $G_n(f)$ of narrowband noise be as shown in Fig. 4.18a (the negative-frequency portion is omitted for simplicity).

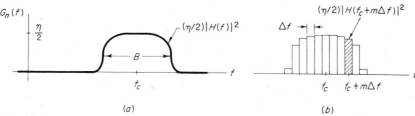

Fig. 4.18 *Narrowband noise.* (a) *Power spectrum;* (b) *partitioned power spectrum.*

The total noise power is $N = \eta B$, and there is no dc component, so

$$\overline{n^2} = \sigma_n{}^2 = N = \eta B$$

The spectrum is then partitioned into very narrow frequency bands of width Δf, centered on $f_c + m\,\Delta f$ (Fig. 4.18b). There are $M = B/\Delta f$ bands. From previous considerations, each band represents (approximately) a sinusoid of frequency $f_c + m\,\Delta f$ and amplitude

$$A_m = \sqrt{2\eta |H(f_c + m\,\Delta f)|^2\,\Delta f}$$

The total narrowband noise signal is the sum of these waves

$$n(t) = \sum_{m=-M/2}^{M/2} A_m \cos\left[2\pi(f_c + m\,\Delta f)t + \theta_m\right]$$

We have included an arbitrary phase term θ_m, since $G_n(f)$ gives no phase information.

Expanding cosine terms, $n(t)$ can be written as

$$n(t) = n_c(t)\cos 2\pi f_c t - n_s(t)\sin 2\pi f_c t \tag{4.57}$$

where

$$n_c(t) = \sum_{m=-M/2}^{M/2} A_m \cos\left(2\pi m\,\Delta f t + \theta_m\right) \tag{4.58a}$$

$$n_s(t) = \sum_{m=-M/2}^{M/2} A_m \sin\left(2\pi m\,\Delta f t + \theta_m\right) \tag{4.58b}$$

Equation (4.57) is called the *quadrature-carrier representation* of narrow-band noise because $\cos 2\pi f_c t$ and $\sin 2\pi f_c t$ are in phase quadrature. More-over, we now have a *phasor* representation for $n(t)$, each term in (4.57) being a phasor with random amplitude, n_c or n_s, rotating at f_c Hz. Recall-ing that $-\sin \alpha = \cos (\alpha + \pi/2)$, the phasor diagram is Fig. 4.19.

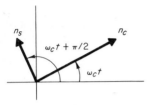

Fig. 4.19   *Phasor interpretation of Eq. (4.57).*

Examining (4.58a) and (4.58b), the highest-frequency component in each is $M\,\Delta f/2 = B/2 \ll f_c$, so the random amplitude variations are quite *slow* compared to the rotational frequency. Furthermore, though $n_c(t)$ and $n_s(t)$ do not exist as physical entities (at least here), a power spectrum can be assigned to each. For example, $G_{n_c}(f)$ will resemble *lowpass* filtered white noise of bandwidth $B/2$ (Fig. 4.20).

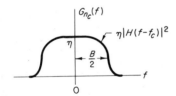

Fig. 4.20   *Power spectrum of $n_c(t)$.*

Further inspection of (4.58) suggests that the central-limit theorem is applicable; n_c and n_s are formed by summing a large number of variates, each of which makes but a small contribution to the sum. The crucial question is the statistical dependence of the individual terms. More sophisticated analysis, e.g., Davenport and Root (1958, chap. 8), reveals that the individual terms are statistically independent. Hence n_c and n_s are *gaussian-distributed* with zero mean and variance $M\eta\,\Delta f = \eta B = N$. It can also be shown that n_c and n_s are themselves statistically inde-pendent, leading to the joint density function

$$p(n_c, n_s) = p(n_c)p(n_s) = \frac{1}{2\pi N} e^{-(n_c{}^2 + n_s{}^2)/2N} \tag{4.59}$$

where

$$\bar{n}_c = \bar{n}_s = 0 \qquad \overline{n_c{}^2} = \overline{n_s{}^2} = N = \eta B$$

For some purposes, (4.57) is not so useful as an *envelope-and-phase representation* of $n(t)$; that is, we would rather have

$$n(t) = r_n(t) \cos [2\pi f_c t + \phi_n(t)] \qquad (4.60)$$

where $r_n(t)$ is the envelope (defined to be nonnegative) and $\phi_n(t)$ the phase. Clearly, r_n and ϕ_n are random variables related to n_c and n_s. In fact,

$$r_n{}^2 = n_c{}^2 + n_s{}^2 \qquad \phi_n = \arctan \frac{n_s}{n_c}$$

as seen from the phasor construction of Fig. 4.21. Conversely, $n_c = r_n \cos \phi_n$, and $n_s = r_n \sin \phi_n$.

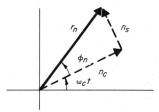

Fig. 4.21 *Phasor diagram for narrowband noise.*

Recalling our derivation of the Rayleigh distribution in Sec. 4.4, it immediately follows that r_n and ϕ_n are *statistically independent*, the former with a *Rayleigh distribution*, the latter *uniformly distributed*. Hence

$$q(r_n,\phi_n) = \frac{r_n}{2\pi N} e^{-r_n{}^2/2N} \qquad \begin{matrix} r_n \geq 0 \\ |\phi_n| \leq \pi \end{matrix} \qquad (4.61)$$

comparable to (4.42). The statistical averages are then

$$\bar{r}_n = \sqrt{\frac{\pi}{2} N} \qquad \bar{\phi}_n = 0$$

$$\overline{r_n{}^2} = 2N \qquad \overline{\phi_n{}^2} = \frac{\pi^2}{3}$$

The fact that $\overline{r_n{}^2} = 2N$ is twice $\overline{n^2} = N$ agrees with $\langle [r_n \cos (\omega_c t + \phi_n)]^2 \rangle = \frac{1}{2}\langle r_n{}^2 \rangle = \langle n^2 \rangle$. The mean value of the envelope, $\bar{r}_n$, is not zero because r_n is never negative.

Summarizing, and referring to Fig. 4.21, we can describe narrowband noise as a *single* phasor of frequency f_c with random amplitude r_n and random phase ϕ_n. The phasor amplitude (envelope) is Rayleigh distributed, while the phase has a uniform distribution. With $B \ll f_c$, the envelope and phase variations are slow compared to the phasor rotation rate. As before, the phasor projection on the real axis is just $n(t)$. A typical sample function of $n(t)$ is illustrated in Fig. 4.22 for $B \approx 0.05 f_c$.

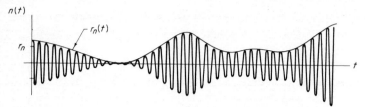

Fig. 4.22 *A sample function of narrowband noise, $B \approx 0.05f_c$.*

Finally, in retrospect, it can be remarked that our analysis is not strictly contingent upon the condition $B \ll f_c$. In fact, Eqs. (4.57) to (4.61) are valid for *wideband* noise as well as for narrowband noise. However, the envelope-and-phase description of wideband noise is less meaningful from a physical viewpoint, since the envelope-and-phase variations are not slow compared to f_c.

PROBLEMS

4.1 In five-card draw poker, what is the probability of being dealt a flush with nothing wild and with deuces wild? Suppose you are dealt three hearts and two other cards; what is the probability of drawing two cards to fill the flush—again, with and without deuces wild?

4.2 In five-card stud poker each player has four cards exposed. Suppose one of your opponents has four hearts showing, while no other hearts are visible. Find the probability that your opponent has a flush if there are five players altogether.

4.3 A binary data system uses two symbols, 0 and 1. They are transmitted with probabilities P_0 and P_1 and with error probabilities p_0 and p_1 at the receiver. Write an expression for the expected number of errors in a sequence of $n \gg 1$ symbols.

4.4 A pair of honest dice is rolled. Totaling the number of spots which come up, the possible outcomes are the integers $x_j = 2, 3, \ldots , 12$. Show that the probabilities are

$$P(x_j) = \begin{cases} \dfrac{x_j - 1}{36} & 2 \le x_j \le 7 \\[2mm] \dfrac{13 - x_j}{36} & 7 \le x_j \le 12 \end{cases}$$

Plot the corresponding distribution function.

4.5 Let x be a discrete variate that is equally likely to be any of the integers 0, 1, $\ldots$, 9. Plot the frequency function, distribution function, and density function.

4.6 Suppose a certain random variable has

$$F(x) = \begin{cases} 0 & x < 0 \\ Kx^2 & 0 \le x \le 10 \\ 100K & x > 10 \end{cases}$$

Determine K and plot $p(x)$.

4.7 Let X and Y be independent random variables with density functions $p(x)$ and $q(y)$. If $Z = X + Y$, show that

$$p_z(z) = \int_{-\infty}^{\infty} p(x)q(z - x)\, dx$$

Hint: $P(Z \leq z) = P(Y \leq z - x)$, etc.

4.8 Calculate the mean and variance of X and Y in Example 4.4.

4.9 Calculate the mean and variance of Z in Example 4.5.

4.10 A large number of independent observations are made on a random process. Designating the outcome of the jth observation as X_j, let

$$Y = \frac{1}{N} \sum_{j=1}^{N} X_j$$

where N is the number of observations. Find the expected range of Y, say $\bar{y} \pm 2\sigma_y$, in terms of $\bar{x}$ and σ_x.

4.11 Prove that $\overline{xy} = \bar{x}\bar{y}$ when X and Y are statistically independent.

4.12 The *characteristic function* associated with a random variable X is defined as the statistical average

$$M_x(jv) = \overline{e^{jvx}}$$

From this definition and Eq. (4.26), show the following relations:

$$|M_x(jv)| \leq M_x(0) = 1$$

$$\overline{x^n} = (-j)^n \frac{d^n M_x(jv)}{dv^n} \bigg|_{v=0}$$

$$p(x) = \frac{1}{2\pi} \int_{-\infty}^{\infty} M_x(jv)e^{-jvx}\, dv$$

4.13 Let $Z = X + Y$, where X and Y are statistically independent. Show that $M_z(jv) = M_x(jv)M_y(jv)$ and therefore

$$p(z) = \frac{1}{2\pi} \int_{-\infty}^{\infty} M_x(jv)M_y(jv)e^{-jvz}\, dv$$

Does this expression agree with Prob. 4.7?

4.14 By applying the binomial distribution, find the probability that there will be fewer than three heads when 10 honest coins are tossed.

4.15 The one-dimensional *random-walk problem* is stated as follows. A man walks along a straight line taking steps of equal length L. At any time he may step forward with probability p or backward with probability q. Describe his position with respect to the starting point after m steps have been taken.

4.16 You have designed a data transmission system with an error probability of 10^{-6} per digit. The customer will test the system by sending a known message of 10^6 digits and checking the received message. If there are more than two errors you will be fired. Calculate the probability of losing your job.

4.17 Let X be uniformly distributed over $|x| \leq 1$. Plot $P(|X| \leq k\sigma_x)$ versus k and the corresponding bound given by Chebyshev's inequality (4.30).

4.18 Consider the random angle Θ. If we say it is uniformly distributed over $[-\pi,\pi]$, then $\bar{\theta} = 0$. If we say it is uniformly distributed over $[0,2\pi]$, then $\bar{\theta} = \pi$. Physically the descriptions should be equivalent. Discuss this apparent contradiction.

4.19 A gaussian noise voltage of zero mean and variance σ^2 is full-wave rectified. Determine the mean-square value of the rectified wave.

4.20 In manufacturing 100-ohm resistors it is found that the actual resistance is approximately gaussian-distributed with $\bar{x} = 100$ and $\sigma_x = 5$. If the actual resistance differs from 100 ohms by more than ± 10 percent, the resistor is not salable.

(a) What percentage of the resistors must be thrown away?

(b) What percentage of the resistors are 1 percent resistors, that is, $99 < X < 101$?

(c) If 1 percent resistors sell for 10 cents, 5 percent resistors sell for 5 cents, 10 percent resistors sell for 2 cents, and the manufacturing cost is 3 cents per resistor, what is the average profit per resistor? Assume that a 5 percent resistor has $95 < X < 99$ or $101 < X < 105$, etc.

4.21 The tabulated *error function* is defined as

$$\text{erf } (u) = \frac{2}{\sqrt{\pi}} \int_0^u e^{-v^2} \, dv$$

Write $F(x)$ for a gaussian variate in terms of erf (u) and relate erf (u) to $\Phi(k)$ as defined in (4.36).

4.22 Derive the approximation (4.39) by starting from (4.36) and integrating by parts.

4.23 Suppose $Z = X + Y$, where X and Y are statistically independent and gaussian-distributed. Find $p(z)$ in terms of m_x, σ_x, m_y, and σ_y.

4.24 In addition to the mean, there are two other measures of the "center" of a probability distribution, namely, the *median* and the *mode*. The median is that value for which $F(x) = \frac{1}{2}$; the mode is that value for which $p(x)$ is maximum. (a) Find the median and mode of the gaussian and Rayleigh distributions, writing them in terms of the mean. (b) Explain why some distributions may not have a meaningful median or mode.

4.25 A certain random signal has a dc component of 2 volts and an rms value of 4 volts. Further measurements indicate that $x(t)$ and $x(t + \tau)$ are independent for $|\tau| \geq 5$ μsec, while $R(\tau)$ decreases linearly with $|\tau|$ for $0 \leq |\tau| \leq 5$ μsec. (a) Plot $R(\tau)$ and fully dimension. (b) Find and plot $G(f)$.

4.26 Let $y(t) = x(t) + x(t - T)$, where T is a constant and $x(t)$ is a random signal from an ergodic source. Find $R_y(\tau)$ and $G_y(f)$ in terms of $R_x(\tau)$ and $G_x(f)$.

4.27 Suppose $R(\tau) = K \exp (-a|\tau|) \cos 2\pi b\tau$. Find and sketch $G(f)$.

4.28 Consider the so-called *random cosine wave* $x(t) = A \cos (2\pi f_c t + \theta)$, where A and f_c are constants and θ is uniformly distributed over $[-\pi, \pi]$. Taking power spectrum as defined by (4.49), show that $G_x(f) = (A^2/4)[\delta(f - f_c) + \delta(f + f_c)]$.

4.29 Show that the system of Fig. P 4.1 is a means of approximately measuring $R_x(\tau)$ (the filter was discussed in Example 3.2). If $G_x(f)$ has negligible content for $|f| > W$, what is the condition on T for accurate results?

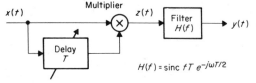

$$H(f) = \text{sinc } fT \, e^{-j\omega T/2}$$

Fig. P 4.1

4.30 Let white noise be the input to a gaussian filter having $H(f) = K \exp (-af^2)$. Find the output power spectrum and autocorrelation function.

4.31 A Butterworth lowpass filter has $|H(f)| = [1 + (f/f_3)^{2n}]^{-\frac{1}{2}}$. Calculate B_N/f_3 for $n = 1, 2, 3,$ and 10.

4.32 A filter is matched to the *deterministic* signal $x(t)$ in the sense of Example 3.3. Suppose $x(t)$ and white noise having density η are applied at the filter's input. Determine the ratio of the maximum output signal voltage to the rms output noise voltage.

4.33 Narrowband noise is processed by a device whose output $y(t)$ is the square of the input envelope. Find the probability density function $p(y)$.

4.34 A certain device has output $y(t)$ related to the input $x(t)$ by

$$y(t) = \begin{cases} 1 & x(t) > A \\ 0 & x(t) < A \end{cases}$$

where $A^2 \ll \overline{x^2}$. If $x(t)$ is narrowband noise, sketch a typical sample of $y(t)$ and write an approximate expression for $p(y)$.

4.35 Consider the signal $y(t) = n(t) \cos(\omega_c t + \theta)$, where $n(t)$ is narrowband noise centered on f_c. Analyze and discuss the properties of $y(t)$, giving attention to its dependence on θ.

SELECTED SUPPLEMENTARY READING

Of the numerous introductory texts on probability theory and random variables, Cramér (1955), Pfeiffer (1965), and Drake (1967) deserve special mention. All three develop the subject from the axiomatic viewpoint, as contrasted with the less rigorous intuitive approach used here, yet they do not sacrifice readability and lively style in the name of mathematical sophistication.

Short treatments of random-signal theory, essentially paralleling that of the present chapter, can be found in Aseltine (1958, chaps. 14 and 15), Schwartz (1959, chap. 7), Hancock (1961, chap. 3), Harman (1963, chap. 3), Downing (1964, chaps. 1 to 3), or Panter (1965, chap. 4). Cooper and McGillem (1967, chaps. 9 to 12) and Beckmann (1967, chaps. 6 and 7) have good discussions of random processes and correlation functions.

More comprehensive and advanced presentations are given by Bendat (1958), Davenport and Root (1958), Lee (1960), and Papoulis (1965). In addition, two excellent papers on noise analysis have appeared in the literature, namely, Rice (1944, 1945) and Bennett (1956). The former is a classic, while the latter is of a tutorial nature.

For references dealing specifically with electrical noise and its physical sources, see the Selected Supplementary Reading for the Appendix.

5
Linear (Amplitude) Modulation

The several purposes of modulation were itemized in Chap. 1, along with a qualitative description of the process. To briefly recapitulate: *modulation* is the systematic alteration of one waveform, called the *carrier wave*, according to the characteristics of another waveform, the modulating signal or message. The fundamental goal is producing an information-bearing modulated wave whose properties are best suited to the given communication task. Often, but not always, this entails frequency translation.

We now launch upon a quantitative discussion and analysis of modulation systems, the *how* and the *why*. This subject forms the core of our study of communication and will be divided into three parts: *linear modulation* (Chap. 5), *exponential modulation* (Chap. 6), and *pulse modulation* (Chap. 7).

The division of material, while convenient, is by no means clear-cut, and some overlapping will be encountered. In particular, both linear and exponential modulation are types of *continuous-wave* (c-w) *modulation;*

i.e., the carrier is a sinusoid.   Consequently, there are inevitable similarities of techniques and instrumentation.   Pulse modulation, for which
the carrier is a periodic pulse train, may resemble linear modulation,
exponential modulation, or neither, depending on the type.   But more
to the point, modern pulse technology has been applied with great success
to the generation and detection of c-w modulation; conversely, many
pulse-modulation systems involve c-w modulation for the final step in
transmission.

As to the specific topic at hand, linear or *amplitude* modulation is
essentially direct frequency translation of the message spectrum; double-
sideband modulation (DSB) is precisely that.   Minor modifications of the
translated spectrum yield conventional amplitude modulation (AM),
single-sideband modulation (SSB), or vestigial-sideband modulation
(VSB).   Each of these variations has its own distinct advantages and
significant practical applications.   Each will be given due consideration
in this chapter, including such matters as waveforms and spectra, detection methods, transmitters and receivers, and system performance in the
face of interference or noise.

Equally important, the concepts and analytic techniques developed
here for linear modulation will serve us in good stead when we come to
other, more complex communication systems.   Thus, the notation and
conventions introduced below will carry over in large part to later
chapters.

5.1 MESSAGE CONVENTIONS

Before plunging into the details of linear modulation, a few conventions
and matters of nomenclature must be disposed of.

Whenever possible, our analyses will be in terms of an *arbitrary
message* (modulating signal) designated by $x(t)$.   Better yet, $x(t)$ can
represent the *ensemble* of all probable messages from a given source.
Though such messages are not strictly bandlimited, it is usually safe to
say that there exists some upper frequency, call it W, above which the
spectral content is negligible and unnecessary for conveying the information in question.   This nominal upper-frequency limit might range from
a few hundred hertz (telegraph signals) to several megahertz (television
video).   Even greater extremes are found in telemetry or computer systems.   At any rate, we shall assume that, to a good approximation, the
message spectrum satisfies

$$X(f) = 0 \qquad \text{for } |f| > W$$

so W can be named the *message bandwidth*.   Figure 5.1 shows a representative spectrum.

Furthermore, for mathematical convenience, let the messages be scaled, or normalized, to have a magnitude not exceeding unity, so that

$$|x(t)| \leq 1$$

This convention has immediate implications on the average *message power;* clearly $\langle x^2(t) \rangle \leq 1$.   Alternately, if each member of the ensemble

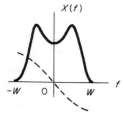

Fig. 5.1 *A representative message spectrum.*

is normalized as above, the ensemble average satisfies

$$\overline{x^2(t)} \leq 1$$

Of course, $\langle x^2(t) \rangle = \overline{x^2(t)}$ if the source is *ergodic,* a not unreasonable assumption in many cases.

Occasionally an analysis with arbitrary $x(t)$ proves difficult, if not impossible.   For such situations it is necessary to resort to a specific modulating signal, usually the simple sinusoid

$$x(t) = A_m \cos 2\pi f_m t \qquad \begin{array}{l} A_m \leq 1 \\ f_m < W \end{array}$$

which is called *tone modulation*. With tone modulation, a positive-frequency line spectrum will suffice for frequency-domain studies.

Although tone modulation may seem unduly simplified, the approach has definite advantages. For one thing, tones are often the only tractable signals in complex problems; they facilitate the calculation of spectra, average powers, etc., that otherwise would be prohibitive. Moreover, if we can find the response of the modulation system to a specific frequency in the message band, we can infer the response for all frequencies in W.

Nonetheless, if superposition is not applicable, one must resort to *multitone* analysis to check on intermodulation effects and so forth.

5.2   DOUBLE-SIDEBAND AMPLITUDE MODULATION: AM AND DSB

Historically, the first type of c-w modulation to evolve was conventional amplitude modulation, the familiar AM of standard radio broadcasting. We shall reserve the term *amplitude modulation* for this specific type, and use *linear modulation* for the general class.   DSB differs from AM by *carrier suppression*, a minor alteration that has major repercussions.

Amplitude modulation (AM)

The unique feature of AM is that the *envelope* of the modulated carrier has the same *shape* as the message waveform.   This is achieved by adding the translated message, appropriately proportioned, to the unmodulated carrier.   Specifically, the modulated signal is

$$
\begin{aligned}
x_c(t) &= A_c \cos \omega_c t + m x(t) A_c \cos \omega_c t \\
&= A_c[1 + m x(t)] \cos \omega_c t
\end{aligned}
\tag{5.1}
$$

where $A_c \cos \omega_c t$ is the unmodulated carrier, $f_c = \omega_c/2\pi$ is the carrier frequency, and the constant m is called the *modulation index*.   Since A_c is the unmodulated carrier amplitude, we can think of the modulated amplitude as being a linear function of the message, namely,

$$
A_c(t) = A_c[1 + m x(t)]
$$

underscoring the meaning of *amplitude* modulation.

Figure 5.2 shows a portion of a typical message and the resulting AM wave for two values of m.   The envelope, shown dashed, has the shape of $x(t)$ providing that: (1) the carrier frequency is much greater than the rate of variation of $x(t)$, otherwise an envelope cannot be visualized; and (2) there are no *phase reversals* in the modulated wave, i.e., the amplitude $A_c[1 + m x(t)]$ does not go negative.   Preserving the desired relationship between envelope and message thus requires

$$
f_c \gg W \qquad \text{and} \qquad m \leq 1
\tag{5.2}
$$

The latter follows since $|x(t)| \leq 1$ by our convention.

The carrier-frequency condition is, of course, in agreement with the frequency-translation aspect of modulation, though it sometimes proves to be detrimental.   The condition $m \leq 1$ sets an upper limit on how heavily the carrier can be modulated.   With $m = 1$, known as 100 percent modulation, the modulated amplitude varies between 0 and $2A_c$. *Overmodulation*, $m > 1$, results in carrier phase reversals and *envelope distortion*, as illustrated in Fig. 5.2c.

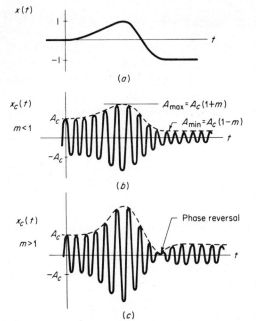

Fig. **5.2** *AM waveforms. (a) $x(t)$;*
(b) $x_c(t)$, $m < 1$; and (c)
$x_c(t)$, $m > 1$.

Turning to the frequency domain, the Fourier transform of an AM wave is easily found to be

$$X_c(f) = \frac{A_c}{2} [\delta(f - f_c) + \delta(f - f_c)]$$

$$+ \frac{mA_c}{2} [X(f - f_c) + X(f + f_c)] \quad (5.3)$$

While Eq. (5.3) may look imposing, the spectrum consists of nothing more than the translated message spectrum, plus a pair of impulses at $\pm f_c$ representing the carrier itself (Fig. 5.3).

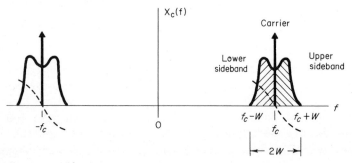

Fig. **5.3** *AM spectrum.*

Two properties of the AM spectrum should be carefully noted. First, there is *symmetry* about the carrier frequency, the amplitude being even and the phase being odd. That portion of the spectrum above f_c is called the *upper sideband*, while that below f_c is the *lower sideband;* hence the designation *double-sideband* amplitude modulation.

Second, the *transmission bandwidth* B_T required for an AM signal is exactly *twice* that of the message bandwidth, so

$$B_T = 2W \tag{5.4}$$

This result points out that AM is not attractive when one must conserve bandwidth—after all, the message could be sent in one-half the AM bandwidth if modulation were omitted.

We shall find that transmission bandwidth is one of the characteristic parameters of modulation types; another important parameter is the *average transmitted power* in the modulated wave, designated as P_T. Forming the mean-square value of $x_c(t)$, per Eq. (5.1), yields

$$P_T = \langle x_c{}^2(t) \rangle = A_c{}^2 \langle [1 + mx(t)]^2 \cos^2 \omega_c t \rangle$$
$$= \frac{A_c{}^2}{2} \{ \langle 1 + 2mx(t) + m^2x^2(t) \rangle + \langle [1 + mx(t)]^2 \cos 2\omega_c t \rangle \}$$

where we have used $\cos^2 \omega_c t = \frac{1}{2} + \frac{1}{2} \cos 2\omega_c t$. Under the condition $f_c \gg W$, the second term averages to zero; then, if the dc component of the message is also zero,[1] we obtain

$$P_T = [1 + m^2 \langle x^2(t) \rangle] \frac{A_c{}^2}{2}$$

Or, if the message source is ergodic,

$$P_T = (1 + m^2 \overline{x^2}) \frac{A_c{}^2}{2} \tag{5.5}$$

where $\overline{x^2}$ is the ensemble average.

Equation (5.5) has an interesting interpretation relative to the AM spectrum. From Fig. 5.3, P_T includes the power in the carrier-frequency component plus two symmetric sidebands. Hence

$$P_T = \frac{A_c{}^2}{2} + m^2 \overline{x^2} \frac{A_c{}^2}{2} = P_c + 2P_{\text{SB}}$$

where the *carrier power* is

$$P_c = \frac{1}{2} A_c{}^2$$

[1] For reasons discussed in Sec. 5.5, messages having $\langle x(t) \rangle \neq 0$ are seldom transmitted via AM.

and the *power per sideband* is

$$P_{SB} = m^2\overline{x^2}\frac{A_c{}^2}{4} = \frac{1}{2}m^2\overline{x^2}P_c \leq \frac{1}{2}P_c$$

The upper limit on P_{SB} follows from $m^2\overline{x^2} \leq 1$.   Thus

$$P_c = P_T - 2P_{SB} \geq \frac{1}{2}P_T \tag{5.6}$$

which says that at least 50 percent of the total transmitted power resides in the carrier.

But, referring back to (5.1), the carrier term alone is seen to be independent of the message and does not contain any of the message information.   We therefore conclude that a substantial portion of the transmitted power P_T is "wasted" in the carrier itself.   It will later be shown that true envelope modulation, and the simplicity of *envelope detection*, depends on this wasted power.

While average powers are convenient for theoretical purposes, practical radio transmitters are often rated in terms of *instantaneous* power.   From (5.1), the maximum voltage is $x_{c_{max}} = 2A_c$, so the *peak* instantaneous power is proportional to $4A_c{}^2$.

AM with tone modulation

Setting $x(t) = A_m \cos 2\pi f_m t$ gives the tone-modulated AM waveform

$$\begin{aligned}
x_c(t) &= A_c(1 + mA_m \cos \omega_m t) \cos \omega_c t \\
&= A_c \cos \omega_c t + \frac{mA_mA_c}{2}[\cos (\omega_c - \omega_m)t + \cos (\omega_c + \omega_m)t] \quad (5.7)
\end{aligned}$$

where we have used the trigonometric expansion for the product of two cosines.   The corresponding positive-frequency line spectrum is shown in Fig. 5.4.

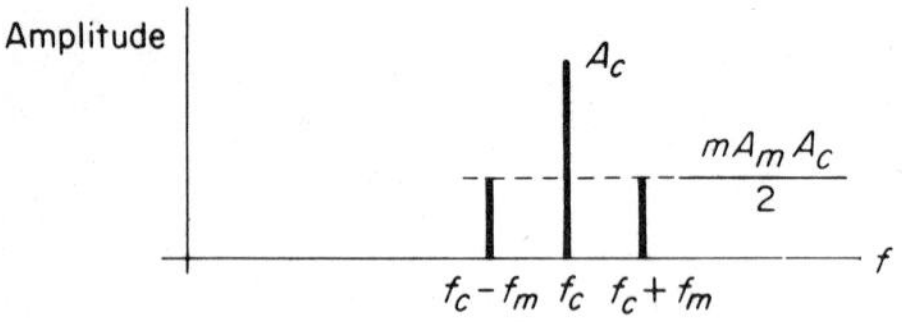

Fig. **5.4** *AM line spectrum: tone modulation.*

Going further, we can nicely bring out the linearity property of AM by examining *multitone modulation*, $x(t) = \sum_i A_i \cos \omega_i t$, for which

$$x_c(t) = A_c \cos \omega_c t + \frac{mA_c}{2}\sum_i A_i[\cos (\omega_c - \omega_i)t + \cos (\omega_c + \omega_i)t]$$

The spectrum then contains one *pair* of sideband lines for each modulating tone. As seen in Fig. 5.5, where there are three tones, the sideband lines are superposed about the carrier as if each pair came from single-tone modulation. Of course, message normalization is required to avoid envelope distortion, that is, $\sum_i A_i \le 1$.

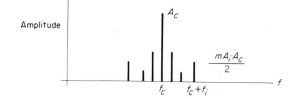

Fig. 5.5 *AM line spectrum: three modulating tones.*

With tone modulation, an AM wave can be treated as a sum of *phasors*, one for each spectral line. An especially informative way of constructing the diagram is shown in Fig. 5.6, where the sideband phasors have been added to the tip of the carrier phasor (a single tone is assumed for simplicity). Since the carrier phasor rotates at f_c Hz, the sideband phasors rotate at speeds of $\pm f_m$ *relative* to the carrier. Note that as long as the sideband lines are equal and of correct phase, the resultant of the sideband phasors is *collinear* with the carrier phasor.

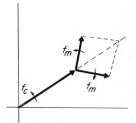

Fig. 5.6 *Phasor diagram for AM with tone modulation.*

As usual, the projection of the phasor sum on the real axis gives $x_c(t)$. But more important, the *magnitude* of the phasor sum is the modulated carrier *envelope*. This observation leads to a simple way of qualitatively studying the effects of transmission imperfections, interference, etc. For example, if the lower sideband line is severely attenuated

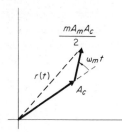

Fig. 5.7

(Fig. 5.7), the resultant envelope is

$$r(t) = \sqrt{\left(A_c + \frac{mA_mA_c}{2}\cos\omega_m t\right)^2 + \left(\frac{mA_mA_c}{2}\sin\omega_m t\right)^2}$$

$$= A_c\sqrt{1 + \left(\frac{mA_m}{2}\right)^2 + mA_m\cos\omega_m t}$$

from which the *envelope distortion* can be determined.

Double-sideband suppressed-carrier modulation (DSB)

Because the carrier-frequency component of AM is independent of the message and represents wasted power, it can just as well be eliminated from the modulated wave. This results in double-sideband suppressed-carrier amplitude modulation—*double-sideband modulation*, or DSB[1] for short.

Dropping the carrier term and the now meaningless modulation index from Eq. (5.1) gives

$$x_c(t) = x(t)A_c\cos\omega_c t \tag{5.8}$$

for the definition of DSB. Note that, unlike AM, the modulated wave is zero in absence of modulation: $x_c(t) = 0$ when $x(t) = 0$. Since the carrier has been suppressed, the average transmitted power is

$$P_T = 2P_{SB} = \tfrac{1}{2}\overline{x^2}A_c^2$$

while the peak power is proportional to A_c^2. The latter is one-fourth that of AM, other things being equal, and can be a significant factor in transmitter design.

The DSB spectrum is simply the translated message spectrum

$$X_c(f) = \frac{A_c}{2}[X(f - f_c) + X(f + f_c)]$$

[1] The abbreviations DSB-SC and DSSC are also used.

sketched in Fig. 5.8.   Of course, the transmission bandwidth is unchanged from the AM case, so

$$B_T = 2W$$

Comparing Figs. 5.8 and 5.3 shows that AM and DSB are quite similar in the frequency domain, but the time-domain picture is another story.   As illustrated by Fig. 5.9, the DSB envelope is not the same

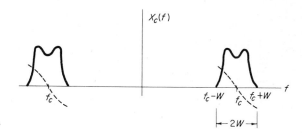

Fig. 5.8   *DSB spectrum.*

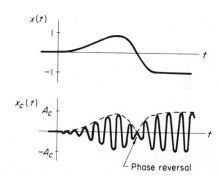

Fig. 5.9   *DSB waveforms.*

shape as the message.   The envelope is, in fact, proportional to $|x(t)|$, and negative values of $x(t)$ are reflected in the carrier *phase*.   So whenever $x(t)$ crosses zero, the modulated wave has a *phase reversal*.   Full recovery of the message therefore entails an awareness of these phase reversals, and an *envelope* detector would not be sufficient.   Stated another way, DSB involves more than just "amplitude" modulation.

This consideration suggests, and later analysis confirms, that there is a trade-off between power efficiency and demodulation methods. DSB conserves power but requires complex detection circuitry; conversely, AM demodulation is simply envelope detection but at the cost of greater transmitted power.

5.3 MODULATORS AND TRANSMITTERS

We have seen that *new* frequencies are created in the modulation process. The device which generates an AM wave, a *modulator*, must therefore be either *time-varying* or *nonlinear*, since linear time-invariant systems cannot produce new frequency components. Figure 5.10 illustrates two possible modulators directly derived from the definition of AM [Eq. (5.1)]. Because a multiplier is required in Fig. 5.10*b*, the system can be termed a *product modulator*. Indeed, recalling that multiplying by cos $\omega_c t$ yields frequency translation, the product operation is basic to linear modulation systems.

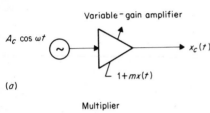

(a)

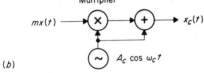

(b)

Fig. **5.10** *AM modulators.* (a) *Variable-gain modulator;* (b) *product modulator.*

In practice, multiplication is achieved with the aid of a nonlinear element and filtering, arranged as in Fig. 5.11, where the nonlinearity may be of the *power-law* or *piecewise-linear* (switching) variety.

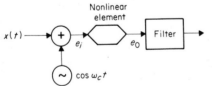

Fig. **5.11**

Power-law modulators

Taking the power-law case, let the nonlinear device have as its transfer characteristic

$$e_0 = a_1 e_i + a_2 e_i^2$$

By omitting higher-order terms, we are assuming a *square-law* device.

In Fig. 5.11, $e_i(t) = x(t) + \cos \omega_c t$, so

$$e_0(t) = a_1 x(t) + a_2 x^2(t) + a_2 \cos^2 \omega_c t + a_1 \left[1 + \frac{2a_2}{a_1} x(t) \right] \cos \omega_c t$$

The last term is the desired AM wave, with $A_c = a_1$ and $m = 2a_2/a_1$, providing it can be separated from the rest.

As to the feasibility of separation, we have already studied the spectrum of $e_0(t)$ in conjunction with nonlinear distortion. For convenience, Fig. 3.17 is repeated here as Fig. 5.12. Clearly, if $f_c > 3W$,

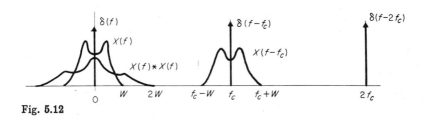

Fig. 5.12

the required separation can be accomplished by a bandpass filter of bandwidth $B_T = 2W$ centered at f_c.

A circuit realization of the complete modulator, employing a square-law diode,[1] is shown in simplified form in Fig. 5.13. The battery serves to bias the diode at a suitable point on its current-voltage curve.

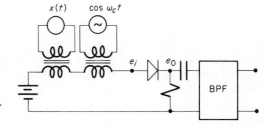

Fig. 5.13 *Square-law diode modulator.*

Because of the heavy filtering required, power-law modulators are used primarily for *low-level* modulation, i.e., at power levels less than the transmitted value. Substantial linear amplification is then necessary to bring the power up to P_T. But RF power amplifiers of the required linearity are not without problems of their own, and it often is better to employ *high-level* modulation if P_T is to be large.

---

[1] Nonlinear resistors (varistors), transistors, or vacuum tubes are just as satisfactory.

Switching modulators

Efficient high-level modulators are arranged so that undesired modulation products never fully develop and need not be filtered out. This is usually accomplished with the aid of a *switching* device, whose detailed analysis is postponed to Chap. 7. However, the basic operation of the supply-voltage modulated class C amplifier is readily understood from its idealized equivalent circuit and waveforms (Fig. 5.14).

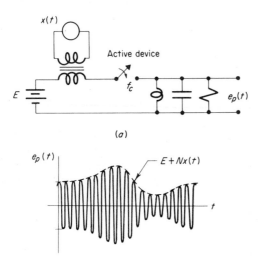

Fig. 5.14 *Supply-voltage modulated class C amplifier. (a) Equivalent circuit; (b) output waveform.*

The active device, a vacuum tube or transistor, serves as a switch driven at the carrier frequency, closing briefly every $1/f_c$ sec. The load, called a *tank* circuit, is tuned to resonate at f_c, so the switching action causes the tank circuit to "ring" sinusoidally. The steady-state load voltage in absence of modulation is then $e_p(t) = E \cos \omega_c t$. Adding the message to the supply voltage, say via transformer, gives $e_p(t) = [E + Nx(t)] \cos \omega_c t$, where N is the transformer turns ratio. If E and N are correctly proportioned, the desired modulation has been accomplished without appreciable generation of undesired components. For additional details, see any standard text on radio electronics, e.g., Terman (1955).

A complete AM transmitter is diagramed schematically in Fig. 5.15 for the case of high-level modulation. The carrier wave is generated by a crystal-controlled oscillator to ensure stability of the carrier frequency. Because high-level modulation demands husky input signals, both the carrier and message are amplified before modulation. The modulated signal is then delivered directly to the antenna.

The performance of such transmitters can be further improved through the use of *negative feedback*. For this purpose, the transmitted signal is demodulated to recover $x_e(t)$, the message as it appears in the

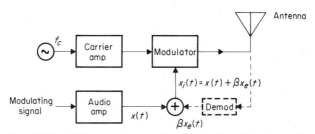

Fig. 5.15 *AM transmitter with high-level modulation and feedback.*

modulated wave. The net input to the modulator is then $x_i(t) = x(t) + \beta x_e(t)$, where β is the feedback factor. Following the usual analysis,[1] the closed-loop gain is

$$A_f = \frac{A_0}{1 + \beta A_0}$$

where A_0 is the open-loop gain of the modulator insofar as the envelope is concerned. Hence, if $|\beta A_0| \gg 1$, then $A_f \approx 1/\beta$ and performance is largely independent of imperfections in the modulation circuitry.

Balanced modulators

DSB modulators differ from AM modulators only by the suppression of the carrier component; in theory, this requires nothing more or less than a *multiplier* [see Eq. (5.8)]. As the reader can verify, the system of Fig. 5.11 will generate DSB if the nonlinear element is a *perfect* square-law device such that $e_0 = ae_i^2$.

Unfortunately, perfect square-law devices are rare, so in practice DSB is obtained using *two* AM modulators arranged in a balanced configuration to cancel out the carrier. Figure 5.16 shows such a *balanced modulator* in block-diagram form. Assuming the AM modulators are identical, save for the reversed sign of one input, the outputs are $A_c[1 + \frac{1}{2}x(t)] \cos \omega_c t$ and $A_c[1 - \frac{1}{2}x(t)] \cos \omega_c t$. Subtracting one from the other yields $x_c(t) = x(t)A_c \cos \omega_c t$, as required. Hence, a balanced

[1] See any text covering feedback amplifiers.

modulator is a *multiplier*.   It should be noted that if the message has a
dc term, that component is *not* canceled out in the modulator, even though
it appears at the carrier frequency in the modulated wave.

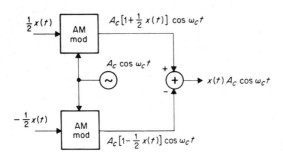

Fig. 5.16   *Balanced modulator.*

A balanced modulator using switching is discussed in Chap. 7 under
the heading of *chopper modulation*.   Other circuit realizations can be
found in the literature.

5.4 SUPPRESSED-SIDEBAND MODULATION: SSB AND VSB

Conventional AM is wasteful of both transmitted power and transmission
bandwidth.   Suppressing the carrier overcomes the former shortcoming;
suppressing one sideband, in whole or part, reduces the latter and leads to
single-sideband modulation (SSB) or vestigial-sideband modulation
(VSB).

Single-sideband modulation (SSB)

The upper and lower sidebands of AM and DSB are uniquely related by
symmetry about the carrier frequency; given the amplitude and phase of
one, we can always construct the other.   Obviously, transmitting both
sidebands is redundant and wasteful of bandwidth.   However, transmission bandwidth can be cut in half if one sideband is suppressed along
with the carrier.

T otal elimination of carrier and one sideband from the AM spectrum
in Fig. 5.3 produces SSB, for which it is obvious that

$$B_T = W \qquad \text{and} \qquad P_T = P_{SB} = \tfrac{1}{4}\overline{x^2}A_c{}^2$$

However, peak-power calculations are quite involved, and will be omitted
here.

SSB is readily visualized in the frequency domain: it is the output of a balanced modulator processed by a *sideband filter* (Fig. 5.17a). The filter is a bandpass circuit passing the upper sideband (SSB_u) or the lower sideband (SSB_l), so the resulting spectra are as given in Fig. 5.17b.

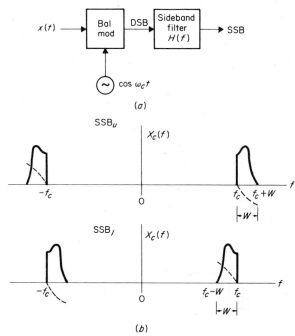

Fig. 5.17 *Single-sideband modulation. (a) Modulator; (b) spectra.*

On the other hand, a time-domain description of the waveform is somewhat more difficult, save for the case of *tone modulation*. With tone modulation $x_c(t)$ is simplicity itself; suppressing the carrier and one sideband line in Fig. 5.4 leaves only the other sideband line, hence

$$x_c(t) = \tfrac{1}{2} A_m A_c \cos (\omega_c \pm \omega_m)t \qquad (5.9)$$

The frequency of the modulated wave is therefore offset from f_c by $\pm f_m$, while the amplitude is constant but proportional to the tone amplitude.

For arbitrary $x(t)$, an analysis given later shows that

$$x_c(t) = \tfrac{1}{2} A_c[x(t) \cos \omega_c t \mp \hat{x}(t) \sin \omega_c t] \qquad (5.10)$$

where the upper sign is taken for SSB_u and vice versa, and where $\hat{x}(t)$ is the *Hilbert transform* of $x(t)$. Mathematically, $\hat{x}(t)$ was defined in Sec.

3.5 as

$$\hat{x}(t) = x(t) * \frac{1}{\pi t} = \frac{1}{\pi} \int_{-\infty}^{\infty} \frac{x(t')}{t - t'} \, dt'$$

Physically, $\hat{x}(t)$ is simply $x(t)$ with its frequency components phase-shifted by $-\pi/2$ rad, i.e., a *quadrature* shift. Checking (5.10) with tone modulation, $\hat{x}(t) = A_m \widehat{\cos \omega_m t} = A_m \cos(\omega_m t - \pi/2) = A_m \sin \omega_m t$, so $x_c(t) = \frac{1}{2}A_c(A_m \cos \omega_m t \cos \omega_c t \mp A_m \sin \omega_m t \sin \omega_c t)$, which reduces to Eq. (5.9).

Further examination of (5.10) reveals an interesting property of SSB, namely, that the modulated wave consists of *two carriers in phase quadrature*, the carriers being double-sideband modulated by $x(t)$ and $\hat{x}(t)$, respectively. Since $x_c(t)$ is *single* sideband, the two DSB waves must be such that their sidebands cancel on one side of the carrier frequency and add on the other side. This observation leads to an alternate method of SSB generation known as the *phase-shift* or *out-phasing* method, to be discussed shortly. For the moment, let us derive (5.10) and examine the envelope of SSB.

SSB waveforms and envelopes ★

To begin the derivation we note from Fig. 5.17a that $x_c(t)$ can be written in terms of the impulse response of the sideband filter. Specifically, $x_c(t)$ is the convolution of $h(t)$ with $x(t)A_c \cos \omega_c t$,

$$x_c(t) = h(t) * [x(t)A_c \cos \omega_c t] = \int_{-\infty}^{\infty} h(t')x(t - t')A_c \cos \omega_c(t - t') \, dt'$$

Expanding $\cos \omega_c(t - t')$ and regrouping gives

$$\begin{aligned}
x_c(t) &= \left[\int_{-\infty}^{\infty} h(t') \cos \omega_c t' \, x(t - t') \, dt' \right] A_c \cos \omega_c t \\
&\quad + \left[\int_{-\infty}^{\infty} h(t') \sin \omega_c t' \, x(t - t') \, dt' \right] A_c \sin \omega_c t \\
&= \xi_c(t) A_c \cos \omega_c t + \xi_s(t) A_c \sin \omega_c t
\end{aligned} \tag{5.11a}$$

where

$$\xi_c(t) = [h(t) \cos \omega_c t] * x(t) \qquad \xi_s(t) = [h(t) \sin \omega_c t] * x(t) \tag{5.11b}$$

Now the bracketed terms in (5.11b) can be thought of as representing a pair of filters related to the sideband filter by

$$H_c(f) = \mathfrak{F}[h(t) \cos \omega_c t] = \frac{1}{2}[H(f - f_c) + H(f + f_c)]$$

$$H_s(f) = \mathfrak{F}[h(t) \sin \omega_c t] = -\frac{j}{2}[H(f - f_c) - H(f + f_c)]$$

as obtained from the frequency-translation theorem and the fact that $\mathfrak{F}[h(t)] = H(f)$. Taking the sideband filter as an ideal BPF passing $f_c - W \leq |f| \leq f_c$ or $f_c \leq |f| \leq f_c + W$, it is a simple matter to show that over the message bandwidth $|f| < W$

$$H_c(f) = \tfrac{1}{2}$$

while

$$H_s(f) = \pm\tfrac{1}{2}j \operatorname{sgn} f = \begin{cases} \pm\tfrac{1}{2}j & |f| > 0 \\ \mp\tfrac{1}{2}j & |f| < 0 \end{cases}$$

where the upper sign corresponds to upper-sideband SSB, etc. Thus, $H_c(f)$ is no filter at all, and $H_s(f)$ is a quadrature phase shifter.

Applying these results to (5.11b), we have immediately

$$\xi_c(t) = [h(t) \cos \omega_c t] * x(t) = \mathfrak{F}^{-1}[H_c(f)X(f)]$$
$$= \mathfrak{F}^{-1}[\tfrac{1}{2}X(f)] = \tfrac{1}{2}x(t)$$

Then, recalling from (3.35a) that $1/\pi t \leftrightarrow -j \operatorname{sgn} f$ gives

$$h(t) \sin \omega_c t = \mathfrak{F}^{-1}[H_s(f)] = \mp\frac{1}{2\pi t}$$

so

$$\xi_s(t) = \left(\mp\frac{1}{2\pi t}\right) * x(t) = \mp\tfrac{1}{2}\hat{x}(t)$$

Finally, inserting in (5.11a) yields for the modulated waveform

$$x_c(t) = \tfrac{1}{2}A_c[x(t) \cos \omega_c t \mp \hat{x}(t) \sin \omega_c t]$$

as previously asserted.

As an exercise, the reader may wish to consider the two components of $x_c(t)$ in the frequency domain and verify that the sidebands either add or cancel. For this purpose the *analytic-signal* representation of Sec. 3.5 is most convenient. (It should now be recognized that Fig. 3.24b is an SSB$_u$ spectrum.)

To complete the discussion, let us find the SSB *envelope* and *phase* for an arbitrary modulating signal. Treating the two terms of (5.10) as quadrature phasors gives directly

$$r(t) = \frac{A_c}{2}\sqrt{[x(t)]^2 + [\hat{x}(t)]^2}$$

$$\phi(t) = \pm \arctan \frac{\hat{x}(t)}{x(t)} \tag{5.12}$$

Clearly, the envelope may bear little resemblance to the message (not that it was expected to).   Furthermore, the phase has a time variation far more complicated than the simple reversals of double sideband.

Illustrative of these points, suppose the message is a rectangular pulse, $x(t) = A\Pi(t/\tau)$.   It is relatively easy to find the Hilbert transform by integration, with the result

$$\hat{x}(t) = \frac{A}{\pi}\ln\left|\frac{t-\tau/2}{t+\tau/2}\right|$$

sketched in Fig. 5.18a.   The corresponding envelope is shown in Fig. 5.18b, while $\phi(t)$ is left to the reader's imagination.   A reasonable conclusion from Fig. 5.18b is that modulating signals having sharp discontinuities will cause the modulated waveform to have large instantaneous

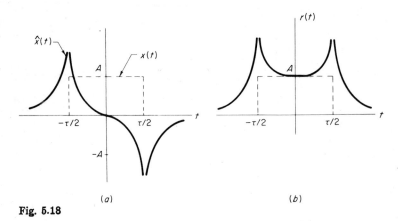

(a)   (b)

Fig. 5.18

peak power, theoretically infinite.   This does not bode well for pulse transmission via SSB.

SSB modulators

The inherent efficiencies of SSB make it quite attractive, particularly when *bandwidth conservation* is an important factor.   However, certain instrumentation problems in SSB are distinct drawbacks, one being the sideband filter included in Fig. 5.17a.

The required sharp cutoff characteristic for $H(f)$ cannot be synthesized exactly, so one must either attenuate a portion of the desired sideband or pass a portion of the undesired sideband.   (Doing both is tantamount to vestigial-sideband modulation.)   Fortunately, many modulating signals of practical interest have little or no low-frequency content, their spectra having "holes" at zero frequency (see Fig. 5.19a). Such spectra are typical of audio signals (voice and music), for example. After translation by the balanced modulator, the zero-frequency hole appears as a vacant space centered about the carrier frequency into which the *transition region* of a practical sideband filter can be fitted (Fig. 5.19b).

As a rule of thumb, the width W_T of the transition region cannot be much smaller than 1 percent of the nominal cutoff frequency, that is, $f_{co} < 100W_T$. Since W_T is constrained by the width of the spectral hole

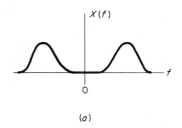

(a)

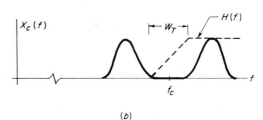

Fig. 5.19   (b)

and f_{co} should equal f_c, it may not be possible to obtain a sufficiently high carrier frequency with a given message spectrum. For these cases the modulation process can be carried out in two or more steps to overcome the limitation (see Prob. 5.9).

An alternate means of SSB generation is suggested by Eq. (5.10) and its interpretation as two DSB waves having quadrature carriers modulated by $x(t)$ and $\hat{x}(t)$. Starting with $x(t)$ and the carrier $\cos \omega_c t$, two balanced modulators and a pair of phase shifters arranged as in Fig. 5.20 will produce SSB_u or SSB_l, depending on whether we add or subtract the outputs. This technique, known as the *phase-shift method*, bypasses the need for sideband filters. However, design of the phase-shift circuitry is not trivial, and imperfections generally result in distortion of the low-frequency modulating components. Thus, the system works best with message spectra of the type previously discussed. (Another phase-shift system is described in Prob. 5.12.)

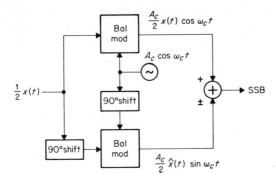

Fig. 5.20 *Phase-shift SSB modulator.*

Vestigial-sideband modulation (VSB)

Consider a modulating signal of very large bandwidth having significant low-frequency content. Principal examples are television video, facsimile, and high-speed data signals. Bandwidth conservation argues for the use of SSB, but practical SSB systems have poor low-frequency response. On the other hand, DSB works quite well for low message frequencies, but the transmission bandwidth is twice that of SSB. Clearly, a compromise modulation scheme is desired; that compromise is VSB.

VSB is derived by filtering DSB (or AM) in such a fashion that one sideband is passed almost completely while just a trace, or *vestige*, of the other sideband is included. The key to VSB is the sideband filter, a typical transfer function being that of Fig. 5.21. While the exact shape

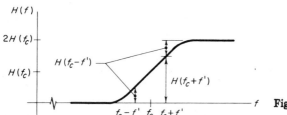

Fig. 5.21 *VSB filter characteristic.*

of the response is not crucial, it must have odd symmetry about the carrier frequency and a relative response of $\frac{1}{2}$ at that point; that is,

$$H(f_c - f') + H(f_c + f') = 2H(f_c) \tag{5.13}$$

over the positive-frequency range $f_c \pm W$.

The VSB sideband filter is thus a *practical* sideband filter with transition width W_T, and a VSB modulator[1] takes the form of Fig. 5.17a. (If carrier suppression is not wanted, the balanced modulator is replaced by an AM modulator.)   Because the width of the partial sideband is one-half the filter transition width, the transmission bandwidth is

$$B_T = W + \tfrac{1}{2}W_T \approx W$$

VSB and SSB spectra are quite similar, particularly when $W_T \ll W$, which is often true.   The similarities exist in the time domain as well, and one can write $x_c(t)$ in a form paralleling (5.11), namely,

$$x_c(t) = \tfrac{1}{2}A_c[x(t) \cos \omega_c t \mp q(t) \sin \omega_c t]$$

where the quadrature component is best expressed in terms of the filter function as

$$q(t) = \mathcal{F}^{-1}\left\{ -\frac{j}{2}[H(f - f_c) - H(f + f_c)]X(f) \right\}$$

If $W_T \ll W$, VSB approximates SSB, and $q(t) \approx \hat{x}(t)$; conversely, for large W_T, VSB approximates DSB, and $q(t) \approx 0$.

   To point up the quadrature content, we take tone modulation with $x(t) = A_m \cos 2\pi f_m t$.   Let $H(f_c) = 0.5$ and $H(f_c + f_m) = 0.5 + a$, so $H(f_c - f_m) = 0.5 - a$.   Applying this filtering to the DSB line spectrum yields

$$
\begin{aligned}
x_c(t) &= \tfrac{1}{2}A_m A_c[(0.5 - a) \cos (\omega_c - \omega_m)t + (0.5 + a) \cos (\omega_c + \omega_m)t] \\
&= \tfrac{1}{2}A_m A_c(\cos \omega_m t \cos \omega_c t - 2a \sin \omega_m t \sin \omega_c t) \qquad (5.14)
\end{aligned}
$$

an expression that also shows the VSB compromise between double side-band and single sideband; for if $a = 0$, we have DSB, while if $a = \pm 0.5$, we have SSB.

   The way in which VSB works, particularly with respect to low modulation frequencies, cannot be fully appreciated without discussing the detection process (Sec. 5.5).   For the moment it can be said that VSB is basically SSB with a finite filter transition and possibly added carrier.

Compatible single sideband ★

On occasion it is desired to combine the envelope modulation of AM with the bandwidth conservation of suppressed sideband.   The purpose is to allow envelope detection without the price of excessive transmission band-width; hence the name *compatible single sideband*.

   Although perfect envelope modulation requires symmetric side-bands, it can be approximated by suppressed-sideband modulation with

---

[1] In practice, the required filter symmetry is achieved primarily at the receiver.

unsuppressed or reinserted carrier. Adding a carrier term and modulation index to Eq. (5.14), the resultant modulated wave is

$$x_c(t) = A_c\{[1 + mx(t)] \cos \omega_c t + mq(t) \sin \omega_c t\} \tag{5.15}$$

where $q(t)$ is the quadrature term that cancels out one sideband in whole or part. If $q(t) = 0$, (5.15) reduces to conventional *amplitude modulation;* if $q(t) = \mp \hat{x}(t)$, we have SSB plus carrier; for VSB plus carrier, $q(t)$ takes on an intermediate value.

The envelope of $x_c(t)$ is found in the usual fashion to be

$$\begin{aligned} r(t) &= A_c \sqrt{[1 + mx(t)]^2 + [mq(t)]^2} \\ &= A_c[1 + mx(t)] \sqrt{1 + \left[\frac{mq(t)}{1 + mx(t)}\right]^2} \end{aligned} \tag{5.16}$$

which is a distorted AM envelope. However, for $|mq(t)| \ll 1$, the envelope is approximately

$$r(t) \approx A_c[1 + mx(t)]$$

as desired. The key to compatible single sideband is therefore keeping the quadrature component small.

If one sideband is *totally* suppressed, i.e., SSB plus carrier, $q(t)$ itself cannot be ignored; effective envelope modulation then requires $m \ll 1$. Under this condition a substantial amount of power is wasted in the carrier, far more than that of AM. Consequently, there is active interest in other schemes for generating compatible single sideband; these are briefly summarized and referenced in Schwartz, Bennett, and Stein (1966, pp. 193–198).

Alternately, if a *vestigial* sideband can be tolerated and its width is not too small, $|q(t)|$ is small compared to $|x(t)|$ most of the time. One can then use a larger modulation index without excessive envelope distortion and thereby reduce the relative carrier power. In fact, it is readily shown from (5.15) that

$$P_T = [1 + m^2(\overline{x^2} + \overline{q^2})] \frac{A_c^2}{2} \approx (1 + m^2\overline{x^2}) \frac{A_c^2}{2} \tag{5.17}$$

essentially the same as AM.

The necessary width of the vestigial sideband must be determined by empirical studies of the envelope with typical program material. In the case of television video signals, which are in fact transmitted as VSB plus carrier, the distortion can be quite sizable without detracting from picture quality, permitting a transmission bandwidth only 30 percent greater than the message bandwidth.

5.5 FREQUENCY CONVERSION, DETECTION, AND RECEIVERS

Linear modulation is primarily direct frequency translation of the message spectrum. The frequency-translation aspect is emphasized by Fig. 5.22, which shows typical modulated spectra for the various types we have discussed.

Demodulation, or *detection*, is the process by which the message is recovered from the modulated wave at the receiver. Therefore, for linear modulation in general, the detection process is basically one of *downward frequency translation*.

Referring to Fig. 5.22, it is seen that if the spectra are shifted down

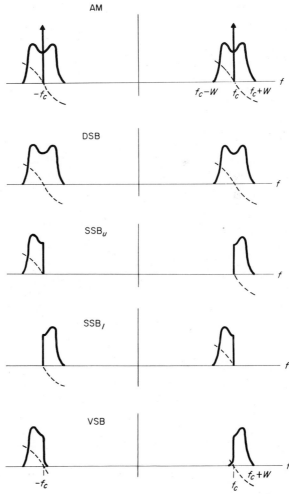

Fig. 5.22 *Typical spectra of linear modulation systems.*

in frequency by f_c units (up by f_c for the negative-frequency components), the original message spectrum is reproduced, plus a possible dc component corresponding to the translated carrier. The reader is urged to try this translation graphically for himself. It is particularly instructive in the case of SSB$_l$ and VSB.

Frequency translation, or *conversion*, is also used to shift a modulated signal to a new carrier frequency (up or down) for amplification or other processing. Thus, translation is a fundamental operation of linear modulation systems and includes modulation and detection as special cases. Before examining detectors, we should look briefly at the general process of frequency conversion.

Frequency conversion

Conversion is accomplished, at least analytically, by multiplication by a sinusoid. Consider, for example, the DSB wave $x(t) \cos \omega_1 t$. Multiplying by $\cos \omega_2 t$, we get

$$x(t) \cos \omega_1 t \cos \omega_2 t = \tfrac{1}{2} x(t) \cos (\omega_1 + \omega_2)t + \tfrac{1}{2} x(t) \cos (\omega_1 - \omega_2)t$$

The product consists of the *sum* and *difference frequencies*, $f_1 + f_2$ and $f_1 - f_2$, each modulated by $x(t)$. [If $f_2 > f_1$, we can write $f_2 - f_1$ for clarity, since $\cos(-\theta) = \cos \theta$.] Assuming $f_2 \neq f_1$, multiplication has translated the signal spectra to *two* new carrier frequencies. With appropriate filtering, the signal is up-converted or down-converted. Devices which carry out this operation are called *frequency converters* or *mixers*. The operation itself is termed *heterodyning* or *mixing*.

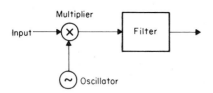

Fig. **5.23** *Frequency converter.*

A generalized frequency converter is diagramed in Fig. 5.23. Its analysis follows directly from the modulation theorem. The multiplier is usually constructed using nonlinear or switching devices, similar to modulators. That this system can perform all the linear modulation operations is illustrated by Table 5.1. However, the specific filter characteristics must be tailored to suit each application.

An ingenious application of the conversion principle is the *regenera-*

Table 5.1

Operation	Input	Oscillator frequency	Filter type
Modulation	$x(t)$	f_c	Bandpass
Detection	$x_c(t)$	f_c	Lowpass
Conversion to f_c'	$x_c(t)$	$|f_c' \pm f_c|$	Bandpass

tive frequency divider (Fig. 5.24). This system takes an input wave of frequency f_0 and produces an output wave of frequency f_0/n, n being a fixed integer. To understand the operation, assume that the output is indeed f_0/n. This is fed back into a harmonic generator that develops the $(n-1)$st harmonic, namely, $(n-1)f_0/n$, which in turn is mixed

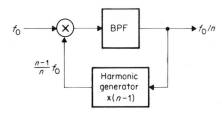

Fig. 5.24 *Regenerative frequency divider.*

with f_0. The mixer output then consists of two frequencies, $f_0 + [(n-1)/n]f_0 = (2 + 1/n)f_0$ and $f_0 - [(n-1)/n]f_0 = f_0/n$. The lower frequency is selected by the bandpass filter, and the cycle is complete.

Other converter applications include beat-frequency oscillators, speech scramblers, spectrum analyzers, etc.

Synchronous detection

All types of linear modulation can be detected by the *product* demodulator of Fig. 5.25. The incoming signal is first multiplied with a locally gener-

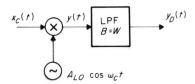

Fig. 5.25 *Synchronous (product) demodulator.*

ated sinusoid and then lowpass-filtered, the filter bandwidth being the same as the message bandwidth W or somewhat larger. It is assumed that the local oscillator (LO) is *exactly synchronized* with the carrier, in both phase and frequency, accounting for the name *synchronous detection*.[1]

For purposes of analysis, let us write the input signal in the generalized form

$$x_c(t) = [K_c + K_m x(t)] \cos \omega_c t + K_m q(t) \sin \omega_c t \qquad (5.18a)$$

which can represent any type of linear modulation with proper identification of K_c, K_m, and $q(t)$—for example, $K_c = 0$ for suppressed carrier, $q(t) = 0$ for double sideband, etc.   The filter input is thus the product

$$x_c(t) A_{LO} \cos \omega_c t = \frac{A_{LO}}{2} \{[K_c + K_m x(t)] + [K_c + K_m x(t)] \cos 2\omega_c t$$
$$+ K_m q(t) \sin 2\omega_c t\}$$

Since $f_c > W$, the double-frequency terms are rejected by the lowpass filter, leaving only the leading term

$$y_D(t) = K_D[K_c + K_m x(t)] \qquad (5.18b)$$

where K_D is the detection constant.   The dc component $K_D K_c$ corresponds to the translated carrier if it is not suppressed in the modulated wave.   This can be removed from the output by a blocking capacitor or transformer; however, any dc term in $x(t)$ will be removed as well. With this minor qualification we can say that the message has been fully recovered from $x_c(t)$.

Although perfectly correct, the above manipulations fail to bring out what goes on in the demodulation of VSB.   This is best seen in the frequency domain with the message spectrum taken to be constant over W (Fig. 5.26a), so the modulated spectrum takes the form of Fig. 5.26b. The downward-translated spectrum at the filter input will then be as shown in Fig. 5.27.   Again, high-frequency terms are eliminated by filtering, while the down-converted sidebands overlap around zero frequency.   Recalling the symmetry property of the vestigial filter, the portion removed from the upper sideband is exactly restored by the corresponding vestige of the lower sideband, so $X(f)$ has been reconstructed at the output and the detected signal is proportional to $x(t)$.

Theoretically, product demodulation borders on the trivial; in practice, it can be rather tricky.   The crux of the problem is *synchronization*—synchronizing an oscillator to a sinusoid that is not even present in the incoming signal if carrier is suppressed.   To facilitate the matter, suppressed-carrier systems may have a small amount of carrier reinserted in $x_c(t)$ at the transmitter.   This *pilot carrier* is picked off at the receiver

---

[1] Also called *coherent detection* because the carrier wave and LO output must be coherent.

by a narrow bandpass filter, amplified, and used in place of an LO. The system, shown in Fig. 5.28, is called *homodyne detection*. (Actually, the amplified pilot more often serves to synchronize a separate oscillator rather than being used directly.)

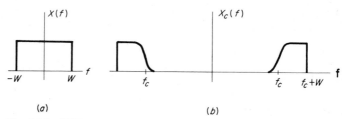

Fig. **5.26** *VSB spectra.* (a) *Message;* (b) *modulated signal.*

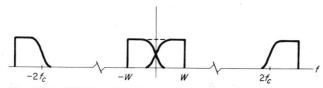

Fig. **5.27** *VSB demodulation.*

A variety of other techniques are possible for synchronization, including the use of highly stable, crystal-controlled oscillators at transmitter and receiver, the oscillators being synchronized periodically rather than on a continuous basis. Another method of synchronizing DSB without pilot carrier is described by Costas (1956).

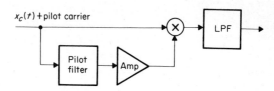

Fig. **5.28** *Homodyne detector.*

Nonetheless, some degree of asynchronism must be expected in synchronous detectors. It is therefore important to investigate the effects of phase and frequency drift in various applications. This we shall do for DSB and SSB in terms of tone modulation.

Let the local oscillator wave be $\cos(\omega_c t + \omega' t + \phi')$, where ω' and

ϕ' represent slowly drifting frequency and phase errors compared to the carrier. For double sideband with tone modulation, the detected signal becomes

$$y_D(t) = K_D \cos \omega_m t \cos (\omega' t + \phi')$$

$$= \begin{cases} \dfrac{K_D}{2} [\cos (\omega_m + \omega')t + \cos (\omega_m - \omega')t] & \phi' = 0 \\ K_D \cos \omega_m t \cos \phi' & \omega' = 0 \end{cases}$$

Similarly, for single sideband, where $x_c(t) = \cos (\omega_c \pm \omega_m)t$,

$$y_D(t) = K_D \cos [\omega_m t \pm (\omega' t + \phi')]$$

$$= \begin{cases} K_D \cos (\omega_m \pm \omega')t & \phi' = 0 \\ K_D \cos (\omega_m t \pm \phi') & \omega' = 0 \end{cases}$$

Clearly, in both DSB and SSB, a frequency drift which is not small compared to W will substantially alter the detected tone. The effect is more severe in DSB since a *pair* of tones, $f_m + f'$ and $f_m - f'$, is produced. While only one tone is produced with SSB, this too can be disturbing, particularly for music transmission. To illustrate, the major triad chord consists of three notes whose frequencies are related as the integers 4, 5, and 6. Frequency error in detection shifts each note by the same absolute amount, destroying the harmonic relationship and giving the music an oriental flavor. (Note that the effect is *not* like playing recorded music at the wrong speed, which preserves the frequency ratios.) For voice transmission, subjective listener tests have shown that frequency drifts of less than ± 10 Hz are tolerable; otherwise, everyone sounds rather like Donald Duck.

As to phase drift, again DSB is more sensitive, for if $\phi' = \pm \pi/2$ (LO and carrier in quadrature), the detected signal vanishes entirely. With slowly varying ϕ', we get an apparent *fading* effect. Phase drift in SSB appears as *delay distortion*, the extreme case being $\phi' = \pm \pi/2$, for which the demodulated signal is $\hat{x}(t)$. However, as was remarked before, the human ear can tolerate sizable delay distortion, so phase drift is not so serious in voice-signal SSB systems.

To summarize, phase and frequency synchronization requirements are rather modest for voice transmission via SSB. But in data, facsimile, and video systems with suppressed carrier, careful synchronization is a necessity. It is primarily for this reason that carrier is not suppressed in television transmission.

Envelope detection

Very little was said above about synchronous demodulation of AM for the simple reason that it is almost never used. True, synchronous detectors work for AM, but so does an *envelope detector*, which is much simpler. Because the envelope of an AM wave has the same shape as the message, independent of carrier frequency and phase, demodulation can be accomplished by extracting the envelope with no worries about synchronization.

A simplified envelope detector and its waveforms are shown in Fig. 5.29, where the diode is piecewise-linear rather than square-law.

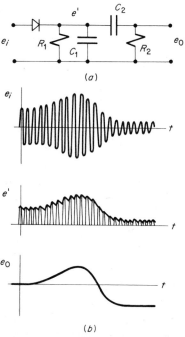

(a)

(b)

Fig. 5.29 *Envelope detection.* (a) *Circuit;*
(b) *waveforms.*

In absence of further circuitry, the voltage e' would be just the half-rectified version of the input e_i. But R_1C_1 acts as a lowpass filter, responding only to variations in the peaks of e_i. This assumes the time constant to be long compared to $1/f_c$ but short compared to the message variation time $1/W$. Thus, as noted earlier, we need $f_c \gg W$ so the envelope is clearly defined. Under these conditions, C_1 discharges only slightly between carrier peaks, and e' is approximately the envelope of e_i. More sophisticated filtering produces further improvement if needed. Finally, R_2C_2 acts as a dc block to remove the bias of the unmodulated carrier component.

The voltage e' may also be filtered to remove the envelope variations and produce a dc voltage proportional to the carrier amplitude. This voltage in turn is fed back to earlier stages of the receiver for *automatic volume control* (AVC) to compensate for carrier fading.

Despite the nonlinear element, Fig. 5.29 is termed a *linear envelope detector;* the output is linearly proportional to the input envelope. Power-law diodes can also be used, but then e' will include terms of the form $e_i{}^2$, $e_i{}^3$, etc., and there may be appreciable second-harmonic distortion unless $m \ll 1$. In fact, the design of linear envelope detectors has several interesting problems not immediately apparent in our discussion [see, for example, Terman (1955, chap. 16)].

Envelope detection is appropriate whenever the wave in question has suitable envelope modulation, e.g., television video transmitted via VSB plus carrier. Furthermore, some suppressed-carrier systems use envelope detectors rather than product demodulators, a large carrier term being inserted ahead of the detector to *reconstruct* the envelope. (This works for SSB as well as for DSB if the local carrier is large enough.) But the procedure does not bypass the need for synchronism, since envelope reconstruction requires the added carrier to be well synchronized.

One final point. The dc blocking capacitor in Fig. 5.29 causes the detector to have poor response to low-frequency message components. Therefore envelope detection may prove unsatisfactory for signals with important dc and slowly varying terms unless additional steps are taken.

Receivers

All that is really required of a receiver is some tuning mechanism, a demodulator, and amplifiers. With sufficiently strong transmitted signal, even the amplifiers may be omitted—witness the historic crystal set. Most radio receivers, however, are the more sophisticated *superheterodyne* type, whose block diagram is given in Fig. 5.30 complete with AVC.

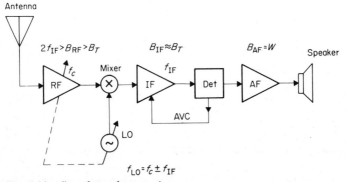

Fig. 5.30   *Superheterodyne receiver.*

There are three types of amplifiers in a superheterodyne: the *radio-frequency* (RF) amplifier, which is tuned to the desired carrier frequency; the *intermediate-frequency* (IF) amplifier, which is fixed-tuned and provides most of the gain and selectivity; and the *audio-frequency* (AF) amplifier, which follows the detector and brings the power level up to that required for the loudspeaker. The mixer is a frequency converter that translates the RF output to the IF band by converting f_c to f_{IF}. The LO provides the mixing frequency and is adjusted in parallel with the RF stage. (The mixer-LO combination is sometimes called the *first detector*, in which case the demodulator is called the *second detector*.)

The operations in a superheterodyne are best understood by considering the relative amplitude of the IF output as a function of input frequency at various points in the receiver (Fig. 5.31). If a constant-amplitude variable-frequency sinusoid is applied to the IF input, the

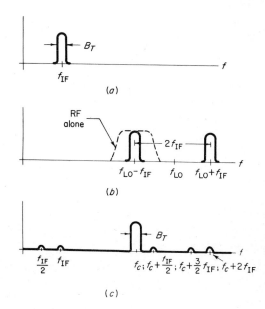

Fig. **5.31** *Superheterodyne-receiver response characteristics. (a) IF alone; (b) mixer and IF; (c) complete receiver with typical spurious responses.*

response (Fig. 5.31a) is just the amplitude response of the IF amplifier itself. Since the IF must pass the modulated signal at the translated carrier frequency, its bandwidth equals or exceeds B_T. Considering the mixer plus LO alone, there are *two* input frequencies that will result in f_{IF} at the output, namely, $f_{LO} \pm f_{IF}$. Thus, the response of the IF and mixer is as shown in Fig. 5.31b. Note that if we are trying to receive a station at $f_c = f_{LO} - f_{IF}$, we will also pick up $f_c' = f_{LO} + f_{IF}$, which is called the *image frequency*. For a given LO setting, the carrier and image frequencies are related by

$$|f_c - f_c'| = 2f_{IF} \tag{5.19}$$

The purpose of the RF stage, whose response is shown dashed on Fig. 5.31*b*, is to reject the image frequency before it gets to the mixer. It can be seen that the RF amplifier need have a bandwidth no narrower than about $2f_{IF}$. On the other hand, the IF bandwidth should be no larger than B_T, for it is up to the IF stage to reject carriers in the immediate vicinity of the desired signal; i.e., the IF provides *adjacent-channel selectivity* while the RF provides *image-channel rejection*. In conventional operation, the RF center frequency is tuned to the desired carrier while the LO is simultaneously adjusted to $f_{LO} = f_c + f_{IF}$ so the correct difference frequency $f_c - f_{LO} = f_{IF}$ is obtained at the mixer output.[1]

Figure 5.31*c* shows the overall receiver response, including some typical *spurious responses*, most of which are due to nonlinearities. For example, when a strong signal of frequency near $\frac{1}{2}f_{IF}$ gets to IF input, its second harmonic may be produced if the first stage of the IF amplifier is nonlinear. This second harmonic, being approximately f_{IF}, will then be amplified by later stages and appear at the detector output as interference, usually a high-frequency audio tone, or whistle.

Other types of receivers used for various purposes include the *heterodyne* receiver, in which the RF stage is omitted, and hence images are a potential problem; the *tuned-RF* (TRF) receiver, in which detection immediately follows the RF stage; and the *double-conversion* receiver,[2] commonly employed for high-quality shortwave AM and single-sideband systems.

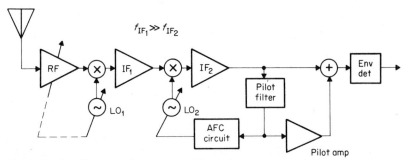

Fig. 5.32   *Double-conversion receiver with envelope reconstruction.*

As the name implies, double-conversion receivers have two mixers and two IF amplifiers (Fig. 5.32). The first IF is fixed-tuned above or

---

[1] Modern superheterodynes have the RF and LO tuning mechanically coupled. But in the early days of radio, the controls were independent, which made tuning a frustrating trial-and-error proposition.

[2] Also called *triple-detection* receivers.

just below the desired carrier-frequency band. The large value of f_{IF_1} maximizes the spacing $2f_{IF_1}$ between a carrier and its image, improving the image rejection of the RF stage. The second IF has a relatively low center frequency to permit much sharper discrimination against adjacent channels than is otherwise possible.

Figure 5.32 also indicates how pilot carrier can be processed for envelope reconstruction and synchronization of suppressed-carrier signals. Similar to homodyne detection, a pilot filter separates the carrier from the modulated wave; the pilot is then amplified and reinserted for envelope reconstruction and detection. The pilot carrier, as translated, is also used to develop an *automatic-frequency-control* (AFC) voltage, which adjusts the second LO frequency to keep the signal centered on the second IF. The RF and first LO are manually tuned, as in conventional superheterodyne receivers.

Spectrum analyzers

Returning to the superheterodyne and Figs. 5.30 and 5.31, if we call the input spectrum $X(f)$ and the IF output spectrum $Y(f)$, then Fig. 5.31c can be thought of as $|Y(f)/X(f)| = |H(f)|$, the amplitude response of the receiver. Assuming that the image and other spurious responses are negligible, the amplitude response takes the form of Fig. 5.33a, where f_{LO} is variable. Thus, a superheterodyne acts as a *tunable bandpass amplifier* of high selectivity.

These observations form the basis of the *scanning spectrum analyzer* (Fig. 5.33b), a most useful laboratory instrument. The LO frequency is

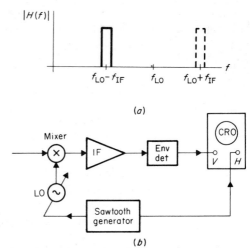

Fig. **5.33** *Scanning spectrum analyzer. (a) Amplitude response; (b) block diagram.*

swept in a sawtooth fashion, scanning $H(f)$ across the input spectrum. If the IF bandwidth is quite narrow, intercep ing only a small range of $X(f)$, the detected output voltage is proportional to $|X(f_{LO} - f_{IF})|$. This voltage is then displayed on an oscilloscope, whose horizontal deflection is derived from the sawtooth wave, the resultant trace being approximately $|X(f)|$. Because it is difficult to sweep electronically, the RF stage is usually omitted from spectrum analyzers. The consequent image problems are disposed of by limiting the sweep *range* to just less than $2f_{IF}$ and incorporating a fixed filter in the front end.

While conceptually simple, experimental spectral analysis entails judicious selection of IF bandwidth, sweep range, and sweep rate for each signal under study. The theory is discussed by Papoulis (1962, chap. 8).

5.6 INTERFERENCE AND NOISE IN LINEAR MODULATION SYSTEMS

Although a clean, virtually noise-free wave may be transmitted, the signal delivered to the demodulator is always accompanied by noise, including that generated in preceding stages of the receiver itself. Furthermore, there may be *interfering* signals in the desired band that are not rejected by the RF or IF amplifiers. Both noise and interference give rise to undesired components at the detector output.

Interference

We begin by considering a very simple case, an unmodulated carrier with an interfering cosine wave (Fig. 5.34). Let the interference have amplitude A_i and frequency $f_c + f_i$. The total signal entering the demodulator is the sum of two sinusoids

$$y(t) = A_c \cos \omega_c t + A_i \cos (\omega_c + \omega_i)t$$
$$= r(t) \cos [\omega_c t + \phi(t)]$$

where

$$r(t) = \sqrt{(A_c + A_i \cos \omega_i t)^2 + (A_i \sin \omega_i t)^2}$$

$$\phi(t) = \arctan \frac{A_i \sin \omega_i t}{A_c + A_i \cos \omega_i t} \tag{5.20}$$

as follows from the phasor construction of Fig. 5.35.

For arbitrary values of A_c and A_i, these expressions cannot be further simplified. However, if the interference is small compared to the carrier, the phasor diagram shows that the resultant envelope is essentially the sum of the in-phase components, while the quadrature compo-

Fig. 5.34   *Line spectrum for inter-*
fering sinusoids.

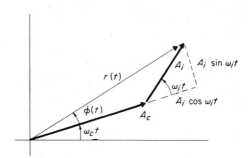

Fig. 5.35

nent determines the phase angle.   That is, if $A_i \ll A_c$, then

$$r(t) \approx A_c + A_i \cos \omega_i t$$

$$\phi(t) \approx \frac{A_i}{A_c} \sin \omega_i t$$

and hence

$$y(t) = A_c(1 + m_i \cos \omega_i t) \cos (\omega_c t + m_i \sin \omega_i t) \tag{5.21}$$

where

$$m_i = \frac{A_i}{A_c} \ll 1$$

The same result is obtained from first-order expansions of (5.20).

   At the other extreme, if $A_i \gg A_c$, the analysis is performed by taking the interference as the reference and decomposing the carrier phasor, which gives $y(t) = A_i(1 + A_c/A_i \cos \omega_i t) \cos [(\omega_c + \omega_i)t - A_c/A_i \sin \omega_i t]$, as might be expected.

   We see from (5.21) that the interfering wave *amplitude-modulates* a carrier just like a modulating tone of frequency f_i with modulation index m_i.[1]   As shown in the next chapter, the time-varying phase corresponds to *frequency modulation* by the same tone.   On the other hand, with strong interference, we can consider the carrier to be modulating the interfering wave.   In either case, the apparent modulation frequency is the difference frequency f_i.

[1] This effect leads to the low-frequency *beat note* heard when two musical instruments, slightly out of tune, are played in unison.

Suppose now that $y(t)$ is the input to an AM receiver. For analysis purposes, we separate the demodulator into three parts, an ideal envelope detector, a lowpass filter, and a dc block (Fig. 5.36). The ideal envelope

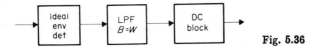

Fig. 5.36

detector is insensitive to carrier phase variations, so its output will be just $K_D r(t) = K_D(A_c + A_i \cos \omega_i t)$. Upon filtering and removal of the dc component, the output signal is

$$y_D(t) = \begin{cases} K_D A_i \cos \omega_i t & |f_i| < W \\ 0 & |f_i| > W \end{cases}$$

For comparison, if the carrier is intentionally modulated at the transmitter, $x_c(t) = A_c[1 + mx(t)] \cos \omega_c t$, the demodulated output is

$$\begin{aligned} y_D(t) &= K_D A_c m x(t) \\ &= K_D A_c m A_m \cos \omega_m t \qquad \text{for tone modulation} \end{aligned}$$

where A_c is carrier amplitude measured at the detector input.

It should be carefully noted that any interference in the frequency band $f_c \pm W$ produces a detected signal whose amplitude depends only on the amplitude of the interference, providing $A_i \ll A_c$. Interference outside $f_c \pm W$ can be rejected by the IF amplifier and the detector filter. Similar results are obtained with synchronous detection, whether or not the carrier is suppressed.

The reader may well wonder why we bother with this simple case, particularly since the results could be obtained directly from application of the frequency-translation concept. The reason is twofold: (1) very few demodulators act strictly as frequency converters, so the above approach is required for complete analysis of other cases (such as when the carrier is modulated); (2) the ideas developed here give insight to the problem of demodulation in the presence of *noise*, examined next.

Modulated signals plus noise

Consider the situation at the input to a demodulator (Fig. 5.37). Emerging from the IF amplifier is the received modulated signal $x_c(t)$, translated and amplified. Invariably the signal is contaminated by undesired noise $n(t)$. Barring nonlinear effects prior to detection, $x_c(t)$ and $n(t)$ are *independent* and *additive*, so the total input becomes

$$y(t) = x_c(t) + n(t) \tag{5.22}$$

We shall assume, not unreasonably, that $n(t)$ is white noise filtered by the IF amplifier and having the power spectrum $G_n(f) = (\eta/2)|H_{IF}(f)|^2$, shown in Fig. 5.38. In this figure η is the *positive-frequency noise density* (power per unit bandwidth), f_0 is the IF center frequency, and $B_{IF} \approx B_T$. Of course $f_0 = f_c$ for double-sideband receivers, f_c being the translated carrier frequency; for single sideband, f_c falls at one edge of the IF pass-band and $B_T = W$, so $f_0 = f_c \pm W/2$.

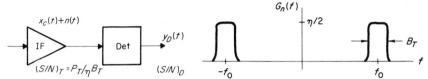

Fig. **5.37** *Parameters for demodulation in the presence of noise.*   Fig. **5.38** *Predetection noise spectrum.*

In general it is also true that $B_T \ll f_0$; hence $n(t)$ is *narrowband noise* centered on f_0 and can be expressed as derived in Sec. 4.7, that is,

$$n(t) = n_c(t) \cos \omega_0 t - n_s(t) \sin \omega_0 t \qquad (5.23)$$

From our study of interference and the notion that narrowband noise is a summation of random sinusoids [see Eqs. (4.57) and (4.58)], we suspect that $n(t)$ will modulate $x_c(t)$ somewhat like a large number of interfering waves. Hence, both signal and noise will appear at the output.

As to the various average powers, let $P_T = \langle x_c^2(t) \rangle$ be the modulated signal power as measured at the detector input, and let $N_T = \eta B_T$ be the noise power at this point. (This assumes the IF noise equivalent bandwidth is approximately B_T.) The *predetection signal-to-noise* ratio is thus

$$\left(\frac{S}{N}\right)_T = \frac{P_T}{N_T} = \frac{P_T}{\eta B_T} \qquad (5.24)$$

We should also recall that the mean-square values of the noise parameters in (5.23) are related by

$$\overline{n^2} = \overline{n_c^2} = \overline{n_s^2} = \eta B_T = N_T$$

since ergodicity is assumed.

With the preliminaries disposed of, the question at hand is this: Given the input $y(t)$, the kind of modulation, and the type of demodulator, what is the detected signal-plus-noise waveform $y_D(t)$? Alternately, if the detected signal and noise are additive, what is the *postdetection signal-to-noise ratio* $(S/N)_D$, say in terms of $(S/N)_T$?

We shall attempt to answer these questions by investigating, in turn, synchronous and envelope detection. But before doing so, it is informative to consider the signal-to-noise ratios for *baseband* message transmission, i.e., direct transmission without carrier modulation. As diagramed in Fig. 5.39, a baseband "detector" need be nothing more than

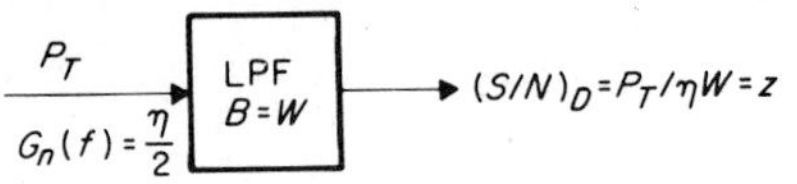

$$\frac{P_T}{G_n(f)=\frac{\eta}{2}} \boxed{\begin{array}{c} LPF \\ B=W \end{array}} \longrightarrow (S/N)_D = P_T/\eta W = z$$

Fig. 5.39 *Baseband "detection."*

a lowpass filter that passes the message bandwidth W and rejects out-of-band noise. For comparison with the modulated case, let the noise be white with $G(f) = \eta/2$ and let P_T be the *message* power at the filter input. The output signal-to-noise ratio is then $(S/N)_D = P_T/\eta W$.

Further thought indicates that this dimensionless quantity is a meaningful parameter of transmission systems in general; it is the ratio of transmitted power to the noise power in the message bandwidth, given the noise density η. Because the ratio will frequently reappear, we give it a symbol of its own by defining

$$z = \frac{P_T}{\eta W} \tag{5.25}$$

In terms of z, the predetection signal-to-noise ratio of (5.24) becomes

$$\left(\frac{S}{N}\right)_T = \frac{W}{B_T}\frac{P_T}{\eta W} = \frac{W}{B_T}z$$

Hence $(S/N)_T = z$ for single sideband, while $(S/N)_T = z/2$ for double sideband. However, the interpretation to keep in mind is that z is the *output* signal-to-noise ratio of baseband transmission.

Synchronous detection

Turning to modulated signals, consider first the synchronous detector of Fig. 5.40 with DSB modulation. For DSB we recall that $x_c(t) = x(t)A_c \cos \omega_c t$, $B_T = 2W$, $P_T = \frac{1}{2}\overline{x^2}A_c^2$, and the noise spectrum is cen-

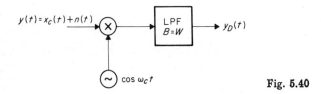

Fig. 5.40

tered on f_c. Hence,

$$y(t) = [x(t)A_c \cos \omega_c t] + [n_c(t) \cos \omega_c t - n_s(t) \sin \omega_c t]$$

and the filter input is proportional to

$$y(t) \cos \omega_c t = \tfrac{1}{2}[A_c x(t) + n_c(t)] + \tfrac{1}{2}[A_c x(t) + n_c(t)] \cos 2\omega_c t$$
$$- \tfrac{1}{2} n_s(t) \sin 2\omega_c t$$

if the LO is perfectly synchronized. Introducing the detection constant K_D, the output is

$$y_D(t) = K_D[A_c x(t) + n_c(t)] \tag{5.26}$$

since all terms but the first are removed by filtering.

Three important conclusions can be gained from Eq. (5.26): (1) the message and noise are additive at the output; (2) the quadrature noise component $n_s(t)$ is completely rejected; and (3) the output-noise power spectrum is $K_D^2 G_{n_c}(f)$ which, as discussed in Sec. 4.7, has the shape of $G_n(f)$ translated to zero frequency. If the IF is relatively flat over B_T, the output noise is essentially white over the message bandwidth W.

As to the postdetection signal-to-noise ratio, taking ensemble averages of the two terms in (5.26) yields

$$\left(\frac{S}{N}\right)_D = \frac{K_D^2 A_c^2 \overline{x^2}}{K_D^2 \overline{n_c^2}} = \frac{2P_T}{\eta B_T} = 2\left(\frac{S}{N}\right)_T$$

Alternately, since $B_T = 2W$,

$$\left(\frac{S}{N}\right)_D = \frac{P_T}{\eta W} = z \qquad \text{DSB} \tag{5.27}$$

Therefore, insofar as noise is concerned, DSB with ideal synchronous detection is equivalent to baseband transmission, even though the transmission bandwidth is twice as great.

From a frequency-translation viewpoint one might have suspected a different result, namely, $(S/N)_D = (S/N)_T = z/2$, since the modulated signal spectrum and the noise spectrum are both shifted down to zero frequency. However, the translated signal sidebands overlap in a *coherent* fashion, whereas the noise sidebands sum *incoherently*. In fact, it is the sideband coherence that counterbalances the doubled predetection noise power $\eta B_T = 2\eta W$ compared to baseband.

The results for synchronously demodulated AM can be readily inferred from (5.26) and (5.27) by recalling that $x_c(t) = A_c[1 + mx(t)] \cos \omega_c t$ and the information-bearing portion of the transmitted power is $2P_{SB} = P_T - P_c = \tfrac{1}{2}m^2 \overline{x^2} A_c^2$. Therefore

$$y_D(t) = K_D[A_c mx(t) + n_c(t)] \tag{5.28}$$

and

$$\left(\frac{S}{N}\right)_D = \frac{2P_{SB}}{\eta W} = \frac{A_c{}^2 m^2 \overline{x^2}}{2\eta W} = \frac{2m^2 \overline{x^2}}{1 + m^2 \overline{x^2}} \left(\frac{S}{N}\right)_T$$

$$= \frac{m^2 \overline{x^2}}{1 + m^2 \overline{x^2}} z \qquad \text{AM} \qquad\qquad (5.29)$$

This ratio is bounded by $(S/N)_D \le z/2$, reflecting our earlier observation that 50 percent or more of the transmitted power is wasted in the AM carrier. Other factors being equal, an AM system must transmit at least twice as much power as a suppressed-carrier system to achieve the same output. Thus, on an average power basis, AM is inferior to DSB by 3 db or more.

Under the best conditions $(S/N)_D = z/2$, corresponding to $m^2 \overline{x^2} = 1$; with full-load tone modulation $m^2 \overline{x^2} = \frac{1}{2}$ and $(S/N)_D = z/3$. More typically, however, $m^2 \overline{x^2} \sim 0.1$, for which $(S/N)_D$ is some 7 db below the maximum value and 10 db below DSB. The consequent reduction of signal-to-noise ratio is a serious problem in AM broadcasting. Hence, special techniques such as volume compression and peak limiting are frequently employed at the transmitter to ensure that the carrier is fully modulated most of the time. These techniques actually distort the recovered message and would be unacceptable for analog data transmission; for audio program material the distortion is more tolerable.

For the case of SSB (or VSB), the DSB analysis must be modified by (1) a quadrature component in $x_c(t)$ and (2) the offset carrier and LO frequency compared to the IF center frequency f_0. With these changes the modulated signal plus noise becomes

$$y(t) = \frac{A_c}{2} [x(t) \cos \omega_c t \mp \hat{x}(t) \sin \omega_c t]$$
$$+ [n_c(t) \cos \omega_0 t - n_s(t) \sin \omega_0 t] \quad (5.30)$$

where $\omega_0 = 2\pi(f_c \pm W/2)$.

Leaving the details to the reader, the demodulated output is

$$y_D(t) = K_D \left[\frac{A_c}{2} x(t) + n_c(t) \cos 2\pi \frac{W}{2} t \pm n_s(t) \sin 2\pi \frac{W}{2} t \right] \quad (5.31)$$

Note that the quadrature signal component $\hat{x}(t)$ is absent, as it should be, but the quadrature noise term now appears at the output. This is because the center frequency of the noise spectrum $G_n(f)$ differs from the LO frequency by $\pm W/2$. Frequency-translation arguments then suggest that the demodulated noise should be "narrowband" noise centered at $W/2$; the last two terms of $y_D(t)$ have precisely this interpretation.

Though Eq. (5.31) perhaps looks more formidable then (5.26), the mean-square values are not hard to find. In particular

$$\overline{\left(n_c \cos 2\pi \frac{W}{2} t \pm n_s \sin 2\pi \frac{W}{2} t \right)^2} = \frac{1}{2}(\overline{n_c^2} + \overline{n_s^2}) = \overline{n^2} = \eta B_T$$

Thus

$$\left(\frac{S}{N}\right)_D = \frac{A_c^2 \overline{x^2}}{4\eta B_T} = \left(\frac{S}{N}\right)_T$$

$$= \frac{P_T}{\eta W} = z \qquad \text{SSB or VSB} \tag{5.32}$$

since $B_T = W$ and $P_T = \frac{1}{4}\overline{x^2}A_c^2$. It can also be shown from (5.31) that the postdetection noise spectrum is essentially constant over W.

Equation (5.32) is equally valid for suppressed-carrier VSB, providing the width of the vestigial band is small compared to the message bandwidth. For the case of VSB plus carrier it is easily shown that

$$\left(\frac{S}{N}\right)_D \approx \frac{m^2\overline{x^2}}{1 + m^2\overline{x^2}} z \qquad \text{VSB plus carrier}$$

which follows from the approximations $B_T \approx W$ and $P_T \approx (1 + m^2\overline{x^2})A_c^2/2$. The reader may wish to consider why this result is the same as for AM, as given by Eq. (5.29).

Reviewing Eqs. (5.26) to (5.32), we can state the following general characteristics of synchronously detected linear modulation.

1. The message and noise are additive at the output if they are additive at the detector input.
2. If the predetection noise spectrum is reasonably flat over the transmission bandwidth, the postdetection noise spectrum is essentially constant over the message bandwidth.
3. Insofar as output signal-to-noise ratios are concerned, suppressed-sideband modulation (SSB and VSB) has no particular advantage over double-sideband modulation (AM and DSB). This is because of the coherence property of double sideband, which compensates for the reduced predetection noise power of single sideband.
4. Making due allowance for the "wasted" power in unsuppressed-carrier systems, all types of linear modulation have the same performance as baseband transmission on the basis of average transmitted power and fixed noise density.

Note that these conclusions are independent of the predetection signal-to-noise ratio $(S/N)_T$.

Up to this point we have compared systems assuming equal values of P_T, the *average* transmitted power. However, it is perhaps more realistic to hold *peak* powers equal, reflecting the peak-power constraint of practical transmitters. Under this condition the postdetection S/N of DSB is four times that of AM (6 db better) since the peak powers are proportional to A_c^2 and $4A_c^2$, respectively. Peak-power calculations for single sideband with representative modulating signals indicate that SSB is 2 to 3 db better than DSB and 8 to 9 db better than AM [see Downing (1964, pp. 77–82)]. But, as might be expected from Fig. 5.18*b*, SSB is *inferior* to DSB if the message has pronounced discontinuities.

Envelope detection—threshold effect

Inasmuch as AM is normally demodulated by an envelope detector, it is necessary to see how this differs from synchronous detection when noise is present. At the detector input we have, as before,

$$y(t) = x_c(t) + n(t)$$
$$= A_c[1 + mx(t)] \cos \omega_c t + [n_c(t) \cos \omega_c t - n_s(t) \sin \omega_c t] \quad (5.33)$$

The phasor construction of Fig. 5.41, showing $x_c(t)$ as a single phasor of

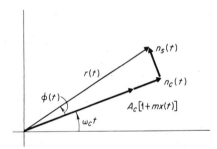

Fig. 5.41

amplitude $A_c[1 + mx(t)]$, gives the *envelope-and-phase* expression

$$y(t) = r(t) \cos [\omega_c t + \phi(t)]$$

where

$$r(t) = \sqrt{\{A_c[1 + mx(t)] + n_c(t)\}^2 + [n_s(t)]^2}$$
$$\phi(t) = \arctan \frac{n_s(t)}{A_c[1 + mx(t)] + n_c(t)} \quad (5.34)$$

Clearly, further analysis calls for some simplifications. Thus, let us assume the signal is either very large or very small.

Taking the signal to dominate, say $A_c^2 \gg \overline{n^2}$, then $A_c[1 + mx(t)]$ will be large compared to $n_c(t)$ and $n_s(t)$, at least most of the time. The envelope can then be approximated by

$$r(t) \approx A_c[1 + mx(t)] + n_c(t)$$

which shows the modulation due to noise, similar to interference modulation. The approximation follows from Fig. 5.41 or expansion of (5.34).

An *ideal* envelope detector is totally insensitive to the phase variation ϕ; hence, upon removal of the dc component $\bar{r} = A_c$, the output is

$$y_D(t) = K_D[r(t) - \bar{r}] = K_D[A_c m x(t) + n_c(t)]$$

identical to that of synchronous detection, Eq. (5.28). The postdetection signal-to-noise ratio is then as previously given in Eq. (5.29), namely,

$$\left(\frac{S}{N}\right)_D = \frac{2m^2\overline{x^2}}{1 + m^2\overline{x^2}} \left(\frac{S}{N}\right)_T = \frac{m^2\overline{x^2}}{1 + m^2\overline{x^2}} z$$

A similar analysis can be made for compatible single-sideband modulation, SSB or VSB plus carrier, with similar results.

But one must bear in mind the condition for these results, namely, $A_c{}^2 \gg \overline{n^2}$. Since $A_c{}^2/\overline{n^2}$ is proportional to $P_T/\eta B_T$, an equivalent requirement is $(S/N)_T \gg 1$. (There is no such condition with synchronous detection.) Thus, providing the predetection signal-to-noise ratio is large, envelope demodulation in the presence of noise has the same performance quality as synchronous demodulation.

At the other extreme, $(S/N)_T \ll 1$, the situation is quite different. For if $A_c{}^2 \ll \overline{n^2}$, the noise dominates in a fashion similar to strong interference, and we can think of $x_c(t)$ as modulating $n(t)$ rather than the reverse. To expedite the analysis, $n(t)$ is represented in envelope-and-phase form according to Eq. (4.60), namely

$$n(t) = r_n(t) \cos [\omega_c t + \phi_n(t)]$$

leading to the phasor diagram of Fig. 5.42. In this figure the noise phasor is the reference because we are taking $n(t)$ to be dominant, and $A_c[1 + mx(t)] \ll r_n$ most of the time. By inspection, the envelope is approximately

$$r(t) \approx r_n(t) + A_c[1 + mx(t)] \cos \phi_n(t)$$

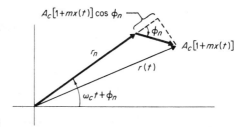

Fig. 5.42

from which

$$y_D(t) = K_D[r_n(t) + A_c m x(t) \cos \phi_n(t) - \overline{r_n}] \tag{5.35}$$

The principal output component is obviously the noise envelope $r_n(t)$, as expected. Furthermore, there is no term in (5.35) strictly proportional to the message; though signal and noise were *additive* at the input, the detected message term is *multiplied* by noise in the form of $\cos \phi_n(t)$, which is random. The message is therefore hopelessly *mutilated*, and its information has been lost. In fact, it can be said that the message *does not exist* at the output since it cannot be recovered. Under these circumstances, an output signal-to-noise ratio is difficult to define, if not meaningless.

The mutilation or loss of message at low predetection signal-to-noise ratios is called *threshold effect*. The name comes about because there is some value of $(S/N)_T$, the *threshold level*, above which mutilation is negligible and below which system performance rapidly deteriorates. Thus, if the transmitted signal fades or the receiver noise increases, the message disappears rather suddenly. [Note carefully that the threshold effect occurs only with envelope detection. With synchronous detection, the output signal and noise are always additive; true, the message is buried in noise if $(S/N)_T \ll 1$, but its identity is preserved.]

Actually, the threshold level for envelope detection is not a unique point unless some convention is established for its definition. Generally speaking, threshold effects are minimal if $A_c \gg r_n$ most of the time. To be more specific we shall define the threshold level as that value of $(S/N)_T$ for which $A_c > r_n$ with probability 0.99, that is, $P(r_n \geq A_c) = 0.01$. It was shown in Chap. 4 that the envelope of narrowband noise is Rayleigh-distributed with $\overline{r_n^2} = 2N_T$. Hence

$$P(r_n \geq A_c) = \int_{A_c}^{\infty} \frac{r_n}{N_T} e^{-r_n^2/2N_T} \, dr_n$$
$$= e^{-A_c^2/2N_T} = e^{-\frac{1}{2}(S/N)_T}$$

where we have assumed $m^2\overline{x^2} \approx 1$, so $A_c^2 \approx P_T$. Solving $e^{-\frac{1}{2}(S/N)_T} = 0.01$ for the threshold level gives

$$\left(\frac{S}{N}\right)_{T_{\text{th}}} = 4 \ln 10 \approx 10 \text{ db} \tag{5.36}$$

or, since $(S/N)_T = z/2$,

$$z_{\text{th}} = 8 \ln 10 \approx 13 \text{ db} \tag{5.37}$$

If $(S/N)_T < (S/N)_{T_{\text{th}}}$ (or $z < z_{\text{th}}$), message mutilation must be expected, along with the consequent loss of information.

Looking at the value of $(S/N)_{T_{\text{th}}}$ and recalling that $(S/N)_D < (S/N)_T$ leads to a significant conclusion: threshold effect is usually not a serious limitation of AM systems. To clarify this assertion, reasonable intelligibility in voice transmission demands a postdetection signal-to-noise ratio of about 30 db or more, $(S/N)_D \geq 1,000$, for which $(S/N)_T$ is well above the threshold level. In other words, additive noise obscures the signal long before multiplicative noise mutilates it. On the other hand, sophisticated processing techniques exist for recovering *digital* signals buried in additive noise. Hence, if AM is used for digital transmission, synchronous detection may be necessary to avoid threshold effects.

Finally, it is informative to consider how an envelope detector can act in synchronous fashion and why this requires large $(S/N)_T$. Referring back to Fig. 5.29 and assuming the input noise is negligible, we see that the diode functions as a switch, closing briefly on the positive carrier peaks; therefore the switching is perfectly synchronized with the carrier. But when noise dominates, the switching is controlled primarily by the noise peaks, so synchronism is lost. The latter effect never occurs in true synchronous detectors, where the locally generated carrier can always be much greater than the noise.

5.7 COMPARISON OF LINEAR MODULATION SYSTEMS

At last we are in position to make a meaningful comparison of the various types of linear modulation systems. Table 5.2 summarizes the points of comparison: transmission bandwidth, postdetection signal-to-noise ratio,

Table 5.2

Type	$\dfrac{B_T}{W}$	$\left(\dfrac{S}{N}\right)_D$	*DC?*	*Complexity*	*Typical applications*
Baseband	1	z	No §	Minor	Short-range telephone, etc. (no modulation)
AM	2	$\dfrac{m^2\overline{x^2}}{1+m^2\overline{x^2}}z$	No	Minor	Broadcast radio
DSB	2	z	Yes	Major	Low-speed data, control systems
SSB	1	z	No	Moderate	Point-to-point voice, multiplexing systems
VSB	1+	z	Yes	Major	High-speed data
VSB with carrier	1+	$\dfrac{m^2\overline{x^2}}{1+m^2\overline{x^2}}z$	Yes ¶	Moderate	Television video, facsimile

§ Baseband systems usually involve transformer coupling, which prevents successful low-frequency transmission.

¶ In TV and facsimile a reference level is transmitted periodically so that the dc component blocked by the envelope detector can be restored by electronic clamping.

dc (or low-frequency) signal transmission, and instrumentation complexity. Also indicated are common applications of the modulation types. As before, $z = P_T/\eta W$, where W is the message bandwidth, P_T is the transmitted power measured at the detector input, and η is the noise density.

Suppressed-carrier modulation is superior to conventional AM on several counts: signal-to-noise ratios are better, and there is no threshold effect. When *bandwidth conservation* is important, single sideband and vestigial sideband are particularly attractive.

But one seldom gets something for nothing in this world, and the price of efficient linear modulation is the increased complexity of instrumentation, especially at the receiver. Synchronous detection, no matter how it is accomplished, requires highly sophisticated circuitry compared to the envelope detector. For *point-to-point* communication (one transmitter, one receiver) the price may be worthwhile. But for *broadcast* systems (one transmitter, *many* receivers) economic considerations tip the balance toward the simplest possible receiver, and hence envelope detection.

From an instrumentation viewpoint AM is the least complex, while suppressed-carrier VSB, with its special sideband filter and synchronization requirements, is the most complex. Between DSB and SSB (in their proper applications) the latter is less difficult to instrument because synchronization is not so critical. In addition, improved filter technology has made the required sideband filters more readily available. Similarly, VSB with carrier is classed as of "moderate" complexity, despite the vestigial filter, since envelope detection is sufficient.

As to the transmission of modulating signals having significant low-frequency components, we have already argued for the superiority of DSB and VSB; this explains their use in data transmission, both analog and digital. For facsimile and TV video, electronic dc restoration makes envelope-detected VSB possible and desirable. AM could also be used in this way, but the bandwidth is prohibitive. Suppressed-carrier single sideband is virtually out of the question.

Not shown in the table is relative system performance in the face of time-varying transmission characteristics, selective fading, multiple-path propagation, etc. An unstable transmission medium has a *multiplicative* effect which is particularly disastrous for envelope detection. Late-night listeners to distant radio stations are familiar with the garbled result. Furthermore, the continually increasing demands for frequency allocations have placed a decided premium on space in the radio spectrum. It is therefore quite likely that the bulk of future radio systems will employ single-sideband modulation. However, economic factors currently preclude the conversion of commercial broadcasting to SSB.

5.8 FREQUENCY-DIVISION MULTIPLEXING

It is often desirable to transmit several messages on one transmission facility, a process called *multiplexing*. Applications of multiplexing range from the vital, if prosaic, telephone network to the glamour of FM stereo and space-probe telemetry systems. There are two basic multiplexing techniques: *frequency-division multiplexing* (FDM), which is treated here, and *time-division multiplexing* (TDM), discussed in Chap. 7.

The principle of FDM is illustrated by Fig. 5.43, where several input messages (three are shown) individually modulate the *subcarriers*

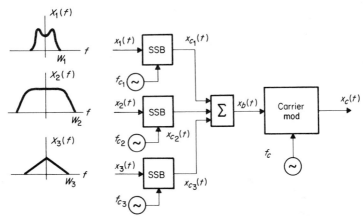

Fig. 5.43 *FDM transmitter.*

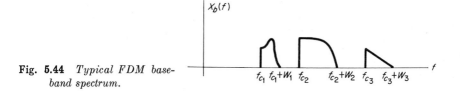

Fig. 5.44 *Typical FDM baseband spectrum.*

f_{c_1}, f_{c_2}, etc. We show the subcarrier modulation as SSB, and it often is; but any of the c-w modulation techniques could be employed, or a mixture of them. The modulated signals are then summed to produce the *baseband signal*, with spectrum $X_b(f)$, as shown in Fig. 5.44. (The designation "baseband" is used here to indicate that final carrier modulation has not yet taken place.) The baseband time function $x_b(t)$ is left to the reader's imagination.

Assuming that the subcarrier frequencies are properly chosen, the multiplexing operation has assigned a slot in the frequency domain for each of the individual messages in modulated form, hence the name *frequency-division* multiplexing. The baseband signal may then be transmitted directly or used to modulate a transmitted carrier of frequency f_c. We are not particularly concerned here with the nature of the final carrier modulation, since it is the baseband spectrum that tells the story.

Message recovery or demodulation of FDM is accomplished in three steps (Fig. 5.45). First, the carrier demodulator reproduces the baseband

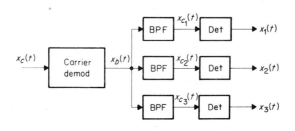

Fig. 5.45 *FDM receiver.*

signal $x_b(t)$. Then the modulated subcarriers are separated by a bank of bandpass filters in parallel, following which the messages are individually detected.

The major practical problem of FDM is *cross talk*, the unwanted coupling of one message into another. Intelligible cross talk (cross modulation) arises primarily because of nonlinearities in the system, which cause one message signal to partially modulate another subcarrier.[1] Equally disturbing is the unintelligible cross talk due to imperfect spectral separation by the filter bank. To reduce this coupling, the modulated message spectra are spaced out in frequency by *guard bands* into which the filter transition regions can be fitted. For example, the guard band between $X_{c_1}(f)$ and $X_{c_2}(f)$ in Fig. 5.44 is of width $f_{c_2} - (f_{c_1} + W_1)$. The net *baseband bandwidth* is therefore the sum of the modulated message bandwidths plus the guard bands.

While the concept of FDM is quite simple, typical systems may be very elaborate. A case in point is the Bell Telephone type L3 carrier system, in which 1,860 voice channels (each with $W = 4$ kHz nominally) are multiplexed together for short-range transmission via coaxial cable. All modulation is single sideband, both SSB_u and SSB_l, and the final baseband spectrum runs from 312 to 8,284 kHz, including pilot carriers and guard bands. To avoid excessive guard-band width at the upper

---

[1] Thus, negative feedback to reduce amplifier nonlinearity is usually a necessity in FDM systems.

end of the baseband spectrum, the multiplexing is done by *groups* in four stages, an arrangement which also facilitates switching and routing of the various channels.[1] (The reader who perchance is not fully convinced of the power of frequency-domain analysis should consider the design of such a system without the aid of spectral concepts.)

To conclude this chapter we briefly describe the FDM used in commercial FM stereophonic broadcasting. Referring to Fig. 5.46a, the

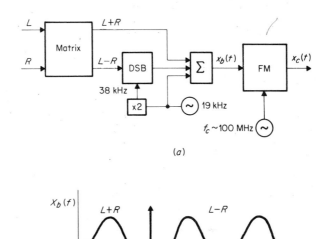

(a)

(b)

Fig. **5.46** *FM stereo multiplexing.* (a) *Transmitter;* (b) *baseband spectrum.*

left-speaker and right-speaker signals are first *matrixed* to produce $L + R$ and $L - R$. ($L + R$ is the signal heard with a monophonic receiver. The matrixing is required so the monaural listener will not be subjected to sound gaps in program material having stereophonic Ping-Pong effects.) The $L + R$ signal is then inserted directly into the baseband, while $L - R$ DSB modulates a 38-kHz subcarrier derived from a 19-kHz supply. Double-sideband modulation is employed to preserve fidelity at low frequencies. The 19-kHz pilot tone is added for receiver synchronization, resulting in the baseband spectrum of Fig. 5.46b.

[1] See *Bell System Tech. J.,* vol. 32, pp. 781–1005, July, 1953, for several articles concerning this system.

We shall have more to say about this system after examining the properties of the carrier modulation, which is *exponential*.

PROBLEMS

5.1 Consider the multitone modulating signal $x(t) = K(2 \cos \omega_m t + \cos 2\omega_m t + 3 \cos 5\omega_m t)$, which is the input to an AM system having $m = 1.0$. (*a*) Find K so that $x(t)$ is properly normalized, and plot the positive-frequency line spectrum of the modulated wave. (*b*) Calculate P_{SB}/P_c and P_c/P_T.

5.2 Suppose the upper sideband of an AM wave is attenuated by a factor of $\frac{1}{2}$. Taking single-tone modulation and $m = 1$, construct the phasor diagram and sketch the resulting envelope.

5.3 Repeat Prob. 5.2 with the upper sideband phase-shifted by 180° but not attenuated.

5.4 Suppose the signal of Prob. 5.1 is the input to a DSB system. Plot the positive-frequency line spectrum, construct the corresponding phasor diagram, and sketch the envelope of the modulated wave.

5.5 Let $x(t)$ be the random telegraph wave of Fig. 4.9, with $A = 1$. Sketch the resulting modulated waveforms if the modulation is AM with $m = 0.5$, AM with $m = 1.0$, and DSB. Referring to these drawings, explain why DSB is sometimes called AM with 200 percent modulation.

5.6 The nonlinear element to be used in an AM modulator (Fig. 5.11) is described by $e_0 = a_1 e_i + a_2 e_i^2 + a_3 e_i^3$. Show that the envelope of the modulated wave contains a term proportional to $x^2(t)$, which represents second-harmonic distortion.

5.7 A vacuum-tube balanced modulator is shown in simplified form in Fig. P 5.1. The tubes are identical and operating in a range where $i_p = g_m e_g + g_{m_2} e_g^2$. (*a*) Write v_3 in terms of v_1 and v_2. (*b*) Taking $v_2(t) = \cos \omega_c t$, find $v_3(t)$ for $v_1(t) = kx(t)$, $v_1(t) = k[1 + x(t)]$, and $v_1(t) = kx(t) \cos \omega_c t$.

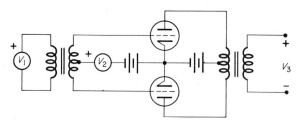

Fig. P 5.1

5.8 Suppose the AM modulators used in the balanced modulator of Fig. 5.16 are constructed with the nonlinear element of Prob. 5.6. Write an expression for the resulting modulated wave in terms of an arbitrary message $x(t)$.

5.9 A practical sideband filter for SSB has its transition region constrained by $W_T \geq f_{co}/100$, where f_{co} is the nominal cutoff frequency. Taking $X(f)$ as shown in Fig. P 5.2a, demonstrate that the system of Fig. P 5.2b produces SSB_u, and find the maximum permitted values of f_{c_1} and f_{c_2}.

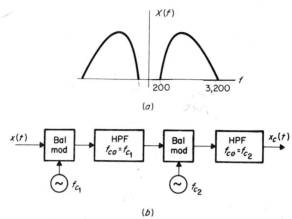

(a)

(b)

Fig. P 5.2

5.10 Taking tone modulation, show that the output of the SSB modulator of Fig. 5.20 is $x_c(t) = \frac{1}{2}A_m A_c \cos (\omega_c \pm \omega_m)t$.

5.11 If $x(t) = A \operatorname{sinc} 2Wt$, find and sketch the SSB envelope $r(t)$ by first determining $\hat{x}(t)$.

5.12 Assuming that $x(t)$ is bandlimited in W, analyze the SSB modulator of Fig. P 5.3 using the analytic signals and their spectra as discussed in Sec. 3.5. Note that this system requires single-frequency phase shifters, rather than the message-band phase shifter of Fig. 5.20.

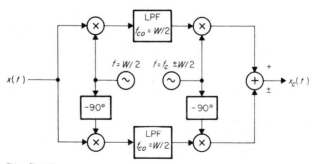

Fig. P 5.3

5.13 Demonstrate the quadrature-component envelope distortion of suppressed-carrier VSB by sketching the envelope of $x_c(t)$ according to Eq. (5.14). Take $a = 0.25$.

5.14 With tone modulation SSB plus carrier can be written as

$$x_c(t) = A_c[\cos \omega_c t + a \cos (\omega_c \pm \omega_m)t]$$

(*a*) Construct the corresponding phasor diagram and write an expression for the envelope $r(t)$.   (*b*) Assuming $a \ll 1$, find the condition on P_{SB}/P_c such that the second-harmonic envelope distortion is less than 5 percent.

5.15 Obtain Eq. (5.17) by starting from (5.15).

5.16 The system shown in Fig. P 5.4 is a simplified *speech scrambler* used to ensure communication privacy and foil wire tapping.   Analyze the system operation by sketching the spectra at points *a*, *b*, *c*, and *d*.   Also show that an identical system will suffice as an *unscrambler*.

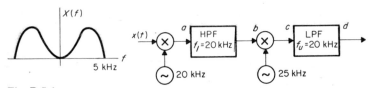

Fig. P 5.4

5.17 The device of Fig. P 5.5 is called a *beat-frequency oscillator* or BFO.   How does it work, and what does it do?   What are its advantages and disadvantages compared to conventional oscillators?

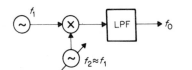

Fig. P 5.5

5.18 Investigate the performance of a synchronous demodulator when the LO *amplitude* drifts slowly with time.

5.19 DSB can be demodulated via *envelope reconstruction* wherein $A_{LO} \cos \omega_c t$ is added to $x_c(t)$ and the resultant is envelope-detected.   Investigate the demodulated signal when the LO is not perfectly synchronized.

5.20 Suppose the diode in Fig. 5.29 has a square-law rather than piecewise-linear characteristic.   Discuss the resulting demodulated signal.

5.21 Referring to the envelope detector of Fig. 5.29, determine suitable upper and lower limits on the time constant $R_1 C_1$ as dictated by the carrier frequency f_c and maximum modulating frequency W.   From these limits find the minimum practical value of f_c/W.   *Hint:* First sketch the rectified wave e_i for the case of tone modulation with $A_m = 1$, $m = 1$, and $f_m = W$.

5.22 In commercial AM the carrier-frequency allocations are spaced by 10 kHz and range from 540 to 1,600 kHz.   The message bandwidth is about 5 kHz.   Usually commercial AM radios are superheterodynes with $f_{IF} = 455$ kHz.   (*a*) Label Figs. 5.30 and 5.31 with numerical values of bandwidths, frequencies, etc.   (*b*)

For what practical reason is $f_{LO} = f_c + f_{IF}$ rather than $f_c - f_{IF}$? *Hint:* Consider the tuning range of the LO.

5.23 Is it possible to design a superheterodyne such that the image frequency always falls outside of the AM broadcast band? If so, what is the necessary f_{IF}?

5.24 Draw and fully label the block diagram of a superheterodyne receiver satisfying the following specifications: DSB modulation with pilot carrier, $W = 15$ kHz, $f_c = 10$ MHz, $f_c' = 4$ MHz. Include a homodyne detector and automatic frequency control.

5.25 The positive-frequency amplitude response of multistage single-tuned amplifiers can be approximated by

$$H(f) = K10^{-(f-f_0)^2/2B^2}$$

where K is the center-frequency voltage gain, f_0 is the center frequency, and B is essentially the 3-db bandwidth. Typically, B is constrained by $0.01 \leq B/f_0 \leq 0.1$. Using this approximation and the information in Prob. 5.22, calculate the adjacent-channel and image-channel rejection ratios (in decibels) for a commercial superheterodyne tuned to $f_c = 1,000$ kHz.

5.26 Repeat the AM interference analysis of Sec. 5.6 with both carriers tone-modulated. Take $A_i \ll A_c$.

5.27 Analyze the characteristics of the output *noise* when the local oscillator in a synchronous detector is drifting in frequency or phase.

5.28 Carry out the omitted details between Eqs. (5.30) and (5.31).

5.29 Investigate the performance of AM envelope detection and AM synchronous detection in the presence of *multipath propagation*; i.e., the received signal is $x_c(t) + ax_c(t - t_0)$, where $|a| < 1$.
Note: If the delayed signal $ax_c(t - t_0)$ is due to a moving reflection point, such as an aircraft, then a and t_0 are functions of time.

5.30 Ten voice signals, each bandlimited to 3 kHz, are to be frequency-divi ion multiplexed with a 1-kHz guard band between channels. The subcarrier modulation is SSB. (*a*) Calculate the baseband bandwidth. (*b*) Sketch the spectrum of the transmitted signal if the carrier modulation is AM.

5.31 Sketch a typical spectrum for an FDM system using AM carrier modulation and AM subcarrier modulation. What are the advantages and disadvantages of this scheme?

5.32 Figure P 5.6 shows two methods for sending two messages on one carrier. Analyze their operation in either the time domain or the frequency domain, as appropriate, and devise block diagrams for the demodulators.

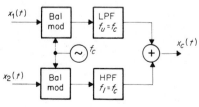
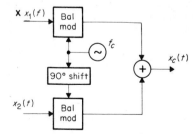

Fig. P 5.6

SELECTED SUPPLEMENTARY READING

For alternate treatments of linear modulation systems at essentially the same level as this chapter, see Schwartz (1959, chaps. 3 and 6), Javid and Brenner (1963, chaps. 7 and 9), Lathi (1965, chap. 11), or Panter (1965, chaps. 5 and 6).   A concise but informative discussion of noise in demodulation is given by Downing (1964, chap. 4).   Frequency-division multiplexing and other topics of particular interest in telephony are considered in Black (1953, chaps. 9 to 11), and Schwartz, Bennett, and Stein (1966, chaps. 3 and 4).

Pappenfus, Bruene, and Schoenike (1964) is devoted entirely to single-sideband modulation, both theory and practice, as is the December, 1956, issue of the *Proceedings of the IRE* (vol. 44).

For the electronic circuits used in transmitters and receivers, see Terman (1955), Henney (1959), or any of the many texts covering radio engineering.

6
Exponential (Angle) Modulation

Three properties of *linear* modulation bear repetition at the outset of this chapter: (1) the modulated spectrum is basically the translated message spectrum; (2) the transmission bandwidth never exceeds twice the message bandwidth; (3) the postdetection signal-to-noise ratio is no better than baseband transmission and, with fixed noise density, can be improved only by increasing the transmitted power. *Exponential* modulation differs on all three counts.

In contrast to linear modulation, exponential modulation is a *nonlinear* process; therefore, it should come as no surprise that the modulated spectrum is not related in a simple fashion to the message spectrum. Moreover, it turns out that the transmission bandwidth is usually much greater than twice the message bandwidth. Compensating for the bandwidth liability is the fact that exponential modulation can provide increased postdetection signal-to-noise ratios without increased transmitted power. Because the price of this improvement is greater bandwidth, the effect is called *wideband noise reduction*.

To gain a qualitative appreciation of this property, consider a form of exponential modulation wherein the instantaneous frequency of the modulated wave is varied in accordance with the message waveform, i.e., *frequency modulation*. The demodulated signal therefore is proportional to the range of frequency variation, called the deviation, and one can increase output signal power by increasing the deviation. Since only the frequency is modulated in this process, the carrier amplitude remains constant, and the improved output is realized without increasing transmitted power. However, as can be imagined, larger deviation does require a greater transmission bandwidth. Thus, with exponential modulation, one can trade bandwidth for signal-to-noise ratio, an intriguing possibility.

Ironically, frequency modulation was first conceived as a means of *bandwidth reduction*, the argument going somewhat as follows: If, instead of modulating the carrier amplitude, we modulate the frequency by swinging it over a range of, say, ± 50 Hz, then the transmission bandwidth will be 100 Hz regardless of the message bandwidth. As we shall soon see, this argument has a serious flaw; specifically, it ignores the distinction between *instantaneous frequency* and *spectral frequency*. Carson (1922) recognized the fallacy of the bandwidth-reduction notion and cleared the air on that score.

Unfortunately, he and many others also felt that exponential modulation had no advantages over linear modulation with respect to noise. It took some time to overcome this belief, but, thanks to Armstrong (1936), the merits of exponential modulation were finally appreciated. Before we can understand them fully, more careful consideration must be given to the fundamental concepts.

6.1 FUNDAMENTALS OF EXPONENTIAL MODULATION

In exponential modulation, the modulated wave in phasor form is an exponential function of the message, that is,

$$x_c(t) = \text{Re} \left[A_c e^{j\theta_c(t)} \right] = A_c \cos \theta_c(t)$$

where $\theta_c(t)$ is a linear function of $x(t)$ and A_c is constant. Since θ_c is the angular position of the phasor, an equally appropriate name for the process is *angle* modulation.

While there are many possible forms of exponential modulation, only two have proved to be practical, namely, *frequency modulation* (FM) and *phase modulation* (PM). These designations suggest *time-varying* frequency or phase, concepts that require special interpretation. This is particularly true of time-varying frequency since frequency implies *periodicity* and time-varying periodicity is meaningless. To clarify the matter, we begin by writing θ_c as

$$\theta_c(t) = 2\pi f_c t + \phi(t) \tag{6.1}$$

so that the carrier frequency f_c is uniquely specified. The second term of (6.1) can then be interpreted as a *relative phase angle*, in the sense that the phasor $e^{j\theta_c}$ differs in angular position from $e^{j\omega_c t}$ by $\phi(t)$. Only in this way can we ascribe meaning to phase as a time-dependent quantity.

Pressing these notions further, it should be recalled that angular frequency (in radians per second) is the time derivative of angular position. Therefore we are led to define *instantaneous frequency* $f(t)$ (in revolutions or cycles per second) by

$$f(t) = \frac{1}{2\pi}\frac{d\theta_c}{dt} = f_c + \frac{1}{2\pi}\frac{d\phi}{dt} \tag{6.2}$$

where the term $(d\phi/dt)/2\pi$ can be interpreted either as the velocity of $e^{j\theta_c}$ compared to $e^{j\omega_c t}$ or as the *instantaneous frequency deviation* relative to f_c. Equation (6.2) also shows that $\theta_c(t)$ is related to $f(t)$ by integration in the form

$$\theta_c(t) = 2\pi \int^t f(t')\,dt' \tag{6.3}$$

The omitted lower limit of integration represents a constant phase term which can be dropped without loss of generality.

With these preliminaries kept in mind, phase modulation is defined as the process whereby the relative phase $\phi(t)$ is proportional to the message, while in frequency modulation the instantaneous frequency deviation $(d\phi/dt)/2\pi$ is proportional to the message.

Specifically, the relative phase of a PM wave is

$$\phi(t) = \phi_d x(t) \tag{6.4}$$

where ϕ_d is the phase-deviation constant, i.e., the maximum phase shift produced by $x(t)$ since $|x(t)| \le 1$. The phasor angle is then

$$\theta_c(t) = 2\pi f_c t + \phi_d x(t)$$

so the modulated waveform becomes

$$x_c(t) = A_c \cos\left[\omega_c t + \phi_d x(t)\right] \tag{6.5}$$

Similarly, the instantaneous frequency of an FM wave is

$$f(t) = f_c + f_d x(t) \tag{6.6}$$

where f_d is the frequency-deviation constant. Substituting (6.6) into

(6.3) yields

$$\theta_c(t) = 2\pi f_c t + 2\pi f_d \int^t x(t')\, dt'$$

and hence

$$x_c(t) = A_c \cos\left[\omega_c t + 2\pi f_d \int^t x(t')\, dt'\right] \tag{6.7}$$

is the modulated waveform.

It is assumed above that the message has no dc component; that is, $\langle x(t)\rangle = 0$. Otherwise the integral in (6.7) would diverge as $t \to \infty$. Physically, a dc term in $x(t)$ produces a constant phase shift of $\phi_d\langle x(t)\rangle$ in phase modulation or a carrier-frequency shift of $f_d\langle x(t)\rangle$ in frequency modulation. Practically, any dc message component is usually blocked in the modulator circuits.

Comparing (6.5) with (6.7), there appears to be little difference between PM and FM, the essential distinction being the integration of the message in FM. Moreover, nomenclature notwithstanding, both FM and PM have both time-varying phase and frequency, as underscored by Table 6.1.

Table 6.1

	Phase modulation	*Frequency modulation*
Relative phase $\phi(t)$	$\phi_d x(t)$	$2\pi f_d \int^t x(t')\, dt'$
Instantaneous frequency $f(t)$	$f_c + \dfrac{\phi_d}{2\pi}\dfrac{dx}{dt}$	$f_c + f_d x(t)$

These relations clearly indicate that, with the help of integrating and differentiating networks, a phase modulator can produce frequency modulation and vice versa. In fact, in the case of tone modulation it is virtually impossible to visually distinguish FM and PM waves.

On the other hand, a comparison of exponential modulation with linear modulation reveals some pronounced differences. For one thing, the amplitude of an FM or PM wave is always constant; therefore, regardless of the message $x(t)$, the average transmitted power is[1]

$$P_T = \tfrac{1}{2}A_c{}^2$$

For another, the *zero crossings* of an exponentially modulated wave are *not periodic*, whereas they are always periodic in linear modulation.

[1] There are theoretical exceptions to this condition, but they seldom occur in practice.

Indeed, because of the constant-amplitude property of FM and PM, it can be said that the message resides in the zero crossings alone, providing the carrier frequency is large.   Finally, since exponential modulation is a nonlinear process, the modulated wave does not look at all like the message waveform.

Figure 6.1 illustrates some of these points by showing typical AM, FM, and PM waves.   As a mental exercise the reader may wish to check

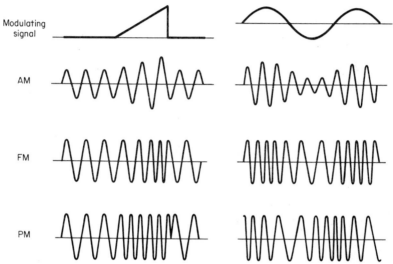

Fig. 6.1   *Illustrative AM, FM, and PM waveforms.*

these waveforms against the corresponding modulating signals.   For FM and PM this is most easily done by considering the instantaneous frequency rather than by substituting $x(t)$ in Eqs. (6.5) and (6.7).

Despite the many similarities of PM and FM, frequency modulation turns out to have superior noise-reduction properties and thus will receive most of our attention.   Fortunately, results and conclusions based on a study of FM are applicable with but minor modifications to all forms of exponential modulation.

6.2 SPECTRAL ANALYSIS OF FREQUENCY MODULATION

The time-domain description of an FM wave with arbitrary message $x(t)$ is provided by Eq. (6.7).   Hence, we begin our study with the frequency-domain description, or spectral analysis, of FM.   Before doing so, it is necessary to observe that *instantaneous frequency $f(t)$* is not the same thing as *spectral frequency f.*   The former is a time-dependent quantity describing $x_c(t)$ in the time domain; the latter is the independent variable of spectral analysis wherein $X_c(f)$ describes $x_c(t)$ in the frequency domain in terms of fixed-frequency sinusoidal components.   Therefore we cannot expect a simple one-to-one correspondence between $f(t)$ and the FM spectrum.

Elaborating on the above point, Fig. 6.2 shows a section of an FM wave whose instantaneous frequency is changing rapidly compared to

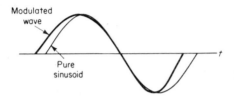

Fig. 6.2   *A wave with changing instantaneous frequency.*

the range of frequency variation.   Under this condition the modulated waveform departs markedly from a simple sinusoid, and, such being the case, instantaneous frequency and spectral frequency must be separate and distinct concepts.   On the other hand, if the variation of $f(t)$ is slow enough, $x_c(t)$ takes the form of an ordinary sinusoidal wave for many cycles, and it is then possible to physically relate the two concepts providing due caution is exercised.

As implied by these considerations, an exact description of FM spectra is difficult save for certain simple modulating signals.   (This of course merely reflects the fact that exponential modulation is a *nonlinear* process.)   Therefore, instead of attempting the analysis with arbitrary $x(t)$, we shall as an alternate tactic examine several specific cases, beginning with tone modulation, and formulate general conclusions based on them.   This approach is admittedly roundabout but necessary.

Tone modulation

With tone modulation, the instantaneous frequency of an FM signal varies in a sinusoidal fashion about the carrier frequency, a typical waveform being illustrated in Fig. 6.1.   Specifically, if $x(t) = A_m \cos \omega_m t$, then

$$f(t) = f_c + f_d A_m \cos \omega_m t$$

and

$$x_c(t) = A_c \cos \left(\omega_c t + 2\pi f_d \int^t A_m \cos \omega_m t' \, dt' \right)$$

$$= A_c \cos \left(\omega_c t + \frac{2\pi f_d A_m}{\omega_m} \sin \omega_m t \right)$$

To simplify the notation let

$$\beta = \frac{2\pi f_d A_m}{\omega_m} = \frac{A_m f_d}{f_m} \tag{6.8}$$

so that

$$x_c(t) = A_c \cos (\omega_c t + \beta \sin \omega_m t) \tag{6.9}$$

This result should be contrasted with Eq. (5.7), the equivalent expression for AM with tone modulation.

The parameter β introduced above is called the FM *modulation index*, and has two rather unusual properties: (1) it is defined only for tone modulation, and (2) it depends on both the amplitude and frequency of the modulating tone.   Physically, β is the maximum *phase deviation* (in radians) produced by the tone in question.   This conclusion follows from inspection of (6.9), which shows that the relative phase of $x_c(t)$ is

$$\phi(t) = \beta \sin \omega_m t = \frac{A_m f_d}{f_m} \sin \omega_m t$$

Thus, different tones having the same amplitude-to-frequency ratio yield the same phase deviation but at different rates.   However, since $f(t) = f_c + f_d A_m \cos \omega_m t$, the frequency deviation depends only on the tone amplitude and f_d, the latter being a property of the modulator.

As to the spectral analysis of $x_c(t)$, we shall not attempt a direct Fourier transformation of Eq. (6.9), for rather obvious reasons.   But it is possible to express $x_c(t)$ as a sum of sinusoids, which then gives us the positive-frequency line spectrum.   For this purpose we first write (6.9) in the form

$$x_c(t) = A_c[\cos \omega_c t \cos (\beta \sin \omega_m t) - \sin \omega_c t \sin (\beta \sin \omega_m t)] \tag{6.10}$$

and observe that, even though $x_c(t)$ itself is not necessarily periodic, $\cos (\beta \sin \omega_m t)$ and $\sin (\beta \sin \omega_m t)$ are periodic in $1/f_m$ and thus can be expanded via Fourier series.   In particular, it is well known in applied mathematics that

$$\cos (\beta \sin \omega_m t) = J_0(\beta) + \sum_{n \text{ even}}^{\infty} 2J_n(\beta) \cos n\omega_m t$$

$$\sin (\beta \sin \omega_m t) = \sum_{n \text{ odd}}^{\infty} 2J_n(\beta) \sin n\omega_m t \tag{6.11}$$

where n is positive and

$$J_n(\beta) = \frac{1}{2\pi} \int_{-\pi}^{\pi} e^{j(\beta \sin u - nu)} \, du \tag{6.12}$$

The coefficients $J_n(\beta)$ are *Bessel functions*[1] of the first kind, of order n and argument β. With the aid of (6.12), the reader should encounter little difficulty in deriving the trigonometric expansions given in (6.11).

Substituting (6.11) into (6.10) and expanding products of sines and cosines finally results in

$$
\begin{aligned}
x_c(t) = \; & A_c J_0(\beta) \cos \omega_c t \\
& + \sum_{n \text{ odd}}^{\infty} A_c J_n(\beta)[\cos (\omega_c + n\omega_m)t - \cos (\omega_c - n\omega_m)t] \\
& + \sum_{n \text{ even}}^{\infty} A_c J_n(\beta)[\cos (\omega_c + n\omega_m)t + \cos (\omega_c - n\omega_m)t] \qquad (6.13a)
\end{aligned}
$$

Alternately, taking advantage of the property that $J_{-n}(\beta) = (-1)^n J_n(\beta)$, we get the more compact but less informative expression

$$
x_c(t) = A_c \sum_{n=-\infty}^{\infty} J_n(\beta) \cos (\omega_c + n\omega_m)t \qquad (6.13b)
$$

In either form, (6.13) is the mathematical representation for a constant-amplitude wave whose instantaneous frequency is varying sinusoidally. A phasor interpretation, to be given shortly, will shed more light on the matter.

Examining (6.13), we see that the FM spectrum consists of a carrier-frequency line plus an *infinite* number of sideband lines at frequencies[2] $f_c \pm n f_m$. As illustrated in the typical spectrum of Fig. 6.3, all lines are equally spaced by the modulating frequency and the odd-order lower sideband lines are reversed in phase compared to the unmodulated carrier. In general, the relative amplitude of a line at $f_c + n f_m$ is given by $J_n(\beta)$,

---

[1] Bessel functions occur in a variety of analyses, notably in boundary-value problems. For a further discussion of their properties, see any text on applied mathematics or guided electromagnetic waves.

[2] We are dealing with a positive-frequency line spectrum, so apparent negative frequencies due to $n f_m > f_c$ are folded back to the positive values $|f_c - n f_m|$. Such components are usually negligible in practice when the carrier frequency is many orders greater than the modulating frequency.

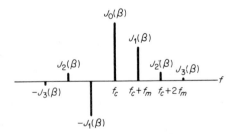

Fig. 6.3 *An FM line spectrum, tone modulation.*

so before we can say more about the spectrum, we must examine the behavior of Bessel functions.

Figure 6.4 shows a few Bessel functions of various order plotted versus the argument β.   Such curves are obtained by series approximations to the defining integral (6.12).   For the record and later use, the approximations with very small and very large β are

$$J_n(\beta) \approx \begin{cases} \dfrac{\beta^n}{2^n n!}\left[1 - \dfrac{\beta^2}{2(n+1)} + \cdots\right] & \beta \ll 1 \\[4mm] \sqrt{\dfrac{2}{\pi\beta}}\,\cos\left(\beta - \dfrac{n\pi}{2} - \dfrac{\pi}{4}\right) & \beta \gg 1 \end{cases} \tag{6.14}$$

The tendency toward damped periodic oscillation when $\beta \gg 1$ is exhibited in the figure by $J_0(\beta)$.

Several important properties can be noted from study of Fig. 6.4. First, the relative amplitude of the carrier line $J_0(\beta)$ varies with the modulation index and hence depends on the modulating signal.   Thus, in contrast to linear modulation, the carrier-frequency component of an FM

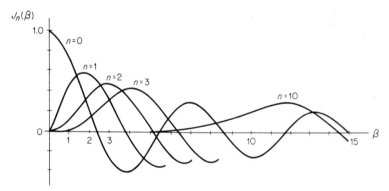

Fig. 6.4   *Bessel functions of fixed order plotted versus the argument β.*

wave "contains" part of the message information.   Nonetheless, there will be spectra in which the carrier line has zero amplitude since $J_0(\beta) = 0$ when $\beta = 2.4$, 5.5, etc.   Second, the number of sideband lines having appreciable relative amplitude also is a function of β.   With $\beta \ll 1$ only J_0 and J_1 are significant, so the spectrum will consist of carrier and two sideband lines, much like AM save for the phase reversal of the lower sideband line.   On the other hand, if $\beta \gg 1$, there will be many sideband lines, giving a spectrum quite unlike linear modulation.   Finally, we observe that large β implies a large bandwidth to accommodate the extensive sideband structure—this in agreement with our physical interpretation of large frequency deviation.

Some of the above points are better illustrated by Fig. 6.5, which gives $J_n(\beta)$ as a function of n/β for various *fixed* values of β. Since FM with tone modulation has constant β, these curves represent the "envelope" of the sideband lines if we multiply the horizontal axis by βf_m to obtain the line position $n f_m$ relative to f_c. Observe in particular that all $J_n(\beta)$ decay monotonically for $n/\beta > 1$ and that $|J_n(\beta)| \ll 1$ if $|n/\beta| \gg 1$.

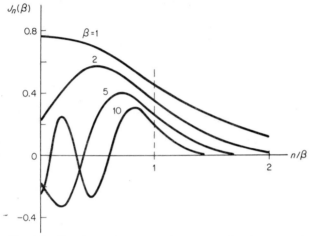

Fig. 6.5 *Bessel functions of fixed argument plotted versus n/β.*

Similar to Fig. 6.5, Table 6.2 lists selected values of $J_n(\beta)$, rounded off at the second decimal place. Blanks in the table correspond to $|J_n(\beta)| < 0.01$. More extensive tabulations can be found in Jahnke and Emde (1945).

Typical line spectra are shown in Fig. 6.6, where reversals of the odd-order lower sideband lines have been omitted for clarity. These spectra should be carefully scrutinized by the reader for the relative influence of modulating amplitude and frequency. Note also the concentration of the spectrum within $f_c \pm \beta f_m$ when β is large.

Phasor interpretation

Because $x_c(t)$ as written in (6.13) is so cumbersome, let us construct the FM phasor diagram to aid physical interpretation.

As a starting point suppose that $\beta \ll 1$, so that, by (6.14), $J_0(\beta) \approx 1$, $J_1(\beta) \approx \beta/2$, and all higher-order lines are negligible. Taking $A_c = 1$ and omitting the rotation of the carrier phasor yields the diagram of

Table 6.2   Selected values of $J_n(\beta)$

n	$J_n(0.1)$	$J_n(0.2)$	$J_n(0.5)$	$J_n(1.0)$	$J_n(2.0)$	$J_n(5.0)$	$J_n(10)$	n
0	1.00	0.99	0.94	0.77	0.22	−0.18	−0.25	0
1	0.05	0.10	0.24	0.44	0.58	−0.33	0.04	1
2			0.03	0.11	0.35	0.05	0.25	2
3				0.02	0.13	0.36	0.06	3
4					0.03	0.39	−0.22	4
5						0.26	−0.23	5
6						0.13	−0.01	6
7						0.05	0.22	7
8						0.02	0.32	8
9							0.29	9
10							0.21	10
11							0.12	11
12							0.06	12
13							0.03	13
14							0.01	14

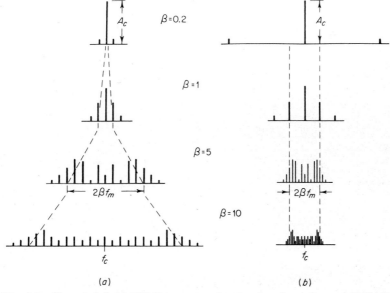

Fig. 6.6   *Tone-modulated FM line spectra showing the effects of tone amplitude and frequency.   (a) f_m fixed, $A_m f_d$ increasing; (b) $A_m f_d$ fixed, f_m decreasing.*

Fig. 6.7a, a diagram that differs from the AM case (Fig. 5.6) only in the phase reversal of the lower sideband line. But because of the phase reversal, the contribution of the sideband pair is perpendicular, or *quadrature*, to the carrier rather than being collinear. This quadrature relationship is precisely what is needed to produce phase or frequency modulation instead of amplitude modulation.

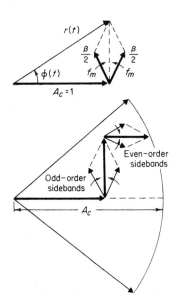

Fig. 6.7 *FM phasor diagrams. (a) Carrier plus first-order sideband pair; (b) carrier plus all sideband pairs.*

Analytically, the envelope and phase of $x_c(t)$ with small β are

$$r(t) \approx \sqrt{1 + \left(\frac{2\beta}{2}\sin \omega_m t\right)^2} \approx 1 + \frac{\beta^2}{4} - \frac{\beta^2}{4}\cos 2\omega_m t$$

$$\phi(t) \approx \arctan\left(\frac{2\beta}{2}\sin \omega_m t\right) \approx \beta \sin \omega_m t$$

Thus the phase variation is approximately as desired, but there is an additional *amplitude* variation at twice the tone frequency. To cancel out the latter we should include the second-order pair of sideband lines that rotate at $\pm 2f_m$ relative to the carrier and whose resultant is collinear with the carrier. While the second-order pair virtually wipes out the undesired amplitude modulation, it also distorts $\phi(t)$. The phase distortion is then corrected by adding the third-order pair, which again introduces amplitude modulation, and so on ad infinitum.

When all spectral lines are included, the odd-order pairs have a resultant in quadrature with the carrier which provides the desired frequency modulation plus unwanted amplitude modulation. The resultant of the even-order pairs, being collinear with the carrier, corrects for the amplitude variations. The net effect is then as illustrated in Fig. 6.7b. The tip of the resultant sweeps through a circular arc, reflecting the constant amplitude A_c.

Multitone modulation

The Fourier series technique used to arrive at (6.13) also can be applied to the case of multitone modulation. For instance, suppose that $x(t) = A_1 \cos \omega_1 t + A_2 \cos \omega_2 t$, where f_1 and f_2 are not harmonically related. The modulated wave is first written as

$$x_c(t) = A_c [\cos \omega_c t (\cos \alpha_1 \cos \alpha_2 - \sin \alpha_1 \sin \alpha_2)$$
$$- \sin \omega_c t (\sin \alpha_1 \cos \alpha_2 + \cos \alpha_1 \sin \alpha_2)]$$

where $\alpha_1 = \beta_1 \sin \omega_1 t$, $\beta_1 = A_1 f_d / f_1$, etc. Terms of the form $\cos \alpha_1$, $\sin \alpha_1$, etc., are then expanded according to Eq. (6.11), and, after some routine manipulations, one arrives at the compact result

$$x_c(t) = A_c \sum_{n=-\infty}^{\infty} \sum_{m=-\infty}^{\infty} J_n(\beta_1) J_m(\beta_2) \cos (\omega_c + n\omega_1 + m\omega_2)t$$

Interpreting this expression in the frequency domain, the spectral lines can be divided into four categories: (1) the carrier line of amplitude $A_c J_0(\beta_1) J_0(\beta_2)$; (2) sideband lines at $f_c \pm n f_1$ due to one tone alone; (3) sideband lines at $f_c \pm m f_2$ due to the other tone alone; and (4) sideband lines at $f_c \pm n f_1 \pm m f_2$ which appear to be beat-frequency modulation at the sum and difference frequencies of the modulating tones and their harmonics. This last category may come as a surprise, for it is unparalleled in linear modulation where simple superposition of sideband lines is the rule. But then we must recall that FM is *nonlinear* modulation, so superposition is not to be expected.

A double-tone FM spectrum showing the various types of spectral lines is given in Fig. 6.8 for $f_1 \ll f_2$ and $\beta_1 \sim \beta_2$. Under these conditions

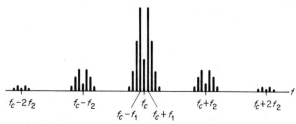

Fig. 6.8 *A double-tone FM spectrum, $f_1 \ll f_2$, $\beta_1 \sim \beta_2$.*

there exists the curious property that each sideband line at $f_c \pm mf_2$ looks like another FM carrier with tone modulation of frequency f_1.

If absolutely necessary, the above technique can be extended to modulation by more than two nonharmonic tones; the procedure is straightforward but exceedingly messy. The results are omitted here since they add nothing of particular importance to our discussion.

However, when the tone frequencies are harmonically related, i.e., when $x(t)$ is a *periodic waveform*, the spectral analysis is greatly facilitated by writing

$$x_c(t) = \text{Re}\,[A_c e^{j\omega_c t} e^{j\phi(t)}]$$

where $\phi(t) = 2\pi f_d \int^t x(t')\,dt'$. Because $x(t)$ is periodic, say in $T_0 = 1/f_0$, $e^{j\phi(t)}$ is also periodic and can be expressed in the usual series form $\sum\limits_n c_n e^{jn\omega_0 t}$.

Therefore

$$x_c(t) = \text{Re}\left[A_c \sum_{n=-\infty}^{\infty} c_n e^{j(\omega_c + n\omega_0)t}\right] \tag{6.15}$$

where the coefficients c_n are given by

$$c_n = \frac{1}{T_0} \int_{-T_0/2}^{T_0/2} \exp\,\{j[\phi(t) - n\omega_0 t]\}\,dt$$

Panter (1965) and Cuccia (1952) give further details and illustrate the method with several interesting examples.

Pulse modulation

Despite the complexities of FM spectra, there are a few modulating signals for which $x_c(t)$ is amenable to direct Fourier transformation. One such signal is the rectangular pulse $x(t) = \Pi(t/\tau)$, an important and informative example.

The key to transformation is the instantaneous frequency, which for pulse modulation is simply

$$f(t) = f_c + f_d \Pi\left(\frac{t}{\tau}\right) = \begin{cases} f_c + f_d & |t| < \dfrac{\tau}{2} \\[2mm] f_c & |t| > \dfrac{\tau}{2} \end{cases}.$$

By inspection or integration of $f(t)$ we have

$$x_c(t) = \begin{cases} A_c \cos\,(\omega_c + \omega_d)t & |t| < \dfrac{\tau}{2} \\[2mm] A_c \cos\,\omega_c t & |t| > \dfrac{\tau}{2} \end{cases}$$

But better suited to our purposes is the form

$$x_c(t) = A_c \left[\cos \omega_c t - \Pi\left(\frac{t}{\tau}\right) \cos \omega_c t + \Pi\left(\frac{t}{\tau}\right) \cos (\omega_c + \omega_d)t \right] \quad (6.16)$$

which says that the modulated signal is a sinusoidal wave of frequency f_c minus an RF pulse of frequency f_c plus an RF pulse of frequency $f_c + f_d$.

Although (6.16) may seem unduly formal, it does lend itself readily to Fourier analysis; indeed, with the result of Example 2.6, Eq. (6.16) can be transformed term by term to give

$$X_c(f) = \frac{A_c}{2}[\delta(f - f_c) + \delta(f + f_c)]$$

$$- \frac{A_c\tau}{2}[\text{sinc } (f - f_c)\tau + \text{sinc } (f + f_c)\tau]$$

$$+ \frac{A_c\tau}{2}[\text{sinc } (f - f_c - f_d)\tau + \text{sinc } (f + f_c + f_d)\tau] \quad (6.17)$$

The positive-frequency portion of $|X_c(f)|$ is sketched in Fig. 6.9 for

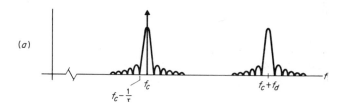

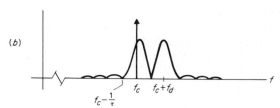

Fig. 6.9 *Pulse-modulated FM spectra.*   (a) $f_d \gg 1/\tau$; (b) $f_d = 2/\tau$.

$f_d \gg 1/\tau$ and $f_d = 2/\tau$.   In both cases we see that the spectrum is *not symmetric*[1] about the carrier frequency and has more content above f_c

[1] This is another difference between exponential and linear modulation, for the spectrum of double-sideband linear modulation always has symmetry about the carrier frequency.

than below, a state of affairs that might have been anticipated in view of the fact that $f(t)$ is never less than f_c. Moreover, even though other frequencies are present, the spectrum is concentrated at the two discrete values of $f(t)$, namely, f_c and $f_c + f_d$, the concentration being quite pronounced when f_d is large compared to $1/\tau$.

The quasi-static approximation ★

To conclude this section we briefly consider an intuitive approach to FM spectra known as the *quasi-static approximation*. This approximation relates *power spectral density* directly to instantaneous frequency and stems from the following physical reasoning.

Suppose the instantaneous frequency of an FM wave changes so slowly that $x_c(t)$ looks very much like an ordinary sinusoid; a two-port network would then respond to $x_c(t)$ as if it were a pure sinusoidal input signal. Furthermore, suppose that $x_c(t)$ is filtered by a narrow bandpass filter having unit gain, center frequency f_0, and bandwidth df. The average power at the filter output will be approximately

$$dP_0 = 2G(f_0)\,df$$

where $G(f_0)$ is the power spectral density of the FM signal at f_0 and the factor of 2 accounts for the negative-frequency portion of $G(f)$. Since the input amplitude is constant, the output power relative to the input power depends only on the relative time that $x_c(t)$ has its instantaneous frequency in the filter passband. Thus

$$dP_0 = \tfrac{1}{2}A_c^2 \times [\text{fractional time } f(t) \text{ is in } f_0 \pm \tfrac{1}{2}df]$$

Recalling the relative-occurrence interpretation of probability, it follows that the fractional time in question is $P[f_0 - \tfrac{1}{2}df < f(t) < f_0 + \tfrac{1}{2}df] \approx p_f(f_0)\,df$, where $p_f(f)$ is the *probability density function* of the instantaneous frequency. Therefore

$$2G(f_0)\,df = \tfrac{1}{2}A_c^2 p_f(f_0)\,df$$

and so the two-sided power spectrum is given by

$$G(f) = \tfrac{1}{4}A_c^2[p_f(f) + p_f(-f)] \tag{6.18}$$

which is the quasi-static approximation for FM power spectral density.

But Eq. (6.18) is only an approximation, and one must determine the conditions for which it is a correct and useful approximation. Generally speaking, the method is valid providing $f(t)$ varies so slowly that $x_c(t)$ looks more or less like a single-frequency waveform; this implies that $f(t) = f_c + f_d x(t)$ must be constant over time intervals large compared to $1/f_d$. Formalizing this condition, if $x(t)$ is essentially bandlimited in W so it does not change significantly over time intervals less than $1/W$, then (6.18) is applicable if $1/W \gg 1/f_d$ or

$$W \ll f_d$$

But even with $W \ll f_d$, the quasi-static approximation gives meaningless results unless $p_f(f)$ is also meaningful in the sense of relative occurrences. Thus, as the reader should convince himself, (6.18) is useless for *time-limited* modulating signals such as a rectangular pulse. Conversely, the approach can be quite useful when dealing with *random* modulating signals.

Illustrating the use of (6.18) and the conditions thereon, let $x(t)$ be a *square wave* of period T_0 such that $f(t)$ takes on the two discrete values $f_c + f_d$ and $f_c - f_d$ with equal probability, and hence

$$p_f(f) = \tfrac{1}{2}[\delta(f - f_c + f_d) + \delta(f - f_c - f_d)]$$

As to the spectrum of the modulating signal, we can say that it has negligible content at frequencies well above the fundamental $1/T_0$. Therefore, providing $1/T_0 \ll f_d$,

$$
\begin{aligned}
G(f) = &\tfrac{1}{8}A_c{}^2[\delta(f - f_c + f_d) + \delta(f - f_c - f_d)] \\
+ &\tfrac{1}{8}A_c{}^2[\delta(f + f_c - f_d) + \delta(f + f_c + f_d)]
\end{aligned}
$$

the positive-frequency portion being as sketched in Fig. 6.10a.   This

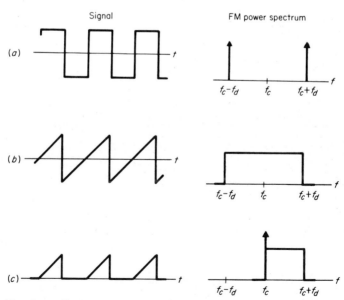

Fig. 6.10 *Periodic modulating signals and the corresponding FM power spectra as obtained from the quasi-static approximation. (a) Square wave; (b) sawtooth; (c) half sawtooth.*

figure reflects the physical notion that with $f(t)$ having but two values, the spectrum should contain only two components. Even so, this notion is strictly correct only in the limit as $f_d T_0 \to \infty$. The exact spectrum for arbitrary $f_d T_0$ can be found by applying Eq. (6.15) [see Panter (1965) for typical results].

Figure 6.10b and 6.10c shows other periodic modulating signals and the corresponding power spectra which result from the quasi-static approximation.

Finally we apply (6.18) to the case of *tone modulation*, that is, $x(t) = A_m \cos \omega_m t$, so the instantaneous frequency can be written as

$$f(t) = f_c + A_m f_d \cos \alpha$$

where $\alpha = 2\pi f_m t$ and $A_m f_d$ is the maximum deviation. Now, insofar as $f(t)$ is concerned, it can be said that α is *uniformly* distributed over $[-\pi, \pi]$ since values of $|\alpha|$ exceeding π yield the same $f(t)$ as values of α within $[-\pi, \pi]$. Hence, drawing upon Example 4.7, one obtains

$$p_f(f) = \begin{cases} \dfrac{1}{\pi A_m f_d \sqrt{1 - [(f - f_c)/A_m f_d]^2}} & |f - f_c| \leq A_m f_d \\ 0 & |f - f_c| > A_m f_d \end{cases}$$

Therefore, if $f_m \ll A_m f_d$, the power spectrum is as shown in Fig. 6.11.

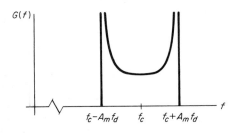

Fig. 6.11 *Tone-modulated FM power spectrum in the limit as $\beta \to \infty$ with f_d held constant.*

Comparing this figure with the line spectra of Fig. 6.6, it is seen to represent the limiting case as $\beta = A_m f_d / f_m \to \infty$ with f_d held constant. The conclusion gained thereby should be obvious, namely, with large values of the modulation index β, the FM spectrum is tightly confined within $f_c \pm A_m f_d$, which is the range of $f(t)$. But again we must emphasize the condition $f_m \ll A_m f_d$ or its equivalent $\beta \gg 1$, otherwise the approximation and conclusions are invalid. In fact, the FM bandwidth-reduction fallacy can be traced directly to incorrect use of this intuitive approach.

6.3 FREQUENCY-MODULATION BANDWIDTHS

We have seen that, in general, an FM spectrum has infinite extent. (The finite spectra stemming from the quasi-static approach are, after all, approximations.) Consequently, the generation and transmission of *pure* FM requires systems of *infinite bandwidth*, whether or not the message is bandlimited. But practical FM systems having finite bandwidth do exist and perform quite well. Their success depends upon the fact that, sufficiently far away from the carrier frequency, the spectral components are quite small and may be discarded. True, omitting any portion of the spectrum will cause *distortion* in the demodulated signal; but the distortion can be minimized by keeping all *significant* spectral components.

Determination of FM transmission bandwidth thus boils down to the question: How much of the modulated signal spectrum is significant? Of course, significance standards are not absolute, being contingent upon the amount of distortion which can be tolerated in a specific application. However, rule-of-thumb criteria based on studies of tone modulation have met with considerable success and lead to useful approximate relations. Our discussion of FM bandwidth requirements therefore begins with the significant sideband lines for tone modulation.

Significant sideband lines

Figure 6.5 indicated that $J_n(\beta)$ falls off rapidly for $|n/\beta| > 1$, particularly if $\beta \gg 1$. Assuming that the modulation index β is large, we can say that $|J_n(\beta)|$ is significant only for $|n| \leq \beta = A_m f_d / f_m$. Therefore, all significant sideband lines are contained in the frequency range

$$f_c \pm n_{max} f_m = f_c \pm \beta f_m = f_c \pm A_m f_d$$

a conclusion which agrees with intuitive reasoning and Fig. 6.11. On the other hand, suppose the modulation index is small; then *all* sideband lines are small compared to the carrier, since $J_0(\beta) \gg J_{n \neq 0}(\beta)$ when $\beta \ll 1$. But we must retain at least the first-order sideband pair, else there would be no frequency modulation at all. Hence, for small β, the significant sideband lines are contained in

$$f_c \pm f_m$$

Summarizing these two extreme cases, the bandwidth for FM with tone modulation is

$$B = \begin{cases} 2A_m f_d & \beta \gg 1 \\ 2f_m & \beta \ll 1 \end{cases}$$

as might have been concluded from the line spectra of Fig. 6.6.

To put the above observations on a quantitative footing, all side-band lines having relative amplitude $|J_n(\beta)| > \epsilon$ are *defined* as being significant, where ϵ ranges from 0.01 to 0.1 according to the application. Then, if $|J_M(\beta)| > \epsilon$ and $|J_{M+1}(\beta)| < \epsilon$, there are M significant sideband *pairs* and $2M + 1$ significant lines all told. The bandwidth is thus

$$B = 2M(\beta)f_m \qquad M \geq 1 \tag{6.19}$$

since the lines are spaced by f_m and M depends on the modulation index β. The condition $M \geq 1$ has been included in (6.19) to account for the fact that B cannot be less than $2f_m$.

Figure 6.12 shows M as a continuous function of β for $\epsilon = 0.01$ and 0.1. Experimental studies indicate that the former is often overly conservative, while the latter may result in small but noticeable distortion.

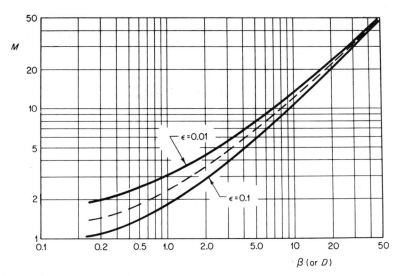

Fig. 6.12 *The number of significant sideband pairs M as a function of β (or D).*

Values of M between these two bounds, as indicated by the dashed line in Fig. 6.12, are acceptable for most purposes and will be used hereafter.

But the bandwidth of (6.19) is not the transmission bandwidth B_T; rather it is the minimum bandwidth necessary for modulation by a tone of specified amplitude and frequency. To illustrate, the maximum frequency deviation f_d of commercial FM is limited by the FCC to 75 kHz, and modulating frequencies typically cover 30 to 15,000 Hz. If a 15-kHz tone has unit amplitude ($A_m = 1$), then $\beta = {}^{75}\!/_{15} = 5$, $M = 7$, and $B = 2 \times 7 \times 15 = 210$ kHz. Had the amplitude been less, the maximum frequency deviation would not be developed, and the bandwidth would be smaller.[1]   Moreover, a lower-frequency tone, say 7.5 kHz, with full amplitude would result in a larger modulation index ($\beta = 10$), a greater number of significant sideband pairs ($M = 12$), but a smaller bandwidth, namely, $B = 2 \times 12 \times 7.5 = 180$ kHz.   In short, bandwidth is determined in a rather complex fashion by both $A_m f_d$ and f_m (or β and f_m), not just β alone.

Pursuing this last point, let us calculate the *maximum* bandwidth required when the tone parameters are constrained by $A_m \leq 1$ and $f_m \leq W$.   For this purpose, the dashed line in Fig. 6.12 can be approximated by

$$M(\beta) \approx \beta + a$$

where a is essentially constant with a value between 1 and 2 (the exact value is immaterial at the moment).   Inserting $M(\beta)$ into (6.19) gives

$$B \approx 2(\beta + a)f_m = 2\left(\frac{A_m f_d}{f_m} + a\right)f_m = 2(A_m f_d + a f_m)$$

Now, bearing in mind that f_d is a property of the modulator, what tone produces the maximum bandwidth?   Clearly, it is the *maximum-amplitude maximum-frequency* tone having $A_m = 1$ and $f_m = W$.   The worst-case tone-modulation bandwidth is then

$$B_{\max} = 2(f_d + aW)$$

Note carefully that the corresponding modulation index $\beta = f_d/W$ is not the maximum value of β but rather the value which, combined with the maximum modulating frequency, yields the maximum bandwidth. Any other tone having $A_m < 1$ or $f_m < W$ will require less bandwidth even though β may be larger.

Transmission bandwidth

We now focus attention on the transmission bandwidth B_T required when $x(t)$ is an *arbitrary modulating signal* having the message bandwidth W and satisfying the normalization convention $|x(t)| \leq 1$.   But we do not turn our backs on the previous conclusions; in fact, we shall estimate B_T directly from the worst-case tone-modulation analysis, assuming that any component in $x(t)$ of smaller amplitude or frequency will require a smaller bandwidth than $B_{\max}$.   Admittedly, this procedure ignores the fact that superposition is not applicable to exponential modulation. However, our investigation of multitone spectra has shown that the beat-

---

[1] It is paradoxical that in *frequency* modulation (with fixed f_d) the bandwidth depends on the tone *amplitude*, whereas in *amplitude* modulation the bandwidth depends on the tone *frequency*.

frequency sideband pairs are contained primarily within the bandwidth of the dominating tone alone, as illustrated by Fig. 6.8.

Therefore, extrapolating tone modulation to an arbitrary modulating signal, we define the *deviation ratio*

$$D = \frac{f_d}{W} \tag{6.20}$$

as the maximum deviation divided by the maximum modulating frequency, analogous to the modulation index of worst-case tone modulation. The transmission bandwidth required for $x(t)$ is then

$$B_T = 2M(D)W \qquad M \geq 1 \tag{6.21}$$

where D is treated just like β to find $M(D)$, say from Fig. 6.12.

Lacking appropriate curves or tables for $M(D)$, there are several approximations to B_T that can be invoked. With extreme values of the deviation ratio we find that

$$B_T = \begin{cases} 2DW = 2f_d & D \gg 1 \\ 2W & D \ll 1 \end{cases}$$

paralleling our results for tone modulation with β very large or very small. Both of these approximations are combined in the convenient relation

$$B_T \approx 2(f_d + W) = 2(D + 1)W \tag{6.22}$$

known as *Carson's rule*. Perversely, the majority of actual FM systems have $2 < D < 10$, for which Carson's rule somewhat underestimates the transmission bandwidth. A better approximation for equipment design is then

$$B_T \approx 2(f_d + 2W) = 2(D + 2)W \qquad D > 2 \tag{6.23}$$

which would be used, for example, to determine the 3-db bandwidths of RF and IF amplifiers.

Applying these relations to commercial FM, $f_d = 75$ kHz and $W = 15$ kHz, so $D = 5$. We have already found that $M = 7$ for $\beta = 5$, hence (6.21) gives $B_T = 210$ kHz. (High-quality FM radios have IF bandwidths of at least 200 kHz.) Carson's rule underestimates B_T by about 10 percent, giving $2(5 + 1) \times 15 = 180$ kHz, whereas (6.23) is right on the mark with $2(5 + 2) \times 15 = 210$ kHz.

The reader is encouraged to review these several approximations and their regions of validity. In deference to most of the literature we shall frequently take B_T as given by Carson's rule, Eq. (6.22), bearing in mind its limitations. However, when $x(t)$ is far from bandlimited, e.g., a rectangular pulse, it is necessary and prudent to carry out the spectral analysis (if possible) and determine the bandwidth therefrom.

Thus, the spectra of Fig. 6.9 require $B_T \geq f_d + 2/\tau$, this being centered not on f_c but on $f_c + \frac{1}{2}f_d$.

Narrowband and wideband frequency modulation (NBFM and WBFM)

Our examinations of spectra and bandwidth point to the conclusion that there are two special cases of FM, namely, when the deviation ratio is very small or very large. These cases have such distinctly different properties that they have been given the characteristic names of *narrowband frequency modulation* (NBFM) and *wideband frequency modulation* (WBFM), respectively.

NBFM is in many ways similar to double-sideband linear modulation. Indeed, with $D \ll 1$,

$$B_T = 2W$$

as was true of AM and DSB. Moreover, the tone-modulated spectrum differs from AM only by the phase reversal of the lower sideband line, as remarked before. Not surprisingly, it also turns out that NBFM has no inherent advantage over linear modulation systems insofar as noise performance is concerned. Consequently, NBFM is not used for broadcast transmission although it often is encountered as an intermediate step in the generation of WBFM.

As to the properties of WBFM, the distinguishing spectral feature is that B_T is large compared to the message bandwidth and *independent* of the message bandwidth, that is,

$$B_T = 2DW = 2f_d$$

when $D \gg 1$. But the important property of WBFM is the wideband noise reduction mentioned at the beginning of this chapter. Indeed, noise reduction is the *raison d'être* of WBFM and will be examined more fully in Sec. 6.6.

6.4 PHASE MODULATION

PM has many similarities to FM. Consequently, our analysis of PM will be quite abbreviated, drawing heavily upon the FM results. Nonetheless, a subtle but significant difference might perhaps be noted at the start.

Recall that PM was defined as having the relative phase

$$\phi(t) = \phi_d x(t)$$

where ϕ_d is the phase deviation constant such that

$$-\phi_d \leq \phi(t) \leq \phi_d$$

since $|x(t)| \leq 1$. Consider now the demodulation of a PM wave: if ambiguities are to be avoided in demodulation, $\phi(t)$ must not exceed the range $\pm\pi$ rad; after all, there is no physical distinction between phase angles of $+3\pi/2$ and $-\pi/2$, for instance. Hence the deviation constant is constrained by

$$\phi_d \leq \pi$$

This restriction is directly analogous to the restriction $m \leq 1$ in AM, and ϕ_d can justly be called the phase modulation index.

A like constraint on the FM deviation constant f_d is not necessary because one can always distinguish $f_c + f_d$ from $f_c - f_d$ providing only that f_d is less than f_c. Therefore the FM deviation can be made as large as desired, going to higher carrier frequencies if necessary. Later it will be shown that the absolute limit on ϕ_d, as contrasted with the relative limit on f_d, accounts in part for the superiority of FM performance in the presence of noise.

Spectra and bandwidth

As before, our spectral analysis of PM begins with tone modulation. However, to facilitate matters, we take the modulating signal to be a *sine* wave rather than a cosine wave. Then, with $x(t) = A_m \sin \omega_m t$, Eq. (6.5) becomes

$$\begin{aligned} x_c(t) &= A_c \cos (\omega_c t + \phi_d A_m \sin \omega_m t) \\ &= A_c \cos (\omega_c t + \beta_p \sin \omega_m t) \end{aligned} \tag{6.24}$$

where

$$\beta_p = A_m \phi_d$$

The reason for using sine-wave modulation is now obvious if one compares Eq. (6.24) with the FM expression, Eq. (6.9); they are identical save that β_p is independent of the tone frequency whereas β depends on both tone amplitude and frequency. Therefore, PM line spectra have the same general characteristics as FM, with the following exception. If the modulating frequency f_m is changed while holding the amplitude A_m fixed, β_p remains constant, and only the line *spacing* is altered. (Both line spacing and relative line amplitude would be affected in the FM case.) Thus, the left-hand side of Fig. 6.6 applies to PM as well as FM, while the right-hand side does not.

In connection with bandwidths, we could repeat the arguments of Sec. 6.3 with appropriate modifications. However, it is only necessary to note that $D = f_d/W$ is the *maximum phase deviation* of an FM wave under worst-case bandwidth conditions. This reveals that D and ϕ_d are equivalent parameters, since ϕ_d is the maximum phase deviation of a PM wave. Therefore, the transmission bandwidth for PM with arbitrary $x(t)$ is given by

$$B_T = 2M(\phi_d)W \qquad M \geq 1 \tag{6.25a}$$

or

$$B_T \approx 2(\phi_d + 1)W \tag{6.25b}$$

which is the approximation equivalent to Carson's rule. These expressions differ from (6.21) and (6.22) in that ϕ_d is independent of W.

Narrowband phase modulation (NBPM)

Because of the constraint $\phi_d \leq \pi$, there is no "wideband" PM in the same sense as WBFM. But NBPM, having $\phi_d \ll \pi$, is both interesting and easy to handle analytically. With small phase deviation the PM wave becomes

$$
\begin{aligned}
x_c(t) &= A_c[\cos \phi_d x(t) \cos \omega_c t - \sin \phi_d x(t) \sin \omega_c t] \\
&\approx A_c \cos \omega_c t - A_c \phi_d x(t) \sin \omega_c t
\end{aligned}
\tag{6.26a}
$$

The spectrum is then

$$X_c(f) = \frac{A_c}{2}[\delta(f - f_c) + \delta(f + f_c)] + \frac{jA_c\phi_d}{2}[X(f - f_c) - X(f + f_c)] \tag{6.26b}$$

so $B_T = 2W$, as expected.

Comparing the NBPM spectrum with that of AM [Eq. (5.3)] shows that the similarities are even more pronounced than those of NBFM. Moreover, it appears from (6.26) that NBPM is a *linear* modulation process, at least approximately, and can be generated using linear modulation devices. Specifically, a balanced modulator together with a quadrature phase shifter can produce NBPM when arranged as diagramed in Fig. 6.13. (Note that the phase shifter need operate only at the carrier frequency.) That Fig. 6.13 is an NBPM modulator is easily verified by inspecting (6.26a).

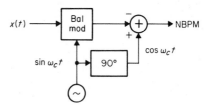

Fig. 6.13 *A narrowband phase modulator constructed using a balanced modulator.*

Because NBPM is generated with relative ease, it forms the core of many FM modulation systems, an integrator being tacked on the input to produce frequency rather than phase modulation. This and other modulators are described in the next section.

6.5 GENERATION AND DETECTION METHODS

In discussing the equipment used for exponential modulation systems, one should keep in mind that the instantaneous frequency of the modulated wave varies as a linear function of the message waveform. Therefore devices are required that produce or are sensitive to frequency variation in a linear fashion, the desired characteristics of an ideal FM modulator and demodulator being shown in Fig. 6.14 for illustration. Needless to

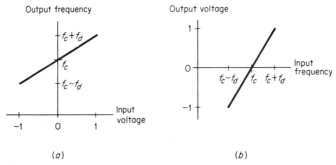

Fig. 6.14 *Idealized FM device characteristics.* (a) *Modulator;* (b) *demodulator.*

say, such idealized characteristics are only approximated in practice, and it is sometimes difficult to obtain a suitably linear voltage-frequency relationship of sufficient range.

On the other hand, the constant-amplitude property of exponential modulation is a definite advantage from the hardware viewpoint. For one thing, the designer need not worry about excessive power dissipation or high-voltage breakdown due to extreme peaks in the waveform. But more important, nonlinear amplitude distortion has virtually no effect on message transmission, since the information resides in the zero crossings of the wave and not in the amplitude. (Phase-shift or delay distortion is of course intolerable.) It likewise follows that any spurious amplitude variations can be eliminated by peak-clipping devices, called *limiters*, without removing the message. Thus, considerable latitude is possible in the design and selection of equipment. As a case in point, the microwave repeater links of long-distance telephone communications use FM primarily because the wideband linear amplifiers required for amplitude modulation are unavailable at microwave frequencies.

Turning specifically to FM, there are two basic generation methods, known as the *direct* and *indirect* systems. We shall examine these one at a time and then take up the subject of FM demodulation.

Direct frequency modulation

Conceptually, direct FM is straightforward and requires nothing more than a *voltage-controlled oscillator* (VCO) whose oscillation frequency has a linear dependence on applied voltage. This is readily implemented in the microwave band ($f_c \geq 1$ GHz), where devices such as the klystron tube have linear VCO characteristics over a substantial frequency range, typically several megahertz. If a lower carrier frequency is desired, the modulated signal can be down-converted by heterodyning with the output of a fixed-frequency oscillator, often another klystron. Some laboratory test generators use precisely this technique.

Alternately, for lower carrier frequencies, it is possible to modulate a conventional tuned-circuit oscillator by introducing a *variable-reactance* element as part of the LC parallel resonant circuit (Fig. 6.15). In the

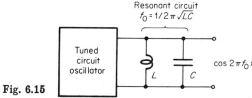

Fig. 6.15

steady state, the oscillation frequency is the usual resonance frequency

$$f_0 = \frac{1}{2\pi \sqrt{LC}}$$

Now suppose the capacitance varies with the modulating signal in the form

$$C(t) = C_0[1 - kx(t)]$$

Providing the fractional variation is small, the oscillation frequency will follow the changing resonance condition, so the instantaneous frequency becomes

$$f(t) = \frac{1}{2\pi \sqrt{LC_0}} [1 - kx(t)]^{-\frac{1}{2}}$$

$$\approx \frac{1}{2\pi \sqrt{LC_0}} [1 + \tfrac{1}{2}kx(t)] \qquad k^2 \ll 1$$

and the desired modulation has been achieved. The condition $k^2 \ll 1$ is equivalent to $f_d \ll f_c$ and usually imposes no hardship.

The variable reactance can be obtained in numerous ways. The workhorse of early direct FM modulators was the *reactance tube*, a pentode biased so that its output impedance contains a capacitive term proportional to grid voltage.[1] Other modulators use Miller effect capacitance, saturable-reactor elements, or the more recently perfected diode reactance of *varactors*.

The principal advantage of direct FM is that large frequency deviations are possible without additional operations. The major disadvantage is that the carrier frequency tends to drift and must be stabilized by rather elaborate feedback frequency control. Because satisfactory stabilization techniques have emerged only lately, many existing FM transmitters are of the indirect type.

Indirect frequency modulation

The heart of indirect FM is a *narrowband phase modulator* whose carrier frequency is supplied by a stable source, usually a crystal-controlled oscillator, to ensure stability. Figure 6.16 shows the parts of a complete

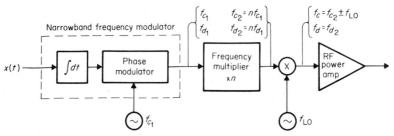

Fig. 6.16 *An indirect FM modulation system.*

indirect system as conceived by Armstrong (1936). The functions are as follows.

Prior to modulation the message is integrated so as to yield frequency rather than phase modulation. The phase modulator itself may be as shown in Fig. 6.13 or a variety of other types, one of which, Day's *serrasoid modulator*, is touched upon in the next chapter. In any case,

---

[1] See Schwartz (1959, chap. 3) for the details.

the resulting NBFM often contains inherent *distortion*, traceable to omitted terms in the approximation of Eq. (6.26a), unless the deviation ratio $D_1 = f_{d_1}/W$ is very small. It is therefore necessary to increase the deviation ratio after modulation, this being accomplished by a chain of frequency doublers and triplers forming a *frequency multiplier*.

An ideal frequency multiplier operates on instantaneous frequency such that if $f_1(t) = f_{c_1} + f_{d_1}x(t)$ is the input, n-fold multiplication produces $f_2(t) = nf_1(t) = f_{c_2} + f_{d_2}x(t)$, where $f_{c_2} = nf_{c_1}$ and $f_{d_2} = nf_{d_1}$. Note that this is a subtle process, affecting the *range* of frequency variation but not the *rate*. Multiplication of a tone-modulated signal, for example, increases the carrier frequency and the modulation index but not the modulating frequency; thus the relative amplitudes of the sideband lines are altered, but the line spacing remains the same.

Returning to Fig. 6.16, n is chosen to give the desired final deviation, namely, $n = f_d/f_{d_1} = D/D_1$. But this multiplication factor usually results in f_{c_2}'s being much higher than the desired carrier frequency. Heterodyning the multiplier output with a second crystal-controlled oscillator translates the spectrum *intact* to the proper location. All these steps are done at low power levels, so the final system component is an RF power amplifier of sufficient bandwidth but whose linearity is not crucial.

By way of illustration with representative values, an early indirect transmitter for commercial FM has $f_{c_1} = 200$ kHz and $f_{d_1} = 25$ Hz. With $W = 15$ kHz, the initial deviation ratio is $D_1 = 25/(15 \times 10^3) \approx 2 \times 10^{-3}$, this minute value being required to minimize distortion and guarantee good fidelity. Since the final deviation is to be $f_d = 75$ kHz, the multiplication factor needed is $n = 75 \times 10^3/25 = 3{,}000$, which entails six triplers and two doublers. After multiplication, the spectrum is located at $f_{c_2} = 3{,}000 \times 200$ kHz $= 600$ MHz, so the second oscillator must have $f_{LO} \sim 600 \pm 100$ MHz to bring the carrier down to the FM band of 88 to 108 MHz.

These calculations indicate that although indirect FM gets around the frequency-stability problem, it is not without difficulties, principally that of the multiplication factor. Indeed, broadband multipliers of the requisite phase characteristics are just as complex as the frequency-control circuits of direct FM, and the latter have been greatly improved by the development of digital counters. Consequently, there is a trend back to direct FM for high-quality large-deviation systems.

FM detection—frequency discriminators

An FM demodulator, or *frequency discriminator*, must produce an output voltage linearly dependent on input frequency. Slightly above or below resonance, a simple tuned circuit plus envelope detector has this property over a limited range. This is called *slope detection* and is illustrated in Fig. 6.17.

For a qualitative analysis[1] of slope detection, suppose $x_c(t)$ is tone-modulated and f_c is less than f_0. Then, as $f(t)$ swings above or below f_c, the amplitude response of the tuned circuit converts the frequency variation to an amplitude variation on top of the FM signal, yielding the waveform $y_c(t)$. Extracting only the amplitude variation with an envelope detector (plus dc block) produces $y_D(t)$ and completes the demodulation.

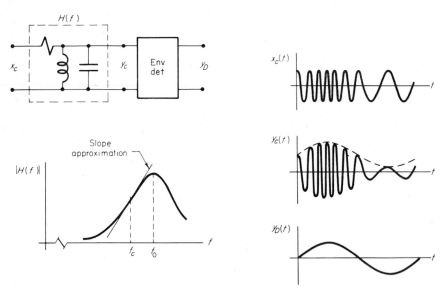

Fig. 6.17 *FM slope-detection circuit and waveforms.*

There are two problems with slope detection as described above: (1) the detector also responds to spurious amplitude variations of the input FM, and (2) the range of linear slope is quite small. A *limiter* preceding the tuned circuit takes care of the former, while extended linearity can be achieved by using the configuration of Fig. 6.18, which is called a *balanced discriminator*.

A balanced discriminator has two resonant circuits, one tuned above f_c and the other below. Thus, as $f(t)$ changes, the amplitude variations are in opposite directions, and taking the difference of these variations gives the frequency-to-voltage characteristic of Fig. 6.19, the well-known

[1] Quantitative analysis of circuit response to time-varying frequency is a difficult matter. Baghdady (1960, chap. 19) presents a concise discussion of the problem and methods of solution.

S curve. Note that the dc component is automatically canceled, bypassing the need for a dc block and thereby improving response to low modulating frequencies. However, an input limiter is still necessary.

Balanced discriminators are often employed as the frequency-sensing element in automatic frequency control, which has many applications aside from exponential modulation systems. They are readily adapted to the microwave band, with resonant cavities serving as tuned circuits and crystal diodes for envelope detectors.

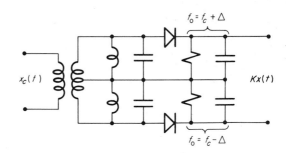

Fig. 6.18 *Schematic diagram of a balanced frequency discriminator.*

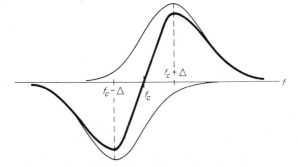

Fig. 6.19 *Frequency-to-voltage characteristics of a balanced discriminator.*

Other common FM demodulators are the *Foster-Seeley phase-shift discriminator* and the *ratio detector*, whose descriptions can be found in texts on radio electronics. The ratio detector is particularly ingenious and economical, for it combines the operations of limiting and demodulating into one unit.

Before leaving the subject of demodulation, it must be pointed out that any device whose output is the *time derivative* of the input will perform an FM-to-AM conversion and therefore can be used to detect FM. That this should be the case follows from the differentiation theorem $dx/dt \leftrightarrow j2\pi f X(f)$, so a differentiator has a linear frequency-to-amplitude characteristic. Specifically, if $x_c(t) = A_c \cos \theta_c(t)$, then $dx_c/dt = -A_c(d\theta_c/dt)$

sin θ_c.　But, from Eqs. (6.2) and (6.6), $d\theta_c/dt = 2\pi f(t) = 2\pi[f_c + f_d x(t)]$, so

$$\frac{dx_c}{dt} = -2\pi A_c[f_c + f_d x(t)] \sin \theta_c(t)$$

which is a waveform having both amplitude and frequency modulation.

FM receivers

Most FM receivers are of the superheterodyne variety.　They differ from Fig. 5.30 in two respects: (1) a limiter-discriminator (or ratio detector) replaces the envelope detector, and (2) automatic frequency control (AFC) is provided to correct for frequency drift of the LO.　Commercial FM radios have a tuning range of 88 to 108 MHz, $f_{IF} = 10.7$ MHz, and an IF bandwidth of 200 to 300 kHz.　Note that the *fractional bandwidth* of the IF amplifier is about 2×10^{-2}, the same as for AM superheterodynes, where $f_{IF} = 455$ kHz and $B_{IF} = 10$ kHz.

6.6 INTERFERENCE AND NOISE IN EXPONENTIAL MODULATION SYSTEMS

A physical argument has been given for the noise-reduction property of exponential modulation.　The time has now arrived to delve into this fascinating subject in greater detail, beginning with a brief study of interference.

Interference

Consider again the sum of an unmodulated carrier plus interfering sinusoid of frequency $f_c + f_i$, as in Sec. 5.6.　Under the small-interference condition[1] $A_i \ll A_c$, we found the resultant to be approximately

$$y(t) = A_c(1 + m_i \cos \omega_i t) \cos (\omega_c t + m_i \sin \omega_i t) \tag{6.27}$$

where

$$m_i = \frac{A_i}{A_c} \ll 1$$

In an exponential system the amplitude variations of (6.27) would be completely removed by the limiter, which is easier said than done, but the time-varying phase corresponds to exponential modulation and causes a detected output.

---

[1] We shall treat only this simplified case.　Discussions of interference with arbitrary A_i/A_c or modulated carriers are found in advanced texts, e.g., Panter (1965, chap. 11).

For the purpose of analysis, we define an ideal phase or frequency demodulator to have the output signal

$$y_D(t) = \begin{cases} K_D \psi(t) & \text{phase demodulator} \\ K_D \dfrac{1}{2\pi} \dfrac{d\psi(t)}{dt} & \text{frequency demodulator} \end{cases} \tag{6.28}$$

where $\psi(t)$ is the relative phase of the resultant input signal with respect to $e^{j\omega_c t}$ and, as usual, K_D is the detection constant. Of course $(d\psi/dt)/2\pi = f(t) - f_c$ is the instantaneous frequency deviation of the input. The carrier amplitude does not appear here, since the ideal demodulator is assumed to have perfect limiting action.

From Eq. (6.27), $\psi(t) = m_i \sin \omega_i t = A_i/A_c \sin \omega_i t$, and the corresponding demodulated signals will be

$$y_D(t) = \begin{cases} K_D \dfrac{A_i}{A_c} \sin \omega_i t & \text{PM} \\ K_D \dfrac{A_i f_i}{A_c} \cos \omega_i t & \text{FM} \end{cases}$$

where $|f_i| \leq B_{\text{IF}}/2 \approx B_T/2$. (Larger values of $|f_i|$ are rejected by pre-detection filtering of the IF amplifier.) Thus, in FM systems, the detected interference is proportional to both the relative *amplitude* and *frequency* of the interfering wave; in PM systems, as in linear modulation, only the amplitude enters the picture.

This difference can be understood with the aid of simple physical considerations. The strength of a detected signal in FM depends on the maximum *frequency deviation*. Interfering waves close to the carrier frequency cannot cause significant change in the frequency of the resultant and therefore produce little effect. The greater the difference between f_c and $f_c + f_i$, the greater the frequency deviation, so we can expect the demodulated output to be proportional to $|f_i|$. But for PM the maximum *phase deviation* depends only on relative amplitudes, as shown by the phasor diagram of Fig. 5.35.

The performance of FM and PM with respect to interference is best displayed by plotting the amplitude of y_D as a function of $|f_i|$ (Fig. 6.20).

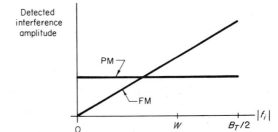

Fig. 6.20 *Detected interference amplitude as a function of $|f_i|$ for an interfering wave at $f_c + f_i$.*

If the interference is due to a *cochannel* (or image-frequency) station, then $f_c + f_i \approx f_c$, $|f_i|$ is small, and FM clearly is less vulnerable.   Conversely, PM has better performance with respect to *adjacent-channel* interference, where $|f_i|$ is relatively large.   In either case, a lowpass filter of bandwidth W should follow the demodulator to eliminate detected interference components which are outside of the message band but not rejected by the IF amplifier, i.e., interference at $W < |f_i| < B_T/2$.   Such postdetection filtering is desirable but not a necessity in linear modulation systems because the transmission bandwidth is no greater than $2W$.

Deemphasis and preemphasis filtering

The fact that detected FM interference is most severe at large values of $|f_i|$ suggests a method for improving system performance with selective postdetection filtering, called *deemphasis filtering*.

Suppose the demodulator is followed by a lowpass filter having an amplitude response that begins to decrease gradually *below W;* this will *deemphasize* the high-frequency portion of the message band and thereby reduce the more serious interference.   A sharp-cutoff (ideal) lowpass filter is still required to remove any residual components above W, so the complete demodulator consists of limiter, discriminator, deemphasis filter, and lowpass filter (Fig. 6.21).

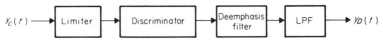

Fig. 6.21   *A complete FM demodulator.*

Of course deemphasis filtering also attenuates the high-frequency components of the message itself, causing distortion of the output signal unless corrective measures are taken.   But it is a simple matter to compensate for deemphasis distortion by *predistorting* or *preemphasizing* the modulating signal at the transmitter before modulation.   The preemphasis and deemphasis filter characteristics should be related by

$$H_{\mathrm{PE}}(f) = \frac{1}{H_{\mathrm{DE}}(f)} \qquad |f| \le W$$

to yield net undistorted transmission.

In essence, we preemphasize the message before modulation (where the interference is absent) so we can deemphasize the interference relative to the message after demodulation.   The technique is by no means restricted to FM systems but can be used to advantage whenever undesired contaminations predominate at certain portions of the desired signal band. For example, high frequencies are commonly preemphasized in sound recording so that high-frequency surface noise can be deemphasized during playback. On the other hand, phase modulation and linear modulation systems profit little from deemphasis because the interference is independent of frequency and there is no *selective* advantage. To be sure, high-frequency deemphasis does yield some improvement, but hardly enough to justify the added complexity.

Returning to FM per se, the deemphasis filter is usually a simple RC network having

$$|H_{\mathrm{DE}}(f)| = \left[1 + \left(\frac{f}{f_3}\right)^2\right]^{-\frac{1}{2}} \qquad f_3 = \frac{1}{2\pi RC}$$

$$\approx \begin{cases} 1 & |f| \ll f_3 \\ \dfrac{f_3}{f} & |f| \gg f_3 \end{cases} \tag{6.29}$$

where the 3-db frequency f_3 is considerably less than the message bandwidth W. Since the interference amplitude increases linearly with $|f_i|$ in the absence of filtering, the deemphasized interference response is $|H_{\mathrm{DE}}(f_i)| \times |f_i|$, as sketched in Fig. 6.22. Note that, like PM, this

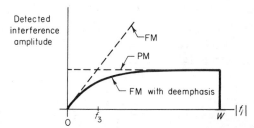

Fig. 6.22 *Detected interference amplitude for FM with deemphasis filtering.*

becomes constant for $|f_i| \gg f_3$. Therefore, FM can be superior to PM for both adjacent-channel and cochannel interference.

At the transmitting end, the corresponding preemphasis filter is

$$|H_{\mathrm{PE}}(f)| = \left[1 + \left(\frac{f}{f_3}\right)^2\right]^{\frac{1}{2}}$$

$$\approx \begin{cases} 1 & |f| \ll f_3 \\ \dfrac{f}{f_3} & |f| \gg f_3 \end{cases}$$

which has little effect on the lower message frequencies. At higher frequencies, however, the filter acts as a *differentiator*, the output spectrum being proportional to $fX(f)$ for $|f| \gg f_3$. But differentiating a signal before frequency modulation is equivalent to *phase modulation!* Hence, preemphasized FM is actually a combination of FM and PM, combining the advantages of both with respect to interference (again see Fig. 6.22). As might be expected, this turns out to be equally effective for reducing *noise*, a topic to which we turn after one final comment.

Referring to $H_{PE}(f)$ as given above, we see that the amplitude of the maximum modulating frequency is increased by a factor of W/f_3, which means that the frequency deviation is increased by this same factor. Generally speaking, the increased deviation requires a greater transmission bandwidth, so the preemphasis-deemphasis improvement is not without price. Fortunately, many modulating signals of interest, particularly audio signals, have relatively little energy in the high-frequency end of the message band, and therefore the higher frequency components do not develop maximum deviation, the transmission bandwidth being dictated by lower components of larger amplitude. Adding high-frequency preemphasis tends to equalize the message spectrum so that all components require the same bandwidth. Under this condition, the transmission bandwidth need not be increased.

Demodulation in the presence of noise

Consider now the demodulation of FM or PM when the modulated signal $x_c(t)$ is contaminated by additive narrowband[1] noise $n(t)$. The situation at the demodulator input is therefore much like that of linear modulation as summarized in Fig. 5.37. In particular, $n(t)$ can be assumed to be white noise filtered by the IF amplifier, so its power spectrum is $G_n(f) = (\eta/2)|H_{IF}(f)|^2$ (see Fig. 5.38) and the predetection noise power is

$$N_T = \overline{n^2} = \eta B_T$$

The modulated signal is $x_c(t) = A_c \cos[\omega_c t + \phi(t)]$, where $\phi(t) = \phi_d x(t)$ for PM or $d\phi/dt = 2\pi f_d x(t)$ for FM. Regardless, the average signal power at the demodulator input is

$$P_T = \tfrac{1}{2}A_c^2$$

giving

$$\left(\frac{S}{N}\right)_T = \frac{P_T}{N_T} = \frac{A_c^2}{2\eta B_T}$$

for the predetection signal-to-noise ratio.

Physically, we expect noise to modulate the carrier in both amplitude and phase. But a demodulator having perfect limiting will respond

---

[1] The assumption of *narrowband* noise is not a necessity, but it is reasonable since normally $B_T/f_c \ll 1$ even with *wideband* FM.

only to the latter. Furthermore, if the rms noise amplitude is less than the carrier amplitude, the noise-induced PM has a small effective phase deviation. Since the desired modulation can have a much larger deviation (without an increase of transmitted power), the noise effects are suppressed. Conversely, if the noise is large, signal suppression must be expected.

Proceeding to the analysis, the demodulator input is

$$
\begin{aligned}
y(t) &= x_c(t) + n(t) \\
&= A_c \cos\left[\omega_c t + \phi(t)\right] + r_n(t) \cos\left[\omega_c t + \phi_n(t)\right]
\end{aligned}
$$

where the envelope-and-phase expression for $n(t)$, as in Sec. 4.7, has been invoked. After a little manipulation, $y(t)$ can be written in the form

$$
y(t) = r(t) \cos\left[\omega_c t + \psi(t)\right]
$$

where

$$
\psi(t) = \arctan \frac{A_c \sin \phi(t) + r_n(t) \sin \phi_n(t)}{A_c \cos \phi(t) + r_n(t) \cos \phi_n(t)}
$$

Now a limiter will remove the envelope variations represented by $r(t)$, so we are concerned only with the relative phase $\psi(t)$. Specifically, from the definitions of (6.28), the demodulated signal $y_D(t)$ is proportional to $\psi(t)$ or its time derivative. But in view of the complexity of $\psi(t)$ as written above, useful expressions for the output call for some simplifying approximations.

Let us therefore assume the signal is either very large or very small compared to the noise, such that $A_c \gg r_n$ or $A_c \ll r_n$ most of the time. Since $\overline{r_n^2} = 2\overline{n^2} = 2N_T$, these are equivalent to $(S/N)_T \gg 1$ or $(S/N)_T \ll 1$. The phasor constructions of Fig. 6.23, where the phase angle $\omega_c t$ common to both $x_c(t)$ and $n(t)$ has been taken as the reference,

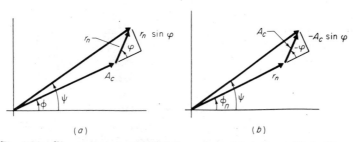

Fig. 6.23 *Phasor diagrams for FM plus noise.* (a) $A_c \gg r_n$; (b) $A_c \ll r_n$.

lead to the approximations

$$\psi(t) \approx \phi(t) + \frac{r_n(t)}{A_c} \sin \varphi(t) \qquad \left(\frac{S}{N}\right)_T \gg 1 \qquad (6.30a)$$

$$\psi(t) \approx \phi_n(t) - \frac{A_c}{r_n(t)} \sin \varphi(t) \qquad \left(\frac{S}{N}\right)_T \ll 1 \qquad (6.30b)$$

where

$$\varphi(t) = \phi_n(t) - \phi(t)$$

Not surprisingly, the leading term is the phase of the dominant component alone, whether signal or noise. But the second term contains both components and is a source of difficulty.

Pursuing this last point, a careful examination of (6.30b) reveals that the message appears only as part of $\sin \varphi(t)$, which also includes $\phi_n(t)$ and in turn is multiplied by the random variable $1/r_n(t)$. This means the message is *mutilated* by noise and cannot be recovered, a phenomenon previously observed in envelope detectors at low predetection signal-to-noise ratios. The resulting *threshold effect* is a serious problem in exponential modulation systems, as we shall discuss shortly. However, further analysis of (6.30b) is exceedingly difficult, so we now focus attention on the large-signal case as given by (6.30a).

If $x_c(t)$ has the phase modulation $\phi(t) = \phi_d x(t)$, the demodulated signal plus noise becomes

$$y_D(t) = K_D \psi(t) = K_D[\phi_d x(t) + n_D(t)]$$

where

$$n_D(t) = \frac{r_n(t)}{A_c} \sin \varphi(t) \qquad \text{PM} \qquad (6.31)$$

Similarly, if the carrier is frequency-modulated, $d\phi/dt = 2\pi f_d x(t)$, and

$$y_D(t) = K_D \frac{1}{2\pi} \frac{d\psi}{dt} = K_D[f_d x(t) + n_D(t)]$$

where

$$n_D(t) = \frac{1}{2\pi A_c} \frac{d}{dt} [r_n(t) \sin \varphi(t)] \qquad \text{FM} \qquad (6.32)$$

It is therefore concluded that, for FM or PM with $(S/N)_T \gg 1$, the message and noise are *additive* at the output, even though the noise term $n_D(t)$ depends in part on the modulating signal. Moreover, the output signal power is

$$S_D = \begin{cases} \phi_d{}^2 \overline{x^2} & \text{PM} \\ f_d{}^2 \overline{x^2} & \text{FM} \end{cases}$$

where we have used ensemble averages and dropped the detector constant K_D, which multiplies both signal and noise.

Despite these conclusions we cannot immediately write down the postdetection signal-to-noise ratios $(S/N)_D$. For one reason, the evaluation of $\overline{n_D{}^2}$ from (6.31) or (6.32) is a more than trivial task; for another, we have not as yet included the effects of postdetection filtering. For both reasons one must go to the frequency domain and examine the postdetection noise power spectra.

Postdetection noise characteristics

A qualitative description of the postdetection noise spectra is obtained by recalling that narrowband noise $n(t)$ can be thought of as a summation of random sinusoids distributed equally over $f_c \pm B_T/2$. Since it was assumed that $(S/N)_T \gg 1$, the amplitudes of these sinusoids are very small compared to A_c, and each will then act as a small interfering signal. Drawing upon our interference studies and Fig. 6.20, we infer that the postdetection noise power spectrum of PM will be constant over $|f| < B_T/2$, i.e., bandlimited white noise. But for FM, where interference amplitude increases linearly with frequency, the power spectrum will go as frequency *squared*. This *parabolic power spectrum* reflects the notion that input noise components close to the carrier frequency cannot cause as much phase deviation as those farther away from f_c.

Going from the qualitative to the quantitative, we shall calculate $G_{n_D}(f)$ from (6.31) and (6.32) by ignoring the carrier modulation $\phi(t)$ and replacing $\varphi(t) = \phi_n(t) - \phi(t)$ by $\phi_n(t)$ alone. The approach is justified because ϕ_n is *uniformly distributed* over $[-\pi,\pi]$; hence, in the sense of ensemble averages, $\phi_n - \phi$ differs from ϕ_n only by a shift of the mean value.[1] With this simplification, $r_n(t) \sin \varphi(t)$ reduces to the input quadrature noise component $n_s(t) = r_n(t) \sin \phi_n(t)$, and the detected noise becomes

$$n_D(t) = \begin{cases} \dfrac{n_s(t)}{A_c} & \text{PM} \\[2ex] \dfrac{1}{2\pi A_c} \dfrac{dn_s(t)}{dt} & \text{FM} \end{cases}$$

That $n_D(t)$ depends on $n_s(t)$ rather than $n_c(t)$ agrees with our phasor interpretation of FM (Fig. 6.7a), wherein it was shown that only those components in quadrature with the carrier cause frequency (or phase) deviation.

The case of PM is quickly disposed of since, by inspection,

$$G_{n_D}(f) = \frac{1}{A_c{}^2} G_{n_s}(f)$$

---

[1] Downing (1964, chap. 5) shows that when $\phi(t)$ is accounted for, there are additional components in $G_{n_D}(f)$ but at frequencies above the message band. Such components are chopped off by the output filter and do not concern us.

where $G_{n_s}(f)$ has the shape of the predetection noise spectrum $G_n(f)$ translated to zero frequency, a conclusion from Sec. 4.7. For an IF amplifier with ideal bandpass (zonal) filter response over $f_c \pm B_T/2$, $G_{n_s}(f) = \eta \Pi(f/B_T)$, and

$$
G_{n_D}(f) = \begin{cases} \dfrac{\eta}{A_c{}^2} & |f| < \dfrac{B_T}{2} \\[2mm] 0 & |f| > \dfrac{B_T}{2} \end{cases} \tag{6.33}
$$

so the output noise is indeed white over $|f| < B_T/2$. But $B_T/2$ is in general greater than the message bandwidth W, which means that out-of-band noise components, like out-of-band interference, should be removed by postdetection filtering. Therefore, assuming an ideal LPF of bandwidth W, the final output noise power is

$$
\begin{aligned}
N_D &= \int_{-W}^{W} G_{n_D}(f)\, df = \frac{\eta}{A_c{}^2} \int_{-W}^{W} df \\
&= \frac{2\eta W}{A_c{}^2} \qquad \text{PM}
\end{aligned} \tag{6.34}
$$

The case of FM is slightly more complicated because of the time derivative in $n_D(t)$. But we can use the differentiation theorem for power spectra, Eq. (2.69), which said that $y(t) = dx(t)/dt$ has $G_y(f) = (2\pi f)^2 G_x(f)$. Hence

$$
\begin{aligned}
G_{n_D}(f) &= \left(\frac{1}{2\pi A_c}\right)^2 (2\pi f)^2 G_{n_s}(f) \\
&= \begin{cases} \dfrac{\eta f^2}{A_c{}^2} & |f| < \dfrac{B_T}{2} \\[2mm] 0 & |f| > \dfrac{B_T}{2} \end{cases}
\end{aligned} \tag{6.35}
$$

again taking a zonal IF response. This spectrum is sketched in Fig. 6.24,

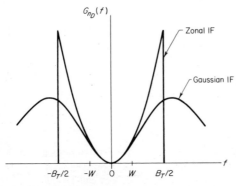

$G_{n_D}(f)$

Zonal IF

Gaussian IF

$-B_T/2$ $-W$ 0 W $B_T/2$ f

Fig. 6.24 *FM postdetection noise power spectrum.*

along with the curve which results from a gaussian-shaped IF amplitude response, the latter being perhaps a more realistic model. In either case the spectrum is essentially parabolic, going as f^2 for $|f| < W$. The final output power in the message band is thus

$$
\begin{aligned}
N_D &= \int_{-W}^{W} \frac{\eta f^2}{A_c^2}\, df \\
&= \frac{2\eta W^3}{3A_c^2} \qquad \text{FM}
\end{aligned}
\tag{6.36}
$$

without deemphasis filtering.

Adding a deemphasis filter modifies the output spectrum to give in general

$$
N_D = \int_{-W}^{W} |H_{\mathrm{DE}}(f)|^2 G_{n_D}(f)\, df
$$

For the simple RC network of Eq. (6.29)

$$
\begin{aligned}
N_D &= \frac{\eta}{A_c^2} \int_{-W}^{W} \frac{f^2\, df}{1 + (f/f_3)^2} \\
&= \frac{2\eta f_3^3}{A_c^2}\left(\frac{W}{f_3} - \arctan\frac{W}{f_3}\right) \qquad \text{deemphasized FM}
\end{aligned}
\tag{6.37}
$$

which is substantially less than (6.36) when $f_3 \ll W$, the usual case. To underscore this improvement we note that if $W/f_3 \gg 1$, then $\arctan (W/f_3) \approx \pi/2 \ll W/f_3$, and hence

$$
N_D \approx \frac{2\eta f_3^3}{A_c^2}\frac{W}{f_3} = 3\left(\frac{f_3}{W}\right)^2 \frac{2\eta W^3}{3A_c^2} \qquad \frac{f_3}{W} \ll 1
$$

for a noise reduction factor of $3(f_3/W)^2$ compared to the output without deemphasis.

A similar analysis for PM with high-frequency deemphasis gives a less impressive factor of approximately $\dfrac{\pi}{2}\dfrac{f_3}{W}$.

Postdetection signal-to-noise ratios

Having found the output noise power for PM, FM, and deemphasized FM, we turn at last to the signal-to-noise ratios and an assessment of wideband noise reduction in exponential modulation systems. The condition $(S/N)_T \gg 1$ is assumed throughout.

As far as output signal power is concerned, the postdetection filtering has no effect; more precisely, any deemphasis filtering is compensated by prior preemphasis filtering. Thus, as previously recorded,

$$
S_D = \begin{cases} \phi_d^2 \overline{x^2} & \text{PM} \\ f_d^2 \overline{x^2} & \text{FM} \end{cases}
$$

whereas the transmitted signal power is $P_T = \frac{1}{2}A_c^2$ at the demodulator input.

Beginning with phase modulation and N_D per (6.34)

$$\left(\frac{S}{N}\right)_D = \frac{\phi_d{}^2\overline{x^2}}{2\eta W/A_c{}^2} = \phi_d{}^2\overline{x^2}\,\frac{P_T}{\eta W}$$

or, introducing the normalized system parameter $z = P_T/\eta W$ from Chap. 5,

$$\left(\frac{S}{N}\right)_D = \phi_d{}^2\overline{x^2}z \qquad \text{PM} \tag{6.38}$$

Since z is the output signal-to-noise ratio for baseband transmission (or suppressed-carrier linear modulation) with power P_T, bandwidth W, and noise density η, it follows that PM gives an improvement over baseband of exactly $\phi_d{}^2\overline{x^2}$. But because of the ambiguity constraint $\phi_d \leq \pi$, the PM improvement is no greater than $\phi_d{}^2\overline{x^2}|_{\max} = \pi^2$, or about 10 db at best. Of course if $\phi_d{}^2\overline{x^2} < 1$, the PM performance is not so good as baseband, while the transmission bandwidth is still $B_T \geq 2W$.

For FM without deemphasis filtering

$$\left(\frac{S}{N}\right)_D = \frac{f_d{}^2\overline{x^2}}{2\eta W^3/3A_c{}^2} = 3\left(\frac{f_d}{W}\right)^2 \overline{x^2}\,\frac{P_T}{\eta W}$$

hence

$$\left(\frac{S}{N}\right)_D = 3D^2\overline{x^2}z \qquad \text{FM} \tag{6.39}$$

where $D = f_d/W$ is the deviation ratio. Adding an RC deemphasis filter with $f_3 \ll W$ further reduces the output noise by about $3(f_3/W)^2$, so

$$\left(\frac{S}{N}\right)_D = \frac{3D^2\overline{x^2}z}{3(f_3/W)^2} = \left(\frac{W}{f_3}\right)^2 D^2\overline{x^2}z$$

$$= \left(\frac{f_d}{f_3}\right)^2 \overline{x^2}z \qquad \begin{array}{l}\text{deemphasized FM}\\ f_3 \ll W\end{array} \tag{6.40}$$

But with or without deemphasis, FM exhibits substantial wideband noise reduction in that $(S/N)_D$ increases with the deviation ratio and the concomitant large transmission bandwidth. In fact, it appears that $(S/N)_D$ can be made *arbitrarily large* by increasing only the deviation, a conclusion which requires further qualification and is reexamined later.

To emphasize the transmission-bandwidth dependence we recall that wideband FM $(D \gg 1)$ has $B_T = 2f_d = 2DW$, so $D^2 = B_T{}^2/4W^2$, and (6.39) becomes

$$\left(\frac{S}{N}\right)_D = \frac{3}{4}\left(\frac{B_T}{W}\right)^2 \overline{x^2}z \qquad \text{WBFM} \tag{6.41}$$

Under this condition the system performance improves as the *square* of the bandwidth ratio B_T/W. With smaller deviations the break-even point compared to baseband occurs at $3D^2\overline{x^2} = 1$; hence

$$D = (3\overline{x^2})^{-\frac{1}{2}} \approx 0.6$$

is sometimes designated as the dividing line between NBFM and WBFM.

When it comes to the question of frequency versus phase modulation, comparing (6.38) and (6.39) on the basis of equal transmission bandwidths, that is, $D = \phi_d$, shows that FM is superior by a factor of 3 (5 db). But this calculation does not reflect the absolute limit on ϕ_d as contrasted to the essentially unlimited FM deviation ratio. Moreover, we have previously observed that deemphasis is far more effective for FM than for PM. Consequently, save in certain special applications,[1] FM is definitely preferred and markedly better than PM insofar as analog message transmission is concerned.

Just how much can be gained with wideband FM is well illustrated by commercial broadcast FM, for which $f_d = 75$ kHz, $W = 15$ kHz, $D = 5$, and the standard deemphasis filter has $f_3 = 2.1$ kHz, corresponding to the time constant $RC = 75$ μsec. Taking $\overline{x^2} = \frac{1}{2}$ as a representative value and excluding deemphasis, $(S/N)_D = (3 \times 5^2 \times \frac{1}{2})z = 37.5z$, or about 16 db better than baseband. Including the deemphasis improvement $(S/N)_D = 640z$ and, other factors being equal, a 1-watt FM system could replace a 640-watt baseband system with no reduction in output signal-to-noise ratio. But bear in mind that the FM transmission bandwidth is 14 times as large as the message bandwidth, underscoring the trade-off of bandwidth for power or signal-to-noise ratio made possible by exponential modulation.

We can now also appreciate why reception may deteriorate when an FM station converts from monaural to stereophonic broadcasting. Referring back to Fig. 5.46b, the maximum modulating frequency becomes $W = 53$ kHz, but the deviation is still fixed at 75 kHz; the final carrier deviation ratio then reduces to a paltry $D = 1.4$. Preemphasis-deemphasis tends to ameliorate the effect [see (6.40), which is independent of D], but a per-channel reduction of output signal-to-noise ratio results unless the effective radiated power is increased proportionately at the time of changeover.

Returning to the FM bandwidth-power exchange, in practice several factors work against full realization of reduced transmitted power at the expense of increased transmission bandwidth. And indeed the goal of commercial FM is not to minimize transmitter power but rather to provide the best possible output signal-to-noise ratio. However, there

[1] One such application is for *digital* systems, in which PM is called *phase-shift keying* (PSK).

are numerous other situations where minimum power is essential; for such applications the condition $(S/N)_T \gg 1$ is a definite hardship, and the FM threshold effect becomes a matter of grave concern.

FM threshold effect

All the above results were based on the approximation of Eq. (6.30a), assuming a large predetection signal-to-noise ratio. We also observed from (6.30b) that if $(S/N)_T \ll 1$, the message is mutilated beyond all recognition by the noise.

Actually, significant mutilation begins to occur when $(S/N)_T \approx 1$, for then $A_c{}^2 \approx \overline{r_n{}^2}$, and the signal and noise *phasors* are of equal length. If they are also of nearly opposite phase $[\phi_n(t) = -\phi(t)]$, the resultant is quite small, and a small change in the phase difference $\phi_n(t) - \phi(t)$ yields a large phase deviation of the resultant, as diagramed in Fig. 6.25.

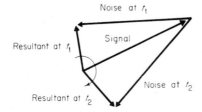

Fig. 6.25

A sudden stepwise phase deviation is equivalent to an *impulsive* frequency shift, so these noise-caused phase deviations appear as sharp pulses or *spikes* at the discriminator output, recognized aurally as a click or crackling sound. As the condition $\phi_n(t) = -\phi(t)$ comes and goes intermittently, the output changes in a sporadic fashion from signal to crackling; hence the name *splutter*.

One also infers from this qualitative picture that the output noise spectrum is no longer parabolic but tends to fill in at dc, the output spikes having appreciable low-frequency content. This conclusion has been verified by the detailed treatments of Rice (1948), Stumpers (1948), and others.

Because the output message may or may not be mutilated, according to the value of $(S/N)_T$, FM suffers from a threshold effect of the same kind as AM with envelope detection. But, unlike the AM case, the FM threshold level usually corresponds to output signal-to-noise ratios sufficiently large to be usable were it not for the mutilation. Moreover, the transition region around the threshold point is relatively narrow, so that small input power variations cause sizable changes in the output signal: one moment it is there, the next moment it has gone.[1] This sudden loss of output is not uncommon in fringe-area FM reception.

As to numerical values for the threshold level, analytic calculations are complicated by three factors: (1) the changing output noise spectrum as $(S/N)_T$ decreases, (2) a modulation-suppression effect at low $(S/N)_T$, and (3) lack of a standard definition for the threshold point. However, experimental studies indicate that mutilation is negligible in most cases of interest if $(S/N)_T \geq 20$ or thereabouts. Hence, we shall define the threshold point to be

$$\left(\frac{S}{N}\right)_{T_{th}} = 20 \approx 13 \text{ db} \tag{6.42}$$

which is somewhat on the conservative side [see Downing (1964, chap. 5) for a tighter definition].

But in using (6.42) one must bear in mind that the predetection noise power $N_T = \eta B_T$ depends on the transmission bandwidth and hence upon the frequency deviation; i.e., widening the IF bandwidth to accommodate a larger deviation also increases the noise entering the demodulator, so it may be necessary to increase P_T along with the deviation to avoid threshold effects. To emphasize this fact we write (6.42) in terms of

$$z = \frac{P_T}{\eta W} = \frac{B_T}{W}\left(\frac{S}{N}\right)_T$$

so that

$$z_{th} = 20\frac{B_T}{W}$$
$$= 40M(D) \approx 40(D + 1) \tag{6.43}$$

where the bandwidth relations of (6.21) and (6.22) have been inserted.

Figure 6.26 summarizes graphically the noise performance of FM relative to baseband, including threshold considerations but excluding deemphasis. The typical message value $\overline{x^2} = \frac{1}{2}$ is assumed. Threshold points calculated from (6.43) are shown as heavy dots, and the approximate performance below threshold is indicated by dashed lines.

The FM improvement above threshold is clearly impressive, and this is without deemphasis. But observe what happens if one attempts to make $(S/N)_D$ arbitrarily large by increasing only the deviation ratio, holding z fixed, say, at 22 db. With $D = 2$ ($B_T \approx 7W$) we are just above

---

[1] A similar phenomenon occurs with cochannel interference of two signals having nearly equal amplitudes at the receiver. Small variations of relative amplitude then cause the stronger of the two to suddenly dominate the situation, displacing the other completely at the output. This has been given the apt name *capture effect*.

threshold, and $(S/N)_D = 29$ db; but with $D = 5$ $(B_T \approx 14W)$ we are below threshold, and the output signal is useless because of mutilation. One therefore cannot achieve an unlimited exchange of bandwidth for signal-to-noise ratio, and system performance may actually deteriorate with increased deviation.

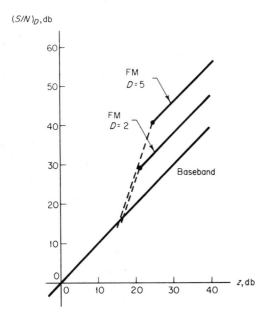

Fig. **6.26** *FM postdetection signal-to-noise ratio as a function of* $z = P_T/\eta W$, *de-emphasis not included.*

Swapping bandwidth in favor of reduced power is likewise restricted. Suppose, for example, a 35-db signal-to-noise ratio is desired with a minimum of transmitted power but the transmission bandwidth can be as large as $B_T = 14W$. Were it not for threshold, we could use FM with $D = 5$ and $z = 19$ db, a power saving compared to baseband of 16 db (a numeric factor of 37.5). But the threshold point for $D = 5$ is $z_{\text{th}} = 24.5$ db, for which $(S/N)_D = 40.5$ db. Thus, the design is dictated by the threshold point rather than the desired signal-to-noise ratio, and the potential power reduction cannot be fully realized.

Fortunately, special techniques have been developed to lower the threshold point for a given deviation ratio and thereby more closely achieve the full benefits of FM. One such *threshold-extension* method is the frequency-following or frequency-compressive feedback receiver originated by Chaffee (1939).

To summarize, the FM threshold effect can be a significant limitation and must be considered in the system design.

6.7 COMPARISON OF EXPONENTIAL AND LINEAR MODULATION SYSTEMS

We have shown that exponential modulation, particularly FM with deemphasis, can provide substantially increased postdetection signal-to-noise ratios compared to those of baseband or linear modulation with equivalent values of $z = P_T/\eta W$. Figure 6.27 illustrates this point in a

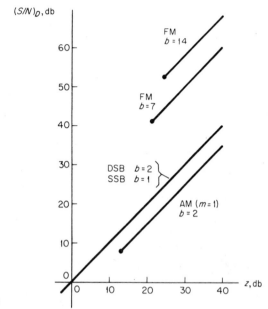

Fig. 6.27

form similar to Fig. 6.26. However, a representative 12-db deemphasis improvement has been added to the FM curves, while performance below threshold is omitted. All curves are labeled with the bandwidth ratio $b = B_T/W$, so the wideband noise reduction of FM is clearly evident.

The penalty for this improvement is of course excessive transmission bandwidth. Therefore, wideband exponential modulation is most appropriate when extremely clean output signals are desired and band-width conservation is a secondary factor. Typical applications include commercial FM and television audio, the latter being transmitted on a separate carrier with a deviation of 25 kHz. At microwave frequencies, both the noise-reduction and constant-amplitude properties are advan-tageous, so that most microwave systems have exponential carrier modulation.

As to power conservation, FM with moderate values of D does offer a saving over linear modulation in some instances, threshold limitations notwithstanding. In this capacity it is used for point-to-point radio, especially mobile systems, where synchronous detection would be a nuisance. Moreover, thanks to threshold-extension techniques, wideband FM is employed in certain satellite and space-probe communication systems, where transmitter power is at a premium.

Another application is in conjunction with signals having very slow time variations and hence substantial dc content. It was briefly noted in Sec. 6.5 that a balanced discriminator has excellent low-frequency response; hence the low-frequency performance of FM can equal that of DSB or VSB, and without troublesome synchronization. Consequently, small-deviation FM or PM subcarrier modulation is often used in telemetry systems for the less active input signals. For similar reasons, high-quality magnetic-tape recorders are equipped with an FM mode in which the input is recorded as a frequency-modulated wave.

But exponential modulation is not the universal solution to all communication problems, nor is any one modulation technique. Significant noise reduction entails very large transmission bandwidths and a received power above the threshold level. Aside from power and bandwidth considerations, other factors such as unstable transmission media and multiple-path propagation often preclude wideband modulation, ionospheric radio being a case in point. In short, the communication engineer must approach each new task with an open mind and a careful review of all available pertinent information.

PROBLEMS

6.1 Suppose that FM were defined as

$$x_c(t) = A_c \cos \omega_c[1 + mx(t)]t$$

in direct analogy with AM. Demonstrate the physical impossibility of this definition by considering the instantaneous frequency.

6.2 With the aid of (6.12), obtain the series expansions of (6.11). *Hint:* First find the exponential Fourier series for exp $(j\beta \sin \omega_m t)$, then equate real and imaginary parts.

6.3 Insert (6.11) into (6.10) and carry out the manipulations to get (6.13a).

6.4 Construct phasor diagrams for tone-modulated FM with $\beta = 1$ when $\omega_m t = 0$, $\pi/4$, and $\pi/2$. Include at least three sideband pairs, and verify that the resultant phase shift is correct and the amplitude is constant.

6.5 Sketch the positive-frequency line spectra for tone-modulated FM having $\beta = 5$ and $f_m = 1$ kHz when $f_c = 20$ and 4 kHz. Justify the dc component in the latter case by sketching the actual modulated waveform.

6.6 The carrier-frequency component of a tone-modulated FM wave is zero whenever $J_0(\beta) = 0$. Using this property, devise an experimental procedure for calibrating the deviation characteristics of an FM modulator.

6.7 Apply (6.15) to FM by a unit-amplitude rectangular pulse train. Sketch the spectrum for $f_d \gg 1/T_0$ and $\tau \ll T_0$. Compare with Figs. 6.9 and 6.10a.

6.8 Use the quasi-static approximation to obtain and sketch FM spectra when the modulating signal is (a) the random telegraph wave of Example 4.8 or (b) band-limited gaussian white noise. Give due attention to the approximation conditions.

6.9 A message has $W = 10$ kHz. Find the transmission bandwidth when $f_d = 0.1$, 1, 10, 100, and 1,000 kHz.

6.10 If $W = 10$ kHz, approximately what percentage of the transmission bandwidth is occupied when the modulating signal is a unit-amplitude tone at $f_m = 100$ Hz, 1 kHz, or 5 kHz and (a) the modulation is FM with $f_d = 30$ kHz, or (b) the modulation is PM with $\phi_d = 3$?

6.11 Frequency modulation is sometimes called *phase-velocity* modulation since $x(t)$ is proportional to $d\phi/dt$. Other types of exponential modulation are then: phase-integral modulation, $x(t) = K \int^t \phi(t') \, dt'$; phase-acceleration modulation, $x(t) = K \, d^2\phi/dt^2$; etc. Discuss the properties of phase-integral and phase-acceleration modulation, giving particular attention to their line spectra and bandwidths.

6.12 When the fractional variation of $f(t)$ is quite small, the response of a linear circuit can be estimated by assuming time-varying impedance of the form $Z(j\omega) = Z[j2\pi f(t)]$, etc. This is called the *quasi-steady-state analysis*. Use this approach to analyze the RC phase modulator of Fig. P 6.1. The modulating signal is introduced via the time-varying resistance $R(t) = R_0 - R_1 x(t)$, where $R_1 \ll R_0$ and the carrier frequency is such that $2\pi f_c R_0 C \gg 1$. Estimate the maximum phase deviation that can be achieved without significant distortion of $\phi(t)$.

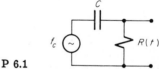

Fig. P 6.1

6.13 By calculating the exact relative phase of the output wave, show that PM generated by the system of Fig. 6.13 contains odd-harmonic distortion.

6.14 An integrator is placed at the input of Fig. 6.13 to make an NBFM modulator. Taking tone modulation, show that the instantaneous frequency contains third-harmonic distortion with relative amplitude of approximately $(f_d/2f_m)^2$. Note that the lowest modulating frequency suffers the most distortion. Why?

6.15 Suppose a multiplexed baseband signal has a spectrum covering 100 to 2,500 kHz and is to be transmitted with FM carrier modulation. Using the information in Prob. 6.14, design an indirect FM modulator with the following specifications: $f_c = 10$ GHz, $D = 2$, third-harmonic distortion less than 10 percent.

6.16 Verify that a nonlinear device having the transfer characteristic

$$y_c(t) = a_1 x_c(t) + a_2 x_c^2(t) + a_3 x_c^3(t) + \cdots$$

produces multiplication of instantaneous frequency. Use your results to explain why, in an indirect system for commercial FM, f_{c_1} should not be much less than 200 kHz.

6.17 Using the quasi-steady-state approach (see Prob. 6.12), obtain an expression for $y_c(t)$ in the slope detector of Fig. 6.17.

6.18 Obtain Eqs. (6.30a) and (6.30b) by suitable analytic approximations of

$$\psi(t) = \arctan \frac{A_c \sin \phi + r_n \sin \phi_n}{A_c \cos \phi + r_n \cos \phi_n}$$

6.19 Calculate and sketch the FM postdetection noise spectrum $G_{n_D}(f)$ if the IF is a simple pentode amplifier with high-Q parallel-tuned load whose 3-db bandwidth equals B_T (see Prob. 3.9).

6.20 Show that the deemphasis noise reduction in phase modulation is a factor of $f_3/W \arctan (W/f_3)$, or approximately $\dfrac{\pi}{2} \dfrac{f_3}{W}$ if $f_3 \ll W$.

6.21 Consider the following communication system. The transmitter has a total radiated power of P_0 watts. At the demodulator input the signal power is 100 db below P_0, and the noise density is $\eta = 10^{-14}$ watt/Hz. The message has $\overline{x^2} = 1$ and is bandlimited to 10 kHz. Calculate the value of P_0 required to give a postdetection signal-to-noise ratio of at least 40 db if the modulation is (*a*) SSB, (*b*) AM with $m = 1$ or $m = 0.1$, (*c*) PM with $\phi_d = \pi$, and (*d*) FM with $f_d = 10$, 50, or 100 kHz. Omit deemphasis in the FM case, but check for threshold.

6.22 Taking the specifications of Prob. 6.21, what value of the FM frequency deviation should be used to *minimize* P_0? Include a deemphasis filter having $f_3 = 1$ kHz.

6.23 Calculate and plot the postdetection signal-to-noise ratio (without deemphasis) as a function of $b = B_T/W$ for FM operated just above threshold, that is, $z \approx z_{\text{th}}$.

6.24 Discuss the advantages and disadvantages of transmitting television video using frequency modulation. The video spectrum extends from 60 Hz to approximately 4.5 MHz.

6.25 A frequency-division multiplexing system uses SSB subcarrier modulation and FM carrier modulation. There are $k \gg 1$ input signals, each bandlimited to W, and negligible guard bands. (*a*) Find the final transmission bandwidth in terms of W and the carrier deviation f_d. (*b*) Recalling that the FM postdetection noise spectrum is parabolic, write an expression for the demultiplexed per-channel signal-to-noise ratio in terms of P_k, the baseband power of the kth channel. How should the P_k's be proportioned to equalize the signal-to-noise ratios?

6.26 Figure P 6.2 shows a portion of an FM receiver with frequency-compressive feedback, the heart of Chaffee's threshold-extension technique. The discriminator output voltage is $y_D(t) = K_1[f_1(t) - f_{\text{IF}}]$, and the instantaneous

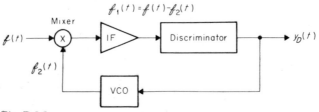

Fig. P 6.2

frequency at the VCO output is $f_2(t) = f_c - f_{\text{IF}} + K_2 y_D(t)$. In practice, $\mu = K_1 K_2 \gg 1$. The input frequency is, of course, $f(t) = f_c + f_d x(t)$. By solving the loop equation for $f_1(t)$, show that $y_D(t)$ is proportional to $f_d x(t)$ and the IF bandwidth need be no greater than $2W$.

SELECTED SUPPLEMENTARY READING

Most of the texts cited for Chap. 5 have material on exponential modulation as well. Deserving special mention here is Panter (1965, chaps. 7 to 16), whose thorough coverage starts with basic principles and goes on to such matters as detailed interference and noise analysis, FM distortion problems, transient response to FM signals, and threshold extension techniques—all supported by abundant literature citations. This is probably the most complete treatment of FM and belongs on the reference shelf of every communication system engineer.

Other useful works are Rowe (1965, chap. 4), for advanced spectral analysis emphasizing random modulating signals, and Downing (1964, chap. 6), for a concise discussion of noise in multiplex systems having FM carrier modulation.

The two classic papers on FM, namely, Carson (1922) and Armstrong (1936), have withstood the test of time and remain lively and informative reading today. More recently, Frutiger (1966) has employed phasor analysis to give a simple treatment of FM threshold effects and frequency-feedback extension.

7
Sampling and Pulse Modulation

Experimental data and mathematical functions are frequently displayed as *continuous* curves even though a finite number of *discrete points* may have been used to construct the graph. If these discrete points, or *samples*, have sufficiently close spacing, a smooth curve is drawn through them, and intermediate values can be interpolated to any reasonable degree of accuracy. It can therefore be said that the continuous display is adequately described by the sample points alone.

In similar fashion, an electric signal satisfying certain requirements can be reproduced entirely from an appropriate set of instantaneous samples. If this is so, and sampling theory will tell us the necessary conditions, we need transmit only the sample values as they occur instead of sending the signal continuously. This is *pulse modulation*.

The key distinction between pulse modulation and c-w modulation is as follows: in c-w modulation, some parameter of the modulated wave varies continuously with the message; in pulse modulation, some parameter of each pulse is modulated by a particular sample value of the message.   Usually the pulses are quite short compared to the time between them, so a pulse-modulated wave is "off" most of the time.

Because of this property, pulse modulation offers two potential advantages over c-w.   First, the transmitted power can be concentrated into short bursts rather than being delivered continuously.   This gives the system engineer added latitude in equipment selection, since certain devices, such as high-power microwave tubes and lasers, are operable only on a pulsed basis.   Second, the time intervals between pulses can be filled with sample values from other messages, thereby permitting the transmission of many messages on one communication system.   Such multiplexing in the time domain is known as *time-division multiplexing* (TDM).

Another distinction between pulse and c-w modulation is that the pulsed wave may contain appreciable dc and low-frequency content. Efficient transmission therefore entails a second operation, namely, c-w modulation, to provide complete frequency translation.   In this light, pulse modulation is a *message-processing* technique rather than modulation in the usual sense.   As a matter of fact, the most common use of pulse modulation is message processing for TDM.

There are two basic types of pulse modulation: *analog*, such as pulse-amplitude or pulse-position modulation, which is in many ways similar to linear or exponential modulation; and *digital* or *coded* pulse modulation, which has no c-w equivalent.   Both types are examined in this chapter, the latter being emphasized because of its unique and desirable properties.

Regardless of type, the key operation for pulsed communication is extracting sample values from the message waveform.   Moreover, the sampling concept plays an equally important role in *information theory*, the subject of Chap. 8, and in the theory of sampled-data control systems. We therefore begin this chapter with a study of sampling, both theoretical and practical.

7.1 SAMPLING THEORY

A simple but highly informative approach to sampling theory is via the switching operation of Fig. 7.1a.   The switch periodically shifts between

(a)   (b)

Fig. 7.1 *A switching sampler.   (a) Circuit; (b) waveforms.*

two contacts at a rate of $f_s = 1/T_s$ Hz, dwelling on the input-signal contact for τ sec and on the grounded contact for the remainder of each period. The output $x_s(t)$ then consists of short segments of the input $x(t)$, as shown in Fig. 7.1b, and can be written as

$$x_s(t) = \begin{cases} x(t) & mT_s - \dfrac{\tau}{2} < t < mT_s + \dfrac{\tau}{2} \quad m = 0, \pm 1, \pm 2, \cdots \\ 0 & \text{otherwise} \end{cases}$$

$$= x(t) \sum_{m=-\infty}^{\infty} \Pi\left(\frac{t - mT_s}{\tau}\right) \tag{7.1}$$

This operation, variously called *single-ended* or *unipolar chopping*, is not instantaneous sampling in the strict sense. Nonetheless, $x_s(t)$ will be designated the sampled wave and f_s the sampling frequency.

We now ask: Are the sampled segments sufficient to describe the original input signal, and, if so, how can $x(t)$ be retrieved from $x_s(t)$? The answer to this question lies in the frequency domain, namely, in the spectrum of the sampled wave.

As a first step toward finding the spectrum, we introduce a *switching function* $S(t)$ such that

$$x_s(t) = x(t)S(t) \tag{7.2}$$

and the sampling operation becomes multiplication by $S(t)$, as indicated schematically in Fig. 7.2a. Clearly, comparing (7.1) and (7.2),

$$S(t) = \sum_{m=-\infty}^{\infty} \Pi\left(\frac{t - mT_s}{\tau}\right)$$

which is a summation of rectangular pulses, having unit amplitude and duration τ, uniformly spaced by T_s. Alternately, $S(t)$ is nothing more than the periodic pulse train of Fig. 7.2b.

Since $S(t)$ is periodic, it can be written as a Fourier series. Using the results of Example 2.1 and inserting the duty cycle $d = \tau/T_s = \tau f_s$,

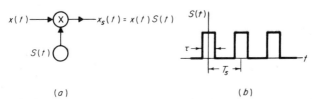

(a)   (b)

Fig. 7.2 *Sampling interpreted as multiplication in the time domain. (a) Operational diagram; (b) switching function $S(t)$.*

we have

$$S(t) = d \sum_{n=-\infty}^{\infty} \text{sinc } nd \; e^{jn\omega_s t} \qquad \omega_s = \frac{2\pi}{T_s}$$

$$= d \left(1 + \sum_{n=1}^{\infty} 2 \text{ sinc } nd \cos n\omega_s t \right) \tag{7.3}$$

Combining (7.3) with (7.2) yields the term-by-term expansion

$$
\begin{aligned}
x_s(t) = d[x(t) &+ (2 \text{ sinc } d)x(t) \cos \omega_s t \\
&+ (2 \text{ sinc } 2d)x(t) \cos 2\omega_s t \\
&+ \cdots]
\end{aligned}
\tag{7.4}
$$

Thus, if the input spectrum is $X(f) = \mathfrak{F}[x(t)]$, the output spectrum is

$$
\begin{aligned}
X_s(f) = d\{X(f) &+ \text{sinc } d \left[X(f - f_s) + X(f + f_s) \right] \\
&+ \text{sinc } 2d \left[X(f - 2f_s) + X(f + 2f_s) \right] \\
&+ \cdots \}
\end{aligned}
\tag{7.5}
$$

which follows directly from the modulation theorem.

While Eq. (7.5) appears rather messy, the spectrum of the sampled wave is readily sketched if the input signal is assumed to be bandlimited. Figure 7.3 shows a convenient $X(f)$ and the corresponding $X_s(f)$. Exam-

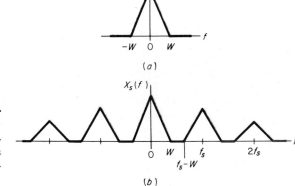

Fig. 7.3 *Spectra for switching sampling. (a) Message spectrum; (b) spectrum of the sampled message.*

ining this figure reveals something quite surprising: the sampling operation has left the message spectrum *intact*, merely repeating it periodically in the frequency domain with a spacing of f_s. We also note that the first term of (7.5) is precisely the message spectrum, attenuated by the duty cycle d.

If sampling preserves the message spectrum, it should be possible to recover or *reconstruct* $x(t)$ from the sampled wave $x_s(t)$. The reconstruction technique is not at all obvious from the time-domain relations, Eqs. (7.2) and (7.4), but referring again to Fig. 7.3, we see that $X(f)$ can be separated from $X_s(f)$ by *lowpass filtering*, providing the components, or sidebands, do not overlap. And if $X(f)$ alone is filtered from $X_s(f)$, we have recovered $x(t)$.

Two conditions obviously are necessary to prevent overlapping sidebands: (1) the message must be *bandlimited*, say in W, and (2) the sampling frequency must be sufficiently great that $f_s - W \geq W$, that is,

$$f_s \geq 2W \qquad \text{or} \qquad T_s \leq \frac{1}{2W} \tag{7.6}$$

The minimum sampling frequency $f_s = 2W$ is called the *Nyquist rate*. When (7.6) is satisfied and $x_s(t)$ is filtered by an ideal LPF, the output signal will be proportional to $x(t)$; message reconstruction from the sampled signal therefore has been achieved. The exact value of the filter bandwidth is unimportant as long as the filter passes $X(f)$ and rejects all other components, i.e., from Fig. 7.3b, the filter bandwidth B must be in the range $W \leq B \leq f_s - W$.

This analysis has shown that if a bandlimited signal is sampled at a frequency greater than the Nyquist rate, it can be *completely reconstructed* from the sampled wave. Reconstruction is accomplished by lowpass filtering. These conclusions may be difficult for the reader to believe at first exposure; they certainly test our faith in spectral analysis. Nonetheless, they are quite correct.

Finally, it should be pointed out that our results are independent of the sample-pulse duration, save as it appears in the duty cycle. If τ is made very small, $x_s(t)$ approaches a string of *instantaneous sample points*, which corresponds to *ideal sampling*. And while the above analysis is representative of practical sampling and is sufficient for many purposes, we shall now devote our attention to ideal sampling, the study of which provides further insight to the how and why of sampling and reconstruction, along with proof of the uniform-sampling theorem.

Ideal sampling and reconstruction

By definition, ideal sampling is *instantaneous* sampling. The switching device of Fig. 7.1a yields instantaneous values only if $\tau \to 0$; but then $d = \tau f_s \to 0$, and so does $x_s(t)$ [see Eq. (7.4)]. Conceptually, this difficulty is overcome by multiplying $x_s(t)$ by $1/\tau$ so that, as $\tau \to 0$ and $1/\tau \to \infty$, the sampled wave becomes a train of *impulses* whose *areas* equal the instantaneous sample values of the input signal. Formally we write

$$x_\delta(t) = \lim_{\tau \to 0} \frac{1}{\tau} x(t) S(t)$$

$$= \lim_{\tau \to 0} x(t) \sum_{m=-\infty}^{\infty} \frac{1}{\tau} \Pi\left(\frac{t - mT_s}{\tau}\right)$$

$$= x(t) \sum_{m=-\infty}^{\infty} \delta(t - mT_s) \qquad (7.7a)$$

having invoked $\delta(u) = \lim_{a \to \infty} a\Pi(au)$. Thus, bringing $x(t)$ inside the summation,

$$x_\delta(t) = \sum_{m=-\infty}^{\infty} x(mT_s)\delta(t - mT_s) \qquad (7.7b)$$

Comparing (7.7) with (7.2) suggests that ideal sampling may be viewed in terms of a switching function that is a train of impulses rather than rectangular pulses. This is in fact the *ideal sampling wave* discussed in Sec. 2.5 and sketched in Fig. 2.22a.

It was also demonstrated in Sec. 2.5 that the transform of $\sum_m \delta(t - mT_s)$ is $f_s \sum_n \delta(f - nf_s)$, a train of impulses in the frequency domain. Hence, applying the convolution theorem to (7.7a),

$$X_\delta(f) = X(f) * [f_s \sum_n \delta(f - nf_s)]$$

But convolving with impulses is a *replication* process such that $g(u) * \delta(u - u_0) = g(u - u_0)$. Therefore

$$X_\delta(f) = f_s \sum_{n=-\infty}^{\infty} X(f - nf_s) \qquad (7.8)$$

or

$$X_\delta(f) = f_sX(f) + f_s[X(f - f_s) + X(f + f_s)]$$
$$+ f_s[X(f - 2f_s) + X(f + 2f_s)]$$
$$+ \cdots$$

which is illustrated in Fig. 7.4 for the message spectrum of Fig. 7.3a.

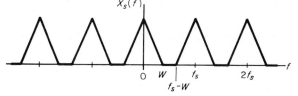

Fig. 7.4 *Spectrum of an ideally sampled message.*

It is immediately apparent that if we invoke the same conditions as before, that is, $x(t)$ bandlimited in W and $f_s \geq 2W$, then a filter of suitable bandwidth will reconstruct $x(t)$ from the ideal sampled wave. Specifically, for an ideal LPF of gain K, time delay t_0, and bandwidth B, the transfer function is

$$H(f) = K\Pi\left(\frac{f}{2B}\right) e^{-j\omega t_0}$$

so filtering $x_s(t)$ produces the output spectrum

$$Y(f) = X_\delta(f)H(f) = Kf_s X(f)e^{-j\omega t_0}$$

assuming B satisfies $W \leq B \leq f_s - W$. Therefore, the output time function is

$$y(t) = \mathcal{F}^{-1}[Y(f)] = Kf_s x(t - t_0) \tag{7.9}$$

which is the original signal amplified by Kf_s and delayed by t_0.

Further confidence in the sampling process can be gained by examining reconstruction in the time domain. The impulse response of the above filter is

$$h(t) = 2BK \text{ sinc } 2B(t - t_0)$$

And, since the input $x_\delta(t)$ is a train of weighted *impulses*, the output is a train of weighted *impulse responses*, namely,

$$y(t) = h(t) * x_\delta(t) = \sum_m x(mT_s)h(t - mT_s)$$

$$= 2BK \sum_{m=-\infty}^{\infty} x(mT_s) \text{ sinc } 2B(t - t_0 - mT_s) \tag{7.10}$$

Equating (7.10) and (7.9) yields

$$Kf_s x(t - t_0) = 2BK \sum_m x(mT_s) \text{ sinc } 2B(t - t_0 - mT_s)$$

or, setting $t_0 = 0$,

$$x(t) = \frac{2B}{f_s} \sum_{m=-\infty}^{\infty} x(mT_s) \text{ sinc } 2B(t - mT_s) \tag{7.11}$$

Equation (7.11) is a series representation of $x(t)$ in terms of $x(mT_s)$, the series being constructed with *sinc* functions.

Now recall that a periodic signal can be expressed as a Fourier series constructed with exponential or trigonometric functions. And just as a *periodic* signal is completely described by its Fourier series coefficients, a *bandlimited* signal is completely described by its instantaneous sample values, if their spacing is $T_s \leq 1/2W$.

Returning to Eq. (7.10), suppose for simplicity that $B = f_s/2$, $K = 1/f_s$, and $t_0 = 0$, so

$$y(t) = \sum_m x(mT_s) \operatorname{sinc} (f_s t - m)$$

We can then carry out the reconstruction process graphically, as shown in Fig. 7.5. Clearly the correct values are reconstructed at the sampling

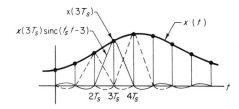

Fig. 7.5 *Ideal reconstruction with sinc functions.*

instants $t = mT_s$, for all sinc functions are zero at these times save one, and that one yields $x(mT_s)$. Between sampling instants $x(t)$ is *interpolated* by summing the precursors and postcursors from *all* the sinc functions. For this reason the LPF is often called an *interpolation filter*, and its impulse response is called the *interpolation function*.

The uniform sampling theorem

The above results are well summarized by stating the important theorem of uniform (periodic) sampling. While there are many variations of this theorem,[1] the following form is best suited to our purposes.

> **Theorem :** If a signal contains no frequency components for $|f| \geq W$, it is completely described by instantaneous sample values uniformly spaced in time with period $T_s \leq 1/2W$.
>
> If a signal has been sampled at the Nyquist rate or greater ($f_s \geq 2W$) and the sample values are represented as weighted impulses, the signal can be exactly reconstructed from its samples by an ideal LPF of bandwidth B, where $W \leq B \leq f_s - W$.

Alternate proof of the sampling theorem ★

As we have stated it, the sampling theorem is couched in the language of electric signals. Moreover, its derivation was based on spectral analysis. But the principle of sampling has much broader applications, and the theorem can be proved on a purely mathematical basis without reference

[1] Black (1953, chap. 4) lists several theorems pertaining to periodic sampling.

to impulses, spectra, filters, and the like. This alternate proof is given below.

To emphasize that sampling is not contingent upon time-domain or frequency-domain interpretations, we introduce two variables u and v such that if $f(u)$ is any real function of u and $F(v)$ is its Fourier transform, then $f(u)$ and $F(v)$ are related by

$$F(v) = \int_{-\infty}^{\infty} f(u)e^{-j2\pi uv}\, du \qquad f(u) = \int_{-\infty}^{\infty} F(v)e^{j2\pi uv}\, dv$$

providing $F(v)$ exists. We shall then show that if

$$F(v) = 0 \qquad |v| \geq V$$

and if U_s, v_s, and Ω are positive constants satisfying

$$U_s = \frac{1}{v_s} \leq \frac{1}{2V} \qquad V \leq \Omega \leq v_s - V$$

then

$$f(u) = \frac{2\Omega}{v_s} \sum_{m=-\infty}^{\infty} f(mU_s)\ \text{sinc}\ 2\Omega(u - mU_s) \tag{7.12}$$

and hence $f(u)$ is defined by its sample values at $u = mU_s$. Equation (7.12) is (7.11) with $x(t) = f(u)$, $T_s = U_s$, etc., and therefore proves the first part of sampling theorem.

As to the proof of (7.12), we begin by defining a new function

$$Z(v) = \sum_{n=-\infty}^{\infty} F(v - nv_s)$$

which is periodic in v with *period* v_s. From the properties of $F(v)$ and the conditions on Ω it follows that $Z(v) = F(v)$ for $|v| \leq \Omega$ and therefore

$$f(u) = \int_{-\Omega}^{\Omega} Z(v)e^{j2\pi uv}\, dv \tag{7.13}$$

But $Z(v)$, being periodic, can also be written as a Fourier series in the v *domain*, namely,

$$Z(v) = \sum_{k=-\infty}^{\infty} c_k e^{j2\pi kU_s v} \qquad U_s = \frac{1}{v_s} \tag{7.14}$$

where

$$c_k = \frac{1}{v_s} \int_{-v_s/2}^{v_s/2} Z(v)e^{-j2\pi kU_s v}\, dv = \frac{1}{v_s} \int_{-\Omega}^{\Omega} Z(v)e^{-j2\pi kU_s v}\, dv \tag{7.15}$$

The change of integration limits is permissible since $Z(v) = 0$ for $V \leq |v| \leq v_s - V$ and both $v_s/2$ and Ω fall in this range.

Next we relate the coefficients c_k to $f(u)$ with the aid of Eq. (7.13). Setting $u = -kU_s$ yields

$$f(-kU_s) = \int_{-\Omega}^{\Omega} Z(v)e^{j2\pi(-kU_s)v}\, dv$$

and comparison with (7.15) shows that

$$c_k = \frac{1}{v_s}f(-kU_s)$$

Substituting this value of c_k in (7.14) and letting $m = -k$ gives

$$Z(v) = \sum_{m=-\infty}^{\infty} \frac{1}{v_s}f(mU_s)e^{-j2\pi mU_s v}$$

Finally $Z(v)$ in series form is inserted into (7.13), the order of summation and integration are interchanged, and the integral is evaluated as

$$f(u) = \int_{-\Omega}^{\Omega} \left[\sum_{m} \frac{1}{v_s}f(mU_s)e^{-j2\pi mU_s v} \right] e^{j2\pi uv}\, dv$$

$$= \frac{1}{v_s} \sum_{m} \left[f(mU_s) \int_{-\Omega}^{\Omega} e^{j2\pi v(u-mU_s)}\, dv \right]$$

$$= \frac{2\Omega}{v_s} \sum_{m} f(mU_s) \operatorname{sinc} 2\Omega(u - mU_s)$$

which completes the proof.

One final comment is appropriate here. If we take $\Omega = v_s/2 = 1/2U_s$, Eq. (7.12) reduces to

$$f(u) = \sum_{m} f(mU_s) \operatorname{sinc}(v_s u - m)$$

Series representations of this form are particularly useful in advanced signal theory by virtue of the fact that the individual terms are *orthogonal*, that is,

$$\int_{-\infty}^{\infty} \operatorname{sinc}(v_s u - m) \operatorname{sinc}(v_s u - n)\, du = \begin{cases} 0 & m \neq n \\ \dfrac{1}{v_s} & m = n \end{cases} \qquad (7.16)$$

paralleling the orthogonality of the trigonometric or exponential functions in ordinary Fourier series. The advantage of sinc functions is the band-limited property of their transforms.

We shall use the above orthogonality relation to analyze recon-struction from sample values which are contaminated by noise.

Reconstruction from "noisy" samples ★

In the process of transmission, sample values may be contaminated by interference, noise, or some other mechanism so that reconstruction must be based on erroneous or noisy samples. Analytically, we can represent the contaminated samples as $x(mT_s) + \epsilon_m$, ϵ_m being the error added to the mth sample. Thus, the input to the reconstruction filter is

$$y_s(t) = \sum_m [x(mT_s) + \epsilon_m]\delta(t - mT_s)$$

Taking the filter to have bandwidth $B = f_s/2$, a reasonable assumption since practical systems use sampling frequencies not too much greater than the Nyquist rate, and assuming zero time delay for simplicity, the filter output will be

$$y(t) = Kf_s \sum_m [x(mT_s) \text{ sinc } (f_s t - m) + \epsilon_m \text{ sinc } (f_s t - m)]$$

$$= Kf_s[x(t) + y_\epsilon(t)]$$

where

$$y_\epsilon(t) = \sum_m \epsilon_m \text{ sinc } (f_s t - m) \tag{7.17}$$

as follows from (7.10) and (7.11). Note that $y(t)$ has the form of the desired signal with *additive* noise or error waveform $y_\epsilon(t)$.

If the contamination is itself a time function such that $\epsilon_m = \epsilon(mT_s)$, and if $\epsilon(t)$ is bandlimited in $W_\epsilon \leq f_s/2$, then $y_\epsilon(t) = \epsilon(t)$. More commonly, we can go no further in the time domain than Eq. (7.17). However, if the statistics of ϵ_m are known, we can find the mean-square value of $y_\epsilon(t)$ as follows.

By definition, the mean square of the error waveform (averaged over time) is $\langle y_\epsilon^2(t) \rangle = \lim_{T \to \infty} \frac{1}{T} \int_{-T/2}^{T/2} y_\epsilon^2(t) \, dt$, which can also be written as

$$\langle y_\epsilon^2(t) \rangle = \lim_{M \to \infty} \frac{1}{2MT_s} \int_{-MT_s}^{MT_s} y_\epsilon^2(t) \, dt$$

Inserting (7.17), interchanging summation and integration, and using the orthogonality relation of (7.16) yields

$$\langle y_\epsilon^2(t) \rangle = \lim_{M \to \infty} \frac{1}{2MT_s} \int_{-MT_s}^{MT_s} \left[\sum_m \epsilon_m \text{ sinc } (f_s t - m) \right]$$

$$\left[\sum_n \epsilon_n \text{ sinc } (f_s t - n) \right] dt$$

$$= \lim_{M \to \infty} \frac{1}{2MT_s} \sum_{m=-M}^{M} \sum_{n=-M}^{M}$$

$$\left[\epsilon_m \epsilon_n \int_{-\infty}^{\infty} \text{ sinc } (f_s t - m) \text{ sinc } (f_s t - n) \, dt \right]$$

$$= \lim_{M \to \infty} \frac{1}{2M} \sum_{m=-M}^{M} \epsilon_m^2$$

which, upon closer inspection, is seen to be just the numerical average of the ϵ_m's.

Now suppose the contamination is due to random noise; then the ϵ_m's are sample values taken from one *sample function* of the ensemble characterizing the noise process. Hence, the value of $\langle y_\epsilon^2(t) \rangle$ may be contingent upon which member of the ensemble is assumed for the calculation. [In essence, $y_\epsilon(t)$ may be *nonstationary* even though $\epsilon(t)$ is stationary.] Under these conditions the best measure of the mean-square error is the ensemble (statistical) average of the time average, that is,

$$\overline{\langle y_\epsilon^2(t) \rangle} = \lim_{M \to \infty} \overline{\left(\frac{1}{2M} \sum_m \epsilon_m^2 \right)} = \lim_{M \to \infty} \frac{1}{2M} \sum_{m=-M}^{M} \overline{\epsilon_m^2} \qquad (7.18)$$

or just the mean-square value of $\epsilon(t)$ in the sense of ensemble averaging, assuming of course the noise process is ergodic.

Correspondingly, we define the signal-to-noise ratio of the reconstructed wave as

$$\frac{S}{N} = \frac{\overline{x^2}}{\overline{\epsilon^2}} \qquad (7.19)$$

That this is a reasonable and meaningful definition is enhanced by noting that sampling and reconstruction are linear operations on both the signal and the contamination.

7.2 PRACTICAL SAMPLING

Having stated and proved the theorem for ideal sampling, we must now examine electric-waveform sampling as it occurs in practice. Practical sampling differs from ideal sampling in three obvious respects:

1. The sampled wave consists of pulses having finite amplitude and duration, rather than impulses.
2. Practical reconstruction filters are not ideal filters.
3. The messages to be sampled are *timelimited* signals and therefore cannot be bandlimited.

We shall treat these differences one at a time and show that only the last one is troublesome.

Pulse-shape effects

While there are several techniques for practical sampling, common to them all is a periodic train of sampling pulses whose amplitudes are modulated by the message. Thus, practical sampling is *pulse-amplitude modulation*. The actual sampling may take place in one of two ways: by *natural* sampling (*exact* scanning), in which the sampling pulse follows the message over the pulse duration, or by *instantaneous*[1] sampling (*square-topped* scanning), in which the amplitude of the sampling pulse equals the instantaneous message sample value. Typical waveforms are shown in Fig. 7.6 assuming rectangular pulses. Note that natural sampling is the same as unipolar chopping.

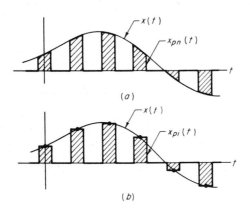

Fig. 7.6 *Practical sampling waveforms. (a) Natural sampling (exact scanning); (b) instantaneous sampling (square-topped scanning).*

Despite Fig. 7.6, perfectly rectangular pulses cannot exist in a finite-bandwidth system, a real system. Moreover, choppers or sampling switches, whether mechanical or electronic, never make an instantaneous transition from off to on, so the rectangular switching function used previously is only an approximation. The distinction between the two types of sampling is therefore not the pulse shape per se but rather the *duration* of the extracted sample.

We can account for nonrectangular sampling pulses by introducing $p(t)$ as the basic *pulse shape*, centered at $t = 0$ and normalized to a peak value of unity. So that adjacent pulses do not overlap, we impose the condition $p(t) = 0$ for $|t| \geq T_s/2$. In terms of $p(t)$, natural sampling may be described by a modified switching function

$$S_p(t) = \sum_m p(t - mT_s)$$

and hence

$$x_{pn}(t) = x(t)S_p(t) = x(t) \sum_m p(t - mT_s) \tag{7.20}$$

---

[1] This is *practical* instantaneous sampling (instantaneous sample values but pulses of finite amplitude and duration) as distinguished from *ideal* sampling.

which of course becomes the same as (7.1) if $p(t) = \Pi(t/\tau)$. With instantaneous sampling the sampled wave is by definition

$$x_{pi}(t) = \sum_m x(mT_s)p(t - mT_s) \tag{7.21}$$

Both types reduce to the ideal sampling if $p(t) = \delta(t)$.

The pulse-shape effect with natural sampling can be quickly disposed of, for, in truth, there is no effect insofar as message recovery is concerned. This conclusion follows from (7.20) and the fact that $S_p(t)$ is periodic in T_s; hence

$$x_{pn}(t) = a_0 x(t) + a_1 x(t) \cos(\omega_s t + \phi_1) + \cdots$$

where a_0 is the dc component of $S_p(t)$, etc. The details are left to the reader.

For the case of practical sampling with instantaneous sample values we must turn to the frequency domain and the spectrum of $x_{pi}(t)$. Again recalling the replication property of convolution with impulses, Eq. (7.21) can be rewritten in the form

$$x_{pi}(t) = p(t) * \left[\sum_m x(mT_s)\delta(t - mT_s) \right]$$

so that

$$X_{pi}(f) = P(f) \left[f_s \sum_n X(f - nf_s) \right]$$
$$= P(f)X_\delta(f) \tag{7.22}$$

where $P(f) = \mathfrak{F}[p(t)]$ is the Fourier transform of the pulse shape and $X_\delta(f)$ is the spectrum of the ideal sampled wave.

Interpreting (7.22), one can think of $P(f)$ as a filter operating on $X_\delta(f)$ and attenuating all frequency components above about $1/\tau$, τ being the nominal duration of $p(t)$. The effect is best illustrated by taking $X(f)$ to be constant over $|f| \leq W$, per Fig. 7.7a, giving $X_{pi}(f)$ as shown in Fig. 7.7b. Generally, the reconstructed signal will be distorted because $P(f)$ attenuates the upper portion of the message spectrum. This loss of high-frequency message components is sometimes called the *aperture effect*—the sampling aperture (pulse duration) is too large—and can be corrected by an *equalizing filter* having $H_{eq}(f) = 1/P(f)$. (Note the similarity to preemphasis-deemphasis filtering.) However, usually $1/\tau \gg W$, so $P(f)$ is essentially constant over the message band, and equalization is not necessary.

We can thus say that pulse-shape effects are relatively inconsequential, and the sampling theorem is valid for nonimpulsive sampled waves.

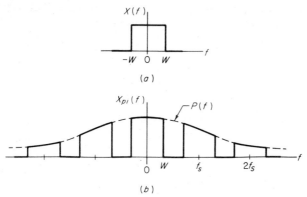

Fig. 7.7 *Aperature effect in practical instantaneous sampling. (a) Message spectrum; (b) spectrum of the sampled message.*

Practical reconstruction filters

The effect of nonideal reconstruction filters is also readily treated in the frequency domain. Consider, for example, a typical filter response superimposed on a sampled-wave spectrum, as in Fig. 7.8. If the filter is reasonably flat over the message band, its output will consist of $x(t)$ plus spurious frequency components at $|f| > f_s - W$, which is outside the message band. Note that these components are considerably attenuated compared to $x(t)$.

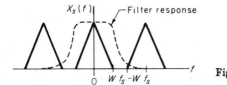

Fig. 7.8

In audio systems such components would sound like high-frequency garble or hissing. However, their strength is proportional to the message, disappearing when $x(t) = 0$, and the message tends to mask their presence, making them easier to tolerate.

Good filter design is obviously the best way to minimize the spurious frequencies. Alternately, for a given filter response, they can be further suppressed by increasing the sampling frequency, which increases $f_s - W$ and produces *guard bands* in the spectrum of the sampled wave. There is an interesting parallel here with FDM: in FDM, guard bands are used to allow message separation by practical bandpass filters with a minimum of *cross talk;* in sampling systems, guard bands are used to allow message reconstruction by practical lowpass filters with a minimum of high-frequency garble.

Nonbandlimited signals—aliasing

Thus far we have ignored the fact that real signals are not strictly band-limited. But a message spectrum, such as Fig. 7.9, is considered to be virtually bandlimited if the frequency content above W is small and presumably unimportant for conveying the information. When such a

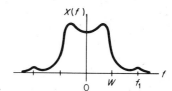

Fig. 7.9 *Nonbandlimited message spectrum.*

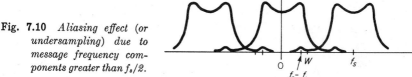

Fig. 7.10 *Aliasing effect (or undersampling) due to message frequency components greater than $f_s/2$.*

message is sampled, there will be unavoidable overlapping of spectral components (Fig. 7.10). In reconstruction, frequencies originally outside the nominal message band will appear at the filter output in the form of much *lower* frequencies. Thus, for example, $f_1 > W$ becomes $f_s - f_1 < W$, as indicated in the figure.

This phenomenon of downward frequency translation occurs whenever a frequency component is *undersampled,* that is, $f_s < 2f_1$, and is given the descriptive name of *aliasing.* The aliasing effect is far more serious than spurious frequencies passed by nonideal reconstruction filters, for the latter fall *outside* the message band, whereas aliased components can fall *within* the message band. Aliasing is combated by filtering the message as much as possible *before* sampling and sampling at much greater than the nominal Nyquist rate if necessary.

As illustration, the average voice spectrum extends well beyond 10 kHz, though most of the energy is concentrated in the range 100 to 600 Hz and a bandwidth of 3 kHz is sufficient for intelligibility. When a voice wave is sampled at $f_s = 8$ kHz, the standard for telephone systems, aliased components are typically 30 db below the desired signal and are virtually unnoticed.

The sampling theorem restated

To summarize our investigation of *practical* sampling as distinguished from *ideal* sampling we can restate the sampling theorem in the following manner:

> **Theorem:** If a signal has negligible spectral content for $|f| > W$, it is adequately described for most purposes by sample values, either instantaneous or of finite duration, uniformly spaced in time with period $T_s \leq 1/2W$.
>
> If a signal has been sampled at the nominal Nyquist rate or greater ($f_s \geq 2W$) and the samples are represented as amplitude-modulated pulses, the signal can be approximately reconstructed from its samples by lowpass filtering.

Admittedly, *perfect* reconstruction, like distortionless transmission, is impossible; but with careful planning, good filter design, and suitable sampling frequency, the recovered wave can be made as close as desired to the original signal for all but the most extreme cases.

Chopper modulation

Before applying sampling theory to pulse-modulation systems, it should be pointed out that practical samplers can be used as *balanced modulators* for linear modulation systems. In particular, returning to the output spectrum of the unipolar chopper (Fig. 7.3*b*), we see that each harmonic of the sampling frequency is *double-sideband-modulated* by the input signal $x(t)$. Bandpass filtering the sampled wave will thus produce DSB, a technique known as *chopper modulation*. The modulation is relatively unaffected by imperfect switching, as shown by our study of pulse-shape effects.

For purposes of suppressed-carrier modulation, switching or chopping methods are often superior to the conventional balanced modulator, which requires more components and carefully matched nonlinear elements. However, the unipolar chopper is not so efficient a modulator as the *bipolar* chopper of Fig. 7.11. This device produces DSB at only the odd harmonics of f_s, and its output contains no dc component. The analysis follows that of unipolar chopping but with a bipolar square-wave switching function.

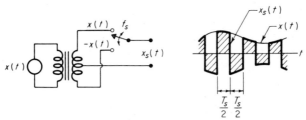

Fig. 7.11   *A bipolar chopper and typical waveforms.*

Aside from modulation for signal transmission, bipolar choppers find application in control systems, where a suppressed-carrier wave is used to drive two-phase servomotors, and for the amplification of slowly varying signals (dc amplifiers).   Figure 7.12 shows a dc amplifier constructed from a pair of choppers and an ac (bandpass) amplifier.   The

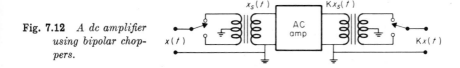

Fig. 7.12   *A dc amplifier using bipolar choppers.*

input chopper translates the signal to the passband of the amplifier, while the output chopper reverses the process.   Note that the output chopper is essentially a *synchronous demodulator* and must be appropriately synchronized.   Problem 7.12 indicates how a bipolar chopper can be modified to achieve the baseband multiplexing for FM stereophonic transmitters.

7.3 ANALOG PULSE MODULATION

If a message is adequately described by its sample values, it can be transmitted via analog pulse modulation, wherein the sample values directly modulate a periodic pulse train with one pulse for each sample. There are numerous varieties of analog pulse modulation, and the terminology has not been standardized.   However, the three types we shall be concerned with are usually designated as *pulse-amplitude modulation* (PAM), *pulse-duration modulation* (PDM), and *pulse-position modulation* (PPM).   PDM and PPM are also lumped together under the general heading of *pulse-time modulation*.

Figure 7.13 shows a typical message and corresponding pulse-modulated waveforms.   For clarity, the pulses are shown as rectangular, and the pulse duration has been grossly exaggerated. Moreover, actual modulated waves are slightly delayed in time compared to the message, since the pulses cannot be generated before the sampling instants.

As shown in the figure, the modulated pulse parameter—amplitude, duration, or relative position—is varied in direct proportion to the sample values. However, in PAM and PDM sample values equal to zero are usually represented by nonzero amplitude or duration. This practice is followed to prevent "missing" pulses and to preserve a constant pulse rate. The latter is particularly important for synchronization purposes in time-division multiplexing.

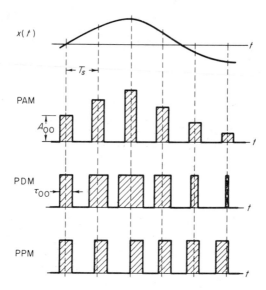

Fig. 7.13 *Types of analog pulse modulation.*

With the aid of Fig. 7.13, certain parallels can be drawn between analog pulse modulation and c-w modulation. Thus, the message information (in the form of sample values) is conveyed by the *amplitude* of a PAM wave, so PAM is analogous to *linear* c-w modulation, particularly AM. In PDM and PPM the information is conveyed by a *time* parameter, namely, the location of the pulse edges. Recalling that instantaneous frequency and phase are also time parameters, it can be said that pulse-time modulation is analogous to *exponential* c-w modulation.

General observations

Before proceeding to more detailed analysis, three general observations about pulse modulation can be stated:

1. Pulse-modulated waves have appreciable dc and low-frequency content, especially near the first few harmonics of $f_s = 1/T_s$. Direct transmission may therefore be difficult, if not impossible.
2. When pulse-modulated waves are transmitted, care must be taken to prevent the pulses from overlapping, for overlapping would destroy the modulation. From the pulse resolution requirements of Sec. 3.4, a bandwidth of at least $1/2\tau$ is necessary, τ being the nominal pulse duration.
3. Pulse-modulated waves can be demodulated via *reconstruction*. Conceptually, the sample values are extracted from the modulated wave, converted into weighted impulses, and lowpass-filtered.

With respect to the first point, short-distance pulse transmission may be feasible over wire circuits or coaxial cable, but efficient radio transmission requires additional frequency translation. Hence, most pulse systems have a carrier-modulation step in which the pulses are converted to RF pulses. Analytically, if $x_p(t)$ is the pulse-modulated wave, the actual transmitted signal is $x_c(t) = x_p(t)A_c \cos \omega_c t$, where $f_c \gg f_s$. Though the carrier modulation is properly classified as DSB, envelope detection can be employed at the receiver if $x_p(t) \geq 0$ and carrier phase reversals do not occur.

A complete pulse-transmission system is diagramed in Fig. 7.14.

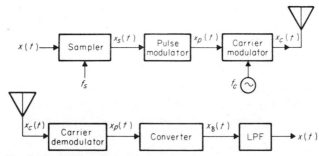

Fig. 7.14 *Pulse-modulation communication system.*

The transmitter consists of a sampler, pulse modulator, and carrier modulator. The receiver includes carrier demodulation followed by a converter which changes the pulse-modulated wave to a train of weighted impulses $x_\delta(t)$ or the realizable equivalent. Message recovery is then by lowpass filtering.

As to transmission bandwidth, the practical advantages of analog pulse modulation depend on the pulse duration's being small compared to the time between pulses, that is, $\tau \ll T_s \leq 1/2W$. Since the carrier modulation *doubles* the baseband bandwidth, we have

$$B_T \geq \frac{1}{\tau} \gg 2W$$

But, with or without carrier modulation, the required bandwidth is large compared to the message bandwidth W.

By analogy with exponential modulation, one suspects that the large bandwidth of analog pulse modulation results in *wideband noise reduction*. And *pulse-time* modulation does indeed have this useful property; moreover, the bandwidth is essentially determined by the desired amount of noise reduction rather than by the message bandwidth per se.

On the other hand, as might be inferred from its similarity to AM, PAM is no better than baseband transmission as far as noise is concerned, and the excessive bandwidth is a definite liability. Because of this shortcoming, PAM is seldom used for single-channel message transmission. (In fact, single-channel pulse communication systems of any type are rare.) However, PAM plays an important role in TDM, in message reconstruction, and in the study of more sophisticated pulse-modulation techniques. A few more words about PAM are thus in order.

Pulse-amplitude modulation (PAM)

The usual PAM waveform consists of nonrectangular unipolar pulses whose peak amplitudes are proportional to instantaneous sample values of the message. Analytically, if $A(mT_s)$ is the amplitude of the mth pulse, then

$$A(mT_s) = A_{00} + A_0 x(mT_s) \qquad A_0 < A_{00} \qquad (7.23)$$

where A_{00} and A_0 are modulation constants. With $x(t)$ normalized according to our convention, the maximum and minimum pulse amplitudes are $A_{00} \pm A_0$. The condition $A_0 < A_{00}$ is imposed to preserve single polarity and prevent missing pulses, as mentioned earlier. However, *bipolar* PAM $(A_{00} = 0)$ is sometimes encountered, usually as an intermediate step in modulation or reconstruction, and strongly resembles DSB as distinguished from AM.

Taking the normalized pulse shape as $p(t)$, the PAM wave is

$$x_p(t) = \sum_m A(mT_s) p(t - mT_s)$$

$$= A_0 \sum_m \left[\frac{A_{00}}{A_0} + x(mT_s) \right] p(t - mT_s) \qquad (7.24)$$

Comparing (7.24) with (7.21) shows that PAM is nothing more than *practical sampling*, the A_{00} term being equivalent to a dc component added to $x(t)$ before sampling.   It then follows that PAM is demodulated by spectral equalization, lowpass filtering, and removal of the dc term, as shown in Fig. 7.15.   Note that the converter of Fig. 7.14 is not required,

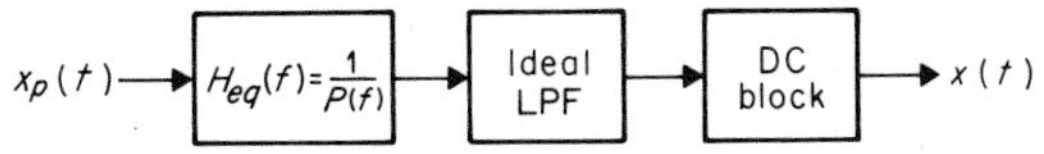

Fig. 7.15 *PAM demodulation.* $x_p(t) \longrightarrow \boxed{H_{eq}(f) = \frac{1}{P(f)}} \longrightarrow \boxed{\begin{matrix}\text{Ideal}\\ \text{LPF}\end{matrix}} \longrightarrow \boxed{\begin{matrix}\text{DC}\\ \text{block}\end{matrix}} \longrightarrow x(t)$

since the PAM wave $x_p(t)$ is the realizable equivalent of the ideal sampled wave $x_\delta(t)$.

Pulse-duration modulation (PDM)

We have remarked that pulse-time modulation is roughly analogous to exponential c-w modulation.   And, not too surprisingly, the spectral calculations turn out to be equally difficult, if not more so, for even the *time-domain* representations of PDM and PPM are in themselves quite clumsy.   One must therefore get involved in some laborious analyses or be content with an approximation.   For PDM we choose the latter course and present below a quasi-static approximation based on Fourier series expansions.   While the approach is lacking in rigor, conclusions gained therefrom are essentially correct.

The approximation stems from two observations about PDM: (1) most messages have their energy concentrated at frequencies well below the nominal maximum frequency W, so adjacent sample values are very nearly equal; (2) in practice, the time slot allowed for the modulated pulse is small compared to the time between pulses, so the variation in pulse duration is much less than T_s.   Under these conditions the PDM waveform is *almost periodic* in the sense that the duration changes but slightly from pulse to pulse.   This viewpoint is similar to the quasi-static treatment of FM, in which the instantaneous frequency was taken to be nearly constant over several cycles.

Assuming rectangular pulses of amplitude A, we expand the waveform as a trigonometric Fourier series

$$x_p(t) = \frac{A\tau}{T} + \sum_{n=1}^{\infty} \frac{A}{\pi n}\left[\sin\frac{\pi n}{T}(2t + \tau) - \sin\frac{\pi n}{T}(2t - \tau)\right]$$

which is a modification of Eq. (7.3) with T interpreted as the pulse *spacing*.   To reflect the modulation, τ and possibly T will be slowly changing in time, proportional to the message.

For the PDM wave of Fig. 7.13 the pulse spacing is constant,

$T = T_s = 1/f_s$, while the duration of the mth pulse is

$$\tau_m = \tau_{00} + t_0 x(mT_s) \qquad t_0 < \tau_{00} \ll T_s \tag{7.25}$$

τ_{00} and t_0 being modulation constants. But in our approximation we will think of the pulse duration as varying continuously such that

$$\tau \approx \tau_{00} + t_0 x(t)$$

Inserting this in the series gives

$$x_p(t) = Af_s\tau_{00} + Af_s t_0 x(t)$$
$$+ \sum_{n=1}^{\infty} \frac{A}{\pi n} \{\sin [n\omega_s t + n\phi(t)] - \sin [n\omega_s t - n\phi(t)]\} \tag{7.26}$$

where $\phi(t) = \pi f_s[\tau_{00} + t_0 x(t)]$.

Equation (7.26) indicates that the PDM wave includes the expected dc component, plus the message itself and other terms which are equivalent to *phase-modulated* carrier waves at all harmonics of f_s. The phase modulation is proportional to $x(t)$; the phase deviation increases with harmonic order n while the carrier amplitude decreases. Thus the sidebands progressively spread out in frequency and diminish in magnitude at the higher "carriers." Without actually calculating the PDM spectrum—virtually impossible for arbitrary $x(t)$—we can make a reasonable sketch of $X_p(f)$, as in Fig. 7.16.

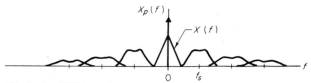

Fig. 7.16 *PDM spectrum.*

This analysis was based on a PDM wave in which both pulse edges are modulated so the pulse spacing is constant. In practice, the leading edge is usually fixed at $t = mT_s$, and only the trailing edge is modulated. This case is somewhat more difficult to handle because the pulse spacing will be variable.[1] However, the results are quite similar to (7.26) when $\tau_{00} \ll T_s$, as is usually true.

A very simple technique for the generation of PDM is illustrated in Fig. 7.17, where the message and a sawtooth waveform are summed, forming the input to a *slicing circuit*. The slicing circuit, essentially a

[1] For which see Black (1953, chap. 17).

clipping and squaring device, produces $+A$ volts whenever the input exceeds the slicing level and has zero output otherwise. The resulting wave is PDM with trailing-edge modulation, and the operations of sampling and modulation have been combined into one step. Reversing the sawtooth produces leading-edge modulation; replacing the sawtooth by a triangular wave produces modulation on both edges. However, as the attentive reader may have noticed, the final pulse duration corresponds to message samples at the time location of the modulated edge,

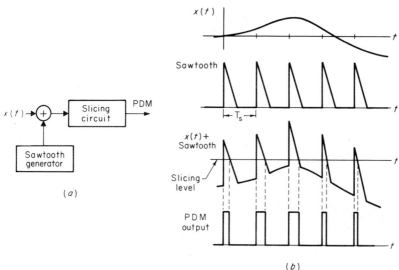

Fig. 7.17 *PDM generation.* (a) *Block diagram;* (b) *waveforms.*

not the apparent sampling time $t = mT_s$. Thus, the sample values are *nonuniformly* spaced, and uniform sampling theory is not strictly applicable. Of course the difference between uniform and nonuniform[1] sampling is insignificant if the pulse duration is small compared to T_s. From the hardware viewpoint, nonuniform sampling is preferred because of its simplified instrumentation.

For demodulation of PDM there are two possibilities. One can convert PDM to $x_\delta(t)$, that is, convert PDM to PAM, and then lowpass-filter, as in the receiver of Fig. 7.14. Alternately, further examination of the PDM spectrum in Fig. 7.16 reveals that direct demodulation by *lowpass filtering* is possible, providing the PM sidebands do not overlap the message band. This rather unexpected conclusion is supported by

[1] Nonuniform sampling for pulse-time modulation is also called *natural* sampling, a designation we shall avoid to prevent possible confusion.

more precise studies taking account of the nonuniform sampling, etc. In fact, nonuniform sampling yields less inherent distortion in the filtered message than uniform sampling, and direct filtering is the conventional means of PDM demodulation.

Pulse-position modulation (PPM)

PDM and PPM are closely allied, a relationship underscored by Fig. 7.18, showing how PPM can be generated from PDM with trailing-edge

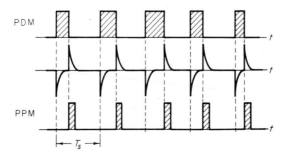

Fig. 7.18

modulation.   The duration-modulated pulses are inverted and differentiated, changing the modulated edges into position-modulated positive spikes.   With a little reshaping these spikes become the desired PPM wave.

As a matter of fact, the principal use of duration modulation is for the generation (and detection) of position modulation because PPM is markedly superior to PDM for message transmission.   To appreciate why this is so, recall that the information resides in the time location of the pulse *edges*, not in the pulses themselves.   Thus, somewhat like the carrier-frequency power of AM, the pulse power of pulse-time modulation is "wasted" power, and it would be more efficient to suppress the pulses and just transmit the edges!   Of course we cannot transmit edges without transmitting pulses to define them.   But we can send very short pulses indicating the position of the edges, a process equivalent to PPM.   The reduced power required for PPM is a fundamental advantage over PDM, an advantage that will be more apparent when we examine the signal-to-noise ratios.

Since PPM with nonuniform sampling is the most efficient type of pulse-time modulation for transmission purposes, we shall analyze its spectrum more carefully than that of PDM.   The analysis shows the presence of PM components, which is expected, as well as some unexpected AM.

To begin let t_m be the time location (center) of the mth pulse.   If the sampling is uniform, the mth pulse carries the sample value at $t = mT_s$, and

$$t_m = mT_s + t_0 x(mT_s)$$

where t_0 is the modulation constant, i.e., the *maximum displacement* relative to $t = mT_s$.   But with nonuniform sampling the sample value is actually extracted at t_m, not mT_s, so

$$t_m = mT_s + t_0 x(t_m) \qquad (7.27)$$

By definition, the PPM wave is a summation of constant-amplitude position-modulated pulses, and can be written as

$$x_p(t) = \sum_m Ap(t - t_m) = Ap(t) * \left[\sum_m \delta(t - t_m) \right]$$

where A is the pulse amplitude and $p(t)$ the pulse shape.   A simplification at this point is made possible by noting that $p(t)$ will (or should) have a very small duration compared to T_s.   Hence, for our purposes, the pulse shape can be taken as impulsive, that is, $p(t) \approx \delta(t)$, so

$$x_p(t) \approx A \sum_m \delta(t - t_m) \qquad (7.28)$$

If desired, Eq. (7.28) can later be convolved with $p(t)$ to account for the nonimpulsive shape.

In their present form, Eqs. (7.27) and (7.28) are unsuited to further manipulation; the trouble is the position term t_m, which cannot be solved for explicitly.   Fortunately, Rowe (1965, chap. 5) has devised a technique whereby t_m can be eliminated entirely, as follows.

Consider any function $g(t)$ having a single first-order zero at $t = \lambda$; that is, $g(\lambda) = 0$, $g(t \neq \lambda) \neq 0$, and $g'(\lambda) = dg/dt|_{t=\lambda} \neq 0$.   Using the properties of impulses, one can show that[1]

$$\delta[g(t)] = \frac{\delta(t - \lambda)}{|g'(t)|}$$

hence

$$\delta(t - \lambda) = |g'(t)|\delta[g(t)] \qquad (7.29)$$

whose right-hand side is independent of λ.   Equation (7.29) can therefore be used to remove t_m from $\delta(t - t_m)$ if we can find a function $g(t)$ which satisfies $g(t_m) = 0$ and the other conditions but does not contain t_m.

Suppose we take $g(t) = t - mT_s - t_0 x(t)$, which is zero at $t = mT_s + t_0 x(t)$.   Now, for a given value of m, there is only one PPM pulse,

[1] See Friedman (1956).

and it occurs at $t_m = mT_s + t_0x(t_m)$. Thus $g(t_m) = t_m - mT_s - t_0x(t_m) = 0$, as desired. Inserting $\lambda = t_m$, $g'(t) = 1 - t_0x'(t)$, etc., into (7.29) gives

$$\delta(t - t_m) = |1 - t_0x'(t)|\delta[t - mT_s - t_0x(t)]$$

and the PPM wave of (7.28) becomes

$$x_p(t) = A[1 - t_0x'(t)]\sum_m \delta[t - t_0x(t) - mT_s]$$

The absolute value is dropped since $|t_0x'(t)| < 1$ for most signals of interest if $t_0 \ll T_s$. We then convert the sum of impulses to a sum of exponentials via $\sum_m \delta(t - mT_s) = f_s \sum_n \exp(jn\omega_s t)$, as derived in Sec. 2.5, to finally obtain

$$x_p(t) = Af_s[1 - t_0x'(t)]\sum_{n=-\infty}^{\infty} e^{jn\omega_s[t-t_0x(t)]}$$

$$= Af_s[1 - t_0x'(t)]\left\{1 + \sum_{n=1}^{\infty} 2\cos[n\omega_s t - n\omega_s t_0 x(t)]\right\} \quad (7.30)$$

Interpreting (7.30), we see that PPM with nonuniform sampling is a combination of linear and exponential carrier modulation, for each harmonic of f_s is phase-modulated by the message $x(t)$ and amplitude-modulated by the derivative $x'(t)$. The spectrum therefore consists of AM and PM sidebands centered at all multiples of f_s, plus a dc component and the spectrum of $x'(t)$. Needless to say, sketching such a spectrum is a tedious exercise even for tone modulation.

The operation of Day's *serrasoid system* for narrowband PM can now be explained with the aid of the above results. The system, diagramed in Fig. 7.19, consists of a PDM generator, a differentiator to convert PDM

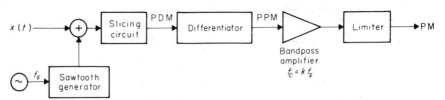

Fig. 7.19 *Serrasoid phase modulation system.*

to PPM, and a bandpass amplifier tuned to the kth harmonic of the sampling frequency followed by an amplitude limiter. The PDM modulator is of the type shown in Fig. 7.17a, whose sawtooth generator is controlled by a crystal oscillator for frequency stability. (*Serra* is Latin for "saw," which explains the name.) Assuming the spectral components of (7.30) are nonoverlapping, the PPM wave after filtering and limiting becomes the PM wave

$$x_c(t) = A_c \cos [\omega_c t + \phi_d x(t)]$$

where $f_c = k f_s$ and $\phi_d = -2\pi k f_s t_0$. The phase deviation is quite small, since $f_s t_0 = t_0/T_s \ll 1$. A message integrator added to the input yields, of course, NBFM, and the serrasoid system was used for some time in commercial FM transmitters of the indirect (Armstrong) type.

Returning to PPM and its demodulation, the leading term of (7.30) suggests that the message can be retrieved by lowpass filtering and *integrating*. However, the integration method does not take full advantage of the noise-reduction properties of PPM, so the usual procedure is conversion to PAM or PDM followed by lowpass filtering.

Signal-to-noise ratios

Regardless of the specific technique employed, the demodulation of a pulse-modulated wave is inherently message *reconstruction* from sample values. This viewpoint allows us to treat added noise in a rather simple fashion: if the received signal plus noise is

$$y(t) = x_p(t) + n(t)$$

then reconstruction must be based on the "noisy" samples

$$K_s x(mT_s) + \epsilon_m$$

where K_s is the modulation constant relating $x(mT_s)$ to $x_p(t)$ and ϵ_m is the error introduced by the presence of $n(t)$. Paralleling Eq. (7.19), the resultant signal-to-noise ratio is

$$\left(\frac{S}{N}\right)_D = K_s^2 \frac{\overline{x^2}}{\overline{\epsilon^2}}$$

$\overline{\epsilon^2}$ being the mean-square value of the error term. In general $\overline{\epsilon^2}$ is proportional to the average noise power $N = \overline{n^2}$.

We shall assume the pulses are transmitted without carrier modulation over a system having bandwidth B_T. Then $n(t)$ is bandlimited white noise of power density η, and $N = \eta B_T$. The average signal power is $P_T = \langle x_p{}^2(t)\rangle$, as usual.

PAM plus noise is readily analyzed with the aid of Fig. 7.20.

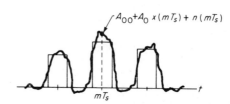

$A_{00} + A_0\, x(mT_s) + n(mT_s)$

Fig. 7.20 *PAM wave pulse noise.*

mT_s

Assuming *bipolar* pulses ($A_{00} = 0$) for simplicity, the sample values are contained in the modulated amplitude $A(mT_s) = A_0 x(mT_s)$, and the noise adds directly to the amplitude. Hence, $K_s = A_0$, $\epsilon_m = n(mT_s)$, $\overline{\epsilon^2} = \overline{n^2} = \eta B_T$, and

$$\left(\frac{S}{N}\right)_D = \frac{A_0{}^2 \overline{x^2}}{\eta B_T}$$

To express this in more familiar terms we note that the average transmitted power is just the duty cycle $d = \tau/T_s$ times the mean-square pulse amplitude, that is,

$$P_T = \frac{\tau}{T_s} \langle A_0{}^2 x^2(mT_s) \rangle = \tau A_0{}^2 \frac{\overline{x^2}}{T_s}$$

so that

$$\left(\frac{S}{N}\right)_D = \frac{T_s P_T}{\tau \eta B_T} = \frac{T_s W}{\tau B_T} \frac{P_T}{\eta W}$$

or

$$\left(\frac{S}{N}\right)_D = \frac{T_s W}{\tau B_T} z \qquad \text{bipolar PAM} \tag{7.31}$$

where $z = P_T/\eta W$.

It was stated earlier that PAM is no better than direct baseband transmission insofar as noise is concerned. Equation (7.31) shows this to be true, for $T_s \leq 1/2W$ and $B_T \geq 1/2\tau$, so $T_s W/\tau B_T \leq 1$ and $(S/N)_D \leq z$. The equality is realized only by sampling at the Nyquist rate and transmitting bipolar PAM. These conditions are seldom achieved in practice; nor are they sought after, for the merit of PAM is its simplicity in multiplexing, not its noise performance.

Turning to pulse-time modulation, we recall that the message samples are contained in the relative position of the pulse edges. If the received pulses were perfectly rectangular, additive noise would have no effect on the edges since the noise is a *vertical* perturbation (Fig. 7.21a). But

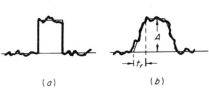

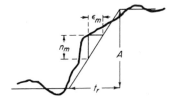

(a)   (b)

Fig. 7.21 *Pulse-time modulation plus noise. (a) Rectangular pulse; (b) finite-bandwidth pulse.*

Fig. 7.22 *Expanded view of noise perturbation.*

rectangular pulses require *infinite* bandwidth; with finite bandwidth the pulses have nonzero rise time $t_r \approx 1/2B_T$, and the noise will alter the edge position according to Fig. 7.21b. The expanded view of Fig. 7.22 shows the error term ϵ_m due to a noise perturbation n_m. From geometry, $\epsilon_m/n_m = t_r/A$, A being the pulse amplitude, so

$$\overline{\epsilon^2} = \left(\frac{t_r}{A}\right)^2 \overline{n^2} = \frac{\eta B_T}{4B_T^2 A^2} = \frac{\eta}{4B_T A^2}$$

Note that the mean-square error *decreases* as the bandwidth *increases*, hence we have *wideband noise reduction* as promised earlier. Figure 7.21a corresponds to the logical extreme, in which $B_T \to \infty$, $t_r \to 0$, and $\overline{\epsilon^2} \to 0$.

As to the signal parameters, the variation in pulse duration or pulse position is $t_0 x(mT_s)$, so $K_s = t_0$. The average transmitted power is $P_T = (\tau/T_s)A^2 = \tau f_s A^2$ since the pulse amplitude is constant. Therefore

$$\left(\frac{S}{N}\right)_D = \frac{t_0^2 \overline{x^2} 4B_T P_T}{\tau \eta f_s} = \frac{4t_0^2 B_T W}{\tau f_s} \overline{x^2 z} \tag{7.32}$$

which applies to both PDM and PPM without carrier modulation. In the case of PDM, τ must be interpreted as the *average* pulse duration, which equals τ_{00} if the message has no dc component.

The only difference between PPM and PDM, as reflected in (7.32), is the smaller pulse duration of PPM. This difference can be made more apparent by considering the best one can do for a given bandwidth B_T. Clearly, maximizing $(S/N)_D$ requires the minimum sampling frequency $(f_s = 2W)$ with maximum t_0 and minimum average pulse duration. In PPM, t_0 and τ are independent and can be treated separately. The minimum possible pulse duration is $\tau \approx 2t_r \approx 1/B_T$, for which the pulse degenerates into a triangular shape. The maximum pulse displacement must be limited so that the *range* of possible pulse positions, $mT_s \pm 2t_0$, does not exceed the time between samples, that is, $t_0 \le T_s/2 = 1/4W$. Taking all values to be optimum gives

$$\left(\frac{S}{N}\right)_{D_{\max}} = \frac{1}{8}\left(\frac{B_T}{W}\right)^2 \overline{x^2 z} \qquad \text{PPM} \tag{7.33}$$

which more clearly exhibits the exchange of bandwidth for signal-to-noise ratio. However, practical systems may fall short of this maximum by an order of magnitude or more.

A similar optimization applied to PDM results in $\tau_{00} \approx t_0 \approx 1/4W$, so

$$\left(\frac{S}{N}\right)_{D_{\max}} = \frac{1}{2}\frac{B_T}{W} \overline{x^2 z} \qquad \text{PDM} \tag{7.34}$$

Note that the signal-to-noise ratio increases linearly with the bandwidth ratio B_T/W rather than as $(B_T/W)^2$. Moreover, to achieve (7.34) the average pulse duration must be one-half the time between samples. Under this condition, the duty cycle is $\tau_{00}/T_s = \frac{1}{2}$, and the practical advantages of pulse modulation with small duty cycle are lost.

It is particularly interesting to compare these results with those of exponential modulation. Recall, for example, that WBFM without deemphasis produced

$$\left(\frac{S}{N}\right)_D = \frac{3}{4}\left(\frac{B_T}{W}\right)^2 \overline{x^2} z \qquad \text{WBFM} \qquad (6.41)$$

Contrasting this with (7.34) shows that PDM falls far short of the noise reduction of FM having the same bandwidth, assuming $B_T/W \gg 1$. PPM is considerably better, giving a signal-to-noise ratio about 8 db below that of FM. However, to make a truly equitable comparison, carrier modulation of the pulsed signal should be included, since it doubles the transmission-bandwidth requirements but halves the transmitted power. With carrier modulation accounted for, PPM falls some 14 db below FM under the best conditions.

It thus appears that, while pulse-time modulation does give wideband noise reduction, the reduction is far less effective than exponential carrier modulation. But certain demodulation refinements do improve the postdetection signal-to-noise ratios beyond those given in (7.33) and (7.34). Moreover, one should also remember that the average power P_T is delivered as short-duration high-power pulses, whereas P_T must be generated continuously in FM. Practical limitations of the carrier supply may favor pulsed operation.

Concluding this discussion, it must be mentioned that the bandwidth ratio cannot be increased without limit, for pulse-time modulation suffers a *threshold effect* similar to that of FM. It is true that $\overline{\epsilon^2} \to 0$ as $B_T \to \infty$. But as $N = \eta B_T$ increases, the rms noise voltage increases, and occasional noise peaks will be mistaken for signal pulses. If such *false pulses* occur often enough, the desired message will be lost completely, for the reconstructed signal will have no relationship to $x(t)$.

As a rule of thumb, false pulses are sufficiently infrequent if $P(n > A) \leq 0.01$. Assuming gaussian noise, the corresponding threshold condition is approximately

$$A \geq 2\sqrt{N}$$

Thus, the pulse amplitude must be strong enough to "lift" the noise by at least twice its rms value. This is the same condition as the *tangential sensitivity* of pulsed radar systems.

7.4 PULSE–CODE MODULATION

All the modulation types covered so far, whether pulse or c-w, have been analog representations of the message. Pulse-code modulation (PCM) is distinctly different in concept; it is digital modulation in which the message is represented by a coded group of digital (discrete-amplitude) pulses. The reasoning behind this procedure is as follows.

In analog modulation, the modulated parameter varies continuously and can take on *any* value corresponding to the range of the message. When the modulated wave is adulterated by noise, there is no way for the receiver to discern the exact transmitted value. Suppose, however, that only a few *discrete* values are allowed for the modulated parameter; if the separation between these values is large compared to the noise perturbations, it will be a simple matter to decide at the receiver precisely which specific value was intended. Thus the effects of random noise can be virtually eliminated, which is the whole idea of PCM.

As a corollary to the discrete-amplitude property, a PCM signal can be periodically *regenerated* along the transmission path. Long-haul communication systems, both radio and cable, often require the use of *repeaters* between terminals to obtain an adequate received signal strength. With analog modulation, a repeater can do no better than amplify both signal and noise; furthermore, it may introduce additional noise of its own, causing the signal-to-noise ratio to progressively decrease with each repeater station. But with PCM, the signal can be regenerated at each repeater—essentially demodulated and remodulated—producing a new signal free from random noise. As a result the noise does not accumulate, and overall system performance is nearly equivalent to that of one repeater-to-repeater link alone.

But the question now arises: How do we represent a continuously varying message waveform by a string of discrete-amplitude pulses? The answer lies in sampling, quantizing, and coding.

Quantizing and coding

The elements of PCM generation are shown in Fig. 7.23. The continuous signal $x(t)$ is first sampled to give $x_s(t)$, as in conventional pulse systems. The sample values are then rounded off to the nearest pre-

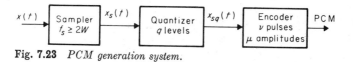

Fig. 7.23 *PCM generation system.*

determined discrete value. This process is called *quantization*, and the discrete values are called *quantum levels*. The sampled and quantized wave $x_{sq}(t)$ which results is discrete in both time (by virtue of the sampling) and amplitude (by virtue of the quantization).

If there are a finite number of quantum levels, say q, each level can be represented by a digital code of finite length; $x_{sq}(t)$ is then *digitized* and *coded*. The encoder converts quantized samples to appropriate code groups, one group for each sample, and generates the corresponding digital pulsed wave forming the baseband PCM signal.

Obviously the parameters of the encoded signal depend upon q, for each code group must uniquely represent one of the possible quantum levels. To ascertain the relationship, let ν be the number of pulses (or digits) in the code group, each having one of μ discrete values. Since there are μ^ν different *combinations* of ν pulses with μ possible amplitudes, we require $\mu^\nu \geq q$ for unique encoding. Therefore, when the parameters are chosen so the equality is realized,

$$\nu = \log_\mu q \tag{7.35}$$

Examining (7.35), we see that if $\mu = q$, then $\nu = 1$, and the quantized signal requires no code translation. In general, however, $\mu < q$ and $\nu > 1$. The most common form of PCM is *binary* PCM, for which $\mu = 2$; the number of quantum levels then is taken as some power of 2, namely, $q = 2^\nu$.

Figure 7.24 illustrates these operations for binary PCM. Eight quantum levels are shown, corresponding to 0, 1, . . . , 7 volts. Note that the quantizing could be performed by a digital voltmeter having a fixed resolution of 1 volt, the spacing between quantum levels. The

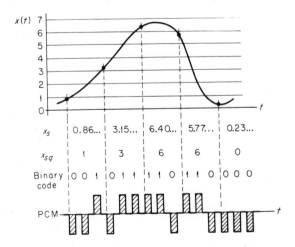

Fig. 7.24

encoder translates quantized voltage to its binary equivalent according to the following table. Three binary digits are necessary since $\nu = \log_2$

Quantized sample	Binary code
0	000
1	001
2	010
3	011
4	100
5	101
6	110
7	111

$8 = \log_2 2^3 = 3$. The binary digits are then represented by *on-off* or *plus-minus* pulses, the latter being shown in the figure.

Because several coded pulses are required for each message sample, it appears that the PCM bandwidth will be much greater than the message bandwidth. An estimate of the bandwidth is obtained as follows. Quantized samples occur at a rate of $f_s \geq 2W$ samples per second, so there must be νf_s coded pulses per second. If there are no "spaces" in the coded signal, the maximum permitted duration of any one pulse is $\tau = 1/\nu f_s$. On the basis of pulse resolution we need a bandwidth of at least $1/2\tau$; therefore

$$B_T \geq \tfrac{1}{2}\nu f_s \geq \nu W \qquad (7.36)$$

The PCM bandwidth is thus a minimum of $\nu = \log_\mu q$ times the message bandwidth.

As the final step of PCM generation, the baseband signal may modulate an RF carrier for transmission purposes. The carrier modulation can be AM (*amplitude-shift keying*), *phase-shift keying* (PSK), or *frequency-shift keying* (FSK). (The latter two are forms of exponential modulation for digital signals and are discussed in Chap. 9.) Of course the carrier modulation results in an even greater transmission bandwidth than that of (7.36), but we shall restrict our attention to the baseband case.

Random noise and error threshold

The purpose of PCM is to eliminate the effects of random electrical noise. To examine the subject in more detail let us consider binary PCM with coded pulses of amplitude $\pm A$. If the code pattern is 010110, for

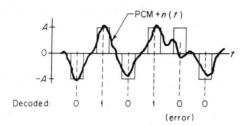

Decoded: 0   1   0   1   0   0   **Fig. 7.25** *PCM wave pulse noise.*
                      (error)

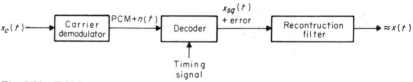

Fig. 7.26 *PCM receiving system.*

example, the signal plus noise at the receiver might look like Fig. 7.25
(after carrier demodulation).   Given this waveform, the receiver must
decode to obtain $x_{sq}(t)$ and reconstruct by lowpass filtering; Fig. 7.26 is
the block diagram of the receiver.

The first task of the decoder is recognition of the binary digits.
This can be done with a *decision circuit* whose crossover level is set at
zero volts; when the waveform is positive, the digit is taken to be a 1, and
vice versa.   Clearly, this decision should be made at the center of each
pulse for maximum reliability.   Hence, synchronizing signals are usually
supplied to trigger the decoder at the optimum times.

As long as the magnitude of the noise is less than the pulse amplitude, things proceed quite smoothly.   But occasionally random-noise
peaks will cause the decoder to make a wrong decision, resulting in
an *error*.   For example, when a 1 is sent, a decoding error occurs if
$A + n(t) < 0$ at the decision time, $n(t)$ being the noise voltage.   Similarly $-A + n(t) > 0$ causes an error when a 0 is sent.   How often such
errors happen depends on the relative strength of the noise and its statistical properties.   The table below lists the probability of decoding errors

$\dfrac{A}{\sqrt{N}}$	Error probability
1	0.16
2	0.023
3	0.00135
4	0.000032

as a function of $A/\sqrt{N}$, assuming the noise to be gaussian with zero mean and $\sigma = \sqrt{N}$. Clearly, the errors decrease rapidly with increasing predetection signal-to-noise ratio, for $A^2/N = (S/N)_T$ is the signal-to-noise ratio at the decoder input for this particular case. With $(S/N)_T = 16$ (12 db) the error probability is less than 10^{-4}, and approximately 1 digit in 10,000 will be erroneously decoded.

Regardless of the specific noise characteristics, we can conclude that "reasonable" signal-to-noise ratios yield negligible decoding errors. Therefore, providing $(S/N)_T$ exceeds some *error threshold*, random noise has virtually no effect on system performance! This is precisely the goal of PCM.

But suppose $(S/N)_T$ drops below the error threshold. When this happens, there is a sharp increase of decoding errors. It is easily appreciated that when errors are frequent, the reconstructed waveform bears little resemblance to the original message; i.e., the message has been *mutilated* by *decoding noise*. Thus PCM suffers from a *threshold effect* similar to that of wideband analog modulation.

The error-threshold level is, of course, somewhat arbitrary, and its calculation is complicated by the fact that the severity of an error depends on where the error occurs; an error in the leading (most significant) digit of a code group is more serious than errors in other digits.[1] As a rule of thumb, it has been found that mutilation is negligible if the error probability is about 10^{-4} or less.

It also should be noted that for a given $(S/N)_T$, *binary* PCM is less vulnerable to decoding errors than μ-*ary* PCM $(\mu > 2)$. With fixed $(S/N)_T$, the *spacing* between levels (relative to the rms noise) decreases with increasing μ, and hence the error probability increases.

Including the dependence on μ, a conservative value for the error-threshold level at the decoder input is

$$\left(\frac{S}{N}\right)_{T_{\text{th}}} = 10(\mu^2 - 1) \tag{7.37}$$

as will be derived in Chap. 8.

Quantization noise

Despite the fact that decoding errors can be ignored, the final output at the receiver will not be identical to the original message—recall that the decoded signal is $x_{sq}(t)$ not $x_s(t)$. When $x_{sq}(t)$ is lowpass-filtered, the reconstructed signal differs from $x(t)$ because the quantized samples are not exact sample values. Furthermore, there is no way of obtaining exact values at the receiver; that information was discarded at the trans-

---

[1] For the interested reader, Downing (1964, chap. 7) and Panter (1965, chap. 21) give some of the details.

mitter in the quantizing process. Therefore, perfect message reconstruction is impossible in pulse-code systems, even when random noise
has negligible influence.

As a result, the quantization effect is a basic limitation of coded
systems, just as random noise is a limitation of conventional analog systems. It is this quantization effect we wish to examine in more detail.

For analysis we decompose $x(t)$ into a stepwise-quantized signal
$x_q(t)$ and a *quantizing-error term* $\epsilon(t)$, such that

$$x_q(t) = x(t) + \epsilon(t)$$

as illustrated in Fig. 7.27. Because the order of sampling and quantizing

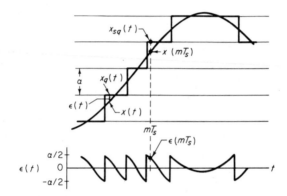

Fig. 7.27 *PCM quantization
error.*

is interchangeable, the quantized sample values as decoded as the receiver
can be written in the form

$$x_{sq}(t) = x_q(mT_s) = x(mT_s) + \epsilon(mT_s)$$

Note that the quantizing error of PCM appears in identical form as noise-
caused errors of analog pulse modulation. Hence, as before, the reconstruction or postdetection signal-to-noise ratio is taken to be

$$\left(\frac{S}{N}\right)_D = \frac{\overline{x^2}}{\overline{\epsilon^2}}$$

where $\overline{\epsilon^2}$ is the mean-square quantizing error, called the *quantization noise*.

The error term is readily evaluated if the quantum levels have
uniform spacing, say α volts, so the error is bounded by $-\alpha/2 \leq \epsilon(t)$
$\leq \alpha/2$ (see Fig. 7.27). Lacking information to the contrary, we assume
that $\epsilon(mT_s)$ is equally likely to be anywhere in this range; that is, ϵ has a
uniform probability density function $p(\epsilon) = 1/\alpha$ over $[-\alpha/2, \alpha/2]$. Hence

$$\overline{\epsilon^2} = \int_{-\alpha/2}^{\alpha/2} \frac{\epsilon^2}{\alpha} \, d\epsilon = \frac{\alpha^2}{12} \tag{7.38a}$$

But a more useful expression is obtained by noting that with q levels of spacing α, the *range* $q\alpha$ must encompass the peak-to-peak swing of the message, which is $-1 \leq x(t) \leq +1$ by our convention. Thus $q\alpha = 2$, $\alpha = 2/q$, and

$$\overline{\epsilon^2} = \frac{1}{3q^2} \tag{7.38b}$$

Therefore

$$\left(\frac{S}{N}\right)_D = 3q^2\overline{x^2} \tag{7.39}$$

is the postdetection signal-to-noise ratio for PCM above threshold.

Clearly, the performance quality increases with q, the number of quantum levels. This merely reiterates that if many quantum levels of small spacing are employed, the quantized samples will closely approximate the exact sample values, and the output can be made as near as desired to $x(t)$. In practice, surprisingly few levels yield satisfactory performance; $q = 128$ is standard for voice telephony.

Several other interesting properties of PCM are contained in (7.39), either explicitly or implicitly: (1) the result is independent of carrier modulation, transmitted power, or random noise, reflecting the fact that decoding errors are ignored; (2) the signal-to-noise ratio is constant and fixed *at the transmitter* by the number of quantum levels; and (3) PCM has the characteristics of *wideband noise reduction* in the sense that quantization noise decreases with increasing q and the bandwidth depends on q. This relationship is made more apparent by the following manipulation.

From Eq. (7.35) we have $q = \mu^\nu$. Now suppose the sampling frequency is close to the Nyquist rate, so the baseband bandwidth is $B_T \approx \nu W$, according to (7.36). Then $q = \mu^{B_T/W}$, and (7.39) can be written in the form

$$\left(\frac{S}{N}\right)_D = 3\mu^{2b}\overline{x^2} \qquad b = \frac{B_T}{W} \tag{7.40}$$

which shows the noise reduction as being an *exponential* exchange of bandwidth for signal-to-noise ratio. This exchange is far more dramatic than that of wideband analog modulation, where $(S/N)_D$ increases linearly or as the square of the bandwidth ratio.

Returning again to quantization and Eq. (7.38a), it should be pointed out that the rms error is fixed at $\alpha/\sqrt{12}$ regardless of the instantaneous value of $x(t)$. Hence, if $|x(t)|$ is small for extended periods of time, the apparent signal-to-noise ratio will be much less than the design value. The effect is particularly acute if the message waveform has a large *crest factor* (the ratio of peak amplitude to rms value), for then $|x(t)| \ll 1$ most of the time, and $\overline{x^2} \ll 1$.

For the transmission of audio signals, typically characterized by large crest factors, it is advantageous to taper the spacing between quantum levels, with small spacing near zero and large spacing at the extremes. A suitable *nonuniform* quantization can result in $\epsilon(t)$ substantially proportional to $x(t)$, thereby masking the noise with the signal insofar as a listener is concerned. In practice, tapered quantization is accomplished with uniformly spaced levels, the message being nonlinearly compressed prior to sampling; a complementary expansion process restores the waveshape at the receiver. The combination of compression and expansion is called *companding* and finds use in analog systems as well, but for different reasons.

Coded versus analog modulation systems

Figure 7.28 illustrates the performance of binary PCM ($\mu = 2$), as described by Eqs. (7.40) and (7.37), for three values of bandwidth ratio. For comparison the corresponding performance of several analog systems is also shown. The PCM and PPM curves are for direct transmission without carrier modulation, and in all cases $\overline{x^2} = \frac{1}{2}$ is assumed. As usual, threshold points are indicated by heavy dots.

Examining the figure reveals that, in the name of efficiency, PCM

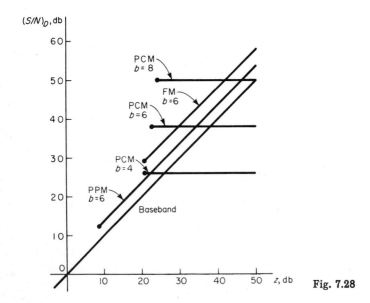

Fig. 7.28

systems should be operated just above threshold, for increasing the transmitted power beyond $(S/N)_{T_{th}}$ causes no improvement in $(S/N)_D$; the latter is dictated only by the number of quantum levels or, equivalently, the bandwidth ratio. But even at threshold, PCM does not show up so well as might have been hoped. Moreover, the instrumentation is considerably more complex and costly than that of uncoded systems. (For precisely this reason, PCM was deemed totally impractical prior to the emergence of high-speed solid-state digital electronics in the late 1950s, some 20 years after PCM's invention.)

Why then is PCM so highly touted? This is an involved and controversial question with several answers, among them:

1. For a given bandwidth ratio, PCM near threshold is superior to all other forms of *pulse* modulation, a significant factor if time-division multiplexing is desired.
2. PCM has a small but definite margin over FM at the lower signal-to-noise ratios. And even a 3-db power reduction, being a factor of 2, may spell the difference between success or failure in minimum-power applications.
3. A PCM system designed for analog message transmission is readily adapted to other input signals, particularly digital data, thereby promoting flexibility and increasing system utilization.[1]
4. By virtue of the *regeneration* capability, PCM is distinctly advantageous for systems having many repeater stations. Indeed, with respect to long-distance telephone, this has been called the real payoff of PCM.

Therefore, PCM should be given due consideration for applications involving TDM, minimum power, a diversity of message types (i.e., analog and digital) or many repeater stations. Because most of these factors are present in long-haul telephone transmission, PCM appears to be the way of the future in telephony; see Reeves (1965) for additional speculation along these lines by the inventor of PCM. However, in more routine applications, the extreme cost of hardware for coded modulation usually proves prohibitive compared to that for analog modulation.[2]

[1] In an interesting survey article, Franklin and Law (1966) discuss the problems of PCM interconnection and compatibility.

[2] In this connection, a PCM variation known as *delta modulation* (DM) has received attention, particularly in Europe. DM has many properties of PCM but with much simpler instrumentation at the cost of even greater transmission bandwidth requirements. Bowers (1957) gives a very readable discussion of DM versus PCM.

7.5 TIME-DIVISION MULTIPLEXING

Time-division multiplexing (TDM) is a technique for transmitting several messages on one facility by dividing the time domain into slots, one slot for each message.

The essentials of TDM are quite simple, as illustrated by Fig. 7.29. The several input signals, all bandlimited in W, are sequentially sampled at the transmitter by a rotary switch, or *commutator*. The

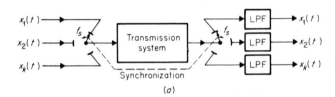

(a)

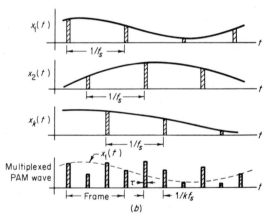

(b)

Fig. 7.29 *TDM system.* (a) *Block diagram;* (b) *waveforms.*

switch makes one complete revolution in $T_s \leq 1/2W$ sec, extracting one sample from each input. Hence, the commutator output is a PAM waveform containing the individual message samples periodically interlaced in time. If there are k inputs, the pulse-to-pulse spacing is $T_s/k = 1/kf_s$, while the spacing between successive samples from any one input is of course T_s. A set of pulses consisting of one sample from each input is called a *frame*.

At the receiver a similar rotary switch, the *decommutator* or *distributor*, separates the samples and distributes them to a bank of lowpass filters, which in turn reconstruct the original messages. The switching action is usually electronic, and synchronizing signals are provided to keep the distributor in step with the commutator. (In fact, synchronization is perhaps the most critical aspect of TDM.)

Within this basic framework are numerous varieties of time-division systems, involving both pulsed and c-w modulation, just as there are many combinations of subcarrier and carrier modulation in frequency-division systems. But before getting into the details, let us pause to contrast the two multiplexing methods.

Clearly TDM and FDM accomplish the same goals, though the means are different. Indeed, they can be visualized as *dual* techniques; for in TDM the signals are separate in the time domain but jumbled together in frequency, whereas in FDM the signals are separate in the frequency domain but jumbled together in time. It is therefore reasonable to ask what advantages, if any, TDM offers compared to FDM. From a theoretical viewpoint there are none. From a practical viewpoint TDM can be superior in two respects.

First, TDM instrumentation is somewhat simpler. Recall that FDM requires subcarrier modulators, bandpass filters, and demodulators for *each* message channel; these are all replaced by the commutator and distributor of TDM. And TDM synchronization is but slightly more demanding than that of FDM with suppressed-carrier modulation.

Second, and equally important, TDM is invulnerable to the usual sources of FDM interchannel *cross talk*, i.e., imperfect channel filtering and cross modulation due to nonlinearities. In fact, there is no cross talk in TDM if the pulses are completely isolated and nonoverlapping, since message separation is achieved by decommutation or gating in time, rather than by filtering. TDM cross-talk immunity is therefore contingent upon a wideband response and the absence of phase-shift (delay) distortion, keeping the pulses short and confined. (Note that phase distortion does not cause cross talk in FDM.)

Actual pulse shapes, having decaying tails, do tend to overlap, of course. However, the resulting cross talk can be effectively reduced by providing *guard times* between pulses, analogous to the guard bands of FDM. Thus, a practical time-division system will have both guard times and guard bands; the former to suppress cross talk, the latter to facilitate message reconstruction with practical filters.

One more point remains to complete our FDM-TDM comparison, namely, the matter of bandwidth conservation. Consider an FDM system with k inputs bandlimited in W. If the guard bands are small compared to W and SSB modulation is used throughout, the transmission bandwidth will be $B_T = kW$, which is obviously the absolute minimum. But what about TDM? It would appear from Fig. 7.29b and our earlier discussion of PAM that the bandwidth of the multiplexed waveform far exceeds the minimum; i.e., the pulse duration is $\tau \ll 1/kf_s$, so transmission as RF pulses requires a bandwidth of $B_T \approx 1/\tau \gg kW$.

But note that the multiplexed wave is nothing more than a series of periodic sample points, albeit from different messages. Reversing the sampling theorem, these points can be completely described by a *continuous* waveform $x_b(t)$, which has no relation to the original messages save that it passes through the correct sample values at the corresponding sample times (Fig. 7.30a). Then, since the points are spaced in time

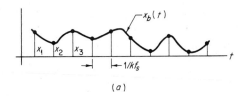

(*a*)

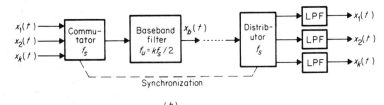

Synchronization

(*b*)

Fig. 7.30 *TDM baseband filtering. (a) Filtered waveform; (b) system diagram.*

by $1/kf_s$, $x_b(t)$ can be bandlimited in $B = kf_s/2$. In fact, $x_b(t)$ is obtained by lowpass (baseband) filtering the multiplexed wave as indicated in Fig. 7.30b. At the receiver, the distributor picks out the original sample points from $x_b(t)$ for reconstruction in the usual fashion.

Thus, if baseband filtering is employed, if the sampling frequency is close to the Nyquist rate ($f_s = 2W$), and if the carrier modulation is SSB, the TDM transmission bandwidth becomes $B_T = kf_s/2 = kW$. Under these conditions TDM can achieve the same minimum bandwidth as FDM, but with loss of cross-talk immunity. TDM systems of this type are designated as PAM-SSB.

This baseband filtering technique is more commonly employed in applications having both TDM and FDM, notably for telemetry systems.[1] The usual arrangement is to combine several slowly varying signals via TDM with baseband filtering, forming a composite wave,

[1] See, for example, Nichols and Rauch (1956).

which is then frequency-division-multiplexed with other signals of comparable bandwidth.

Returning to TDM per se, the interlaced sample values can be directly converted (no baseband filtering) to PDM, PPM, or PCM and transmitted with or without carrier modulation. In fact, it is time-division-multiplexed PCM that holds the most promise for telephony. For noise-reduction purposes, PPM-AM, PCM-FM, etc., are particularly attractive. Note, however, that guard times are quite important in multiplexed PDM and PPM, where the modulated pulses "move around" in their portion of the frame. Another complication is that certain systems, e.g., PCM-FM, have *two* threshold conditions which must be satisfied.

PROBLEMS

7.1 Using Eq. (7.5), describe what will happen to the spectrum $X_s(f)$ in Fig. 7.3b when the duty cycle of the switching function is $d = 1/k$, where k is an integer. Give particular attention to the case $k = 1$.

7.2 The signal $x(t) = \cos 2\pi(W/2)t$ is sampled at $t = 0$, $\pm 1/2W$, $\pm 2/2W$, ... Plot the sample points and convince yourself that no other waveform, bandlimited in W, can be interpolated from these points.

7.3 Graphically determine the sample values when $\sin 2\pi(7W/8)t$ is sampled at $t = 0$, $\pm 1/2W$, $\pm 2/2W$, ... Repeat for $\sin 2\pi Wt$. Explain why the sampling theorem apparently fails in the second case.

7.4 The signal $x(t) = \text{sinc } 2Bt$ is ideally sampled at $t = 0$, $\pm 1/f_s$, $\pm 2/f_s$, ..., where $f_s = 2B + \epsilon$ and $\epsilon \ll B$. Sketch $x_\delta(t)$ and $X_\delta(f)$ and discuss what happens as $\epsilon \to 0$.

7.5 Suppose $x(t)$ is *timelimited* such that $x(t) = 0$ for $|t| \geq T$. Show that its *spectrum* is completely described by specifying the sample values $X(mf_0)$, where $f_0 \leq 1/2T$. *Hint:* See Eq. (7.12) et seq.

7.6 Prove the orthogonality of sinc functions as expressed in (7.16).

7.7 Show that the mean square of the sample points, that is,

$$\overline{x_m^2} = \lim_{M \to \infty} \frac{1}{2M} \sum_{m=-M}^{M} x^2(mT_s)$$

is related to the average power by

$$\overline{x_m^2} = \langle x^2(t) \rangle + \left\langle \sum_{n=1}^{\infty} 2x^2(t) \cos n\omega_s t \right\rangle$$

Hint: Write $x^2(mT_s)$ as $\int_{mT_s-\epsilon}^{mT_s+\epsilon} x^2(t) \xi_m(t)\, dt$, where

$$\xi_m(t) = \lim_{\tau \to 0} \frac{1}{\tau} \Pi\left(\frac{t - mT_s}{\tau}\right)$$

and expand $\sum_{m=-\infty}^{\infty} \xi_m(t)$ in a series of the form of Eq. (7.3).

does $\overline{x_m^2}$ equal $\langle x^2(t) \rangle$?

7.8   Starting with Eq. (7.20), write $S_p(t)$ as an exponential Fourier series in terms of $P(nf_s)$, where $P(f) = \mathfrak{F}[p(t)]$, and thereby show that

$$X_{pn}(f) = f_s \sum_n P(nf_s)X(f - nf_s)$$

Sketch this spectrum, taking $X(f)$ as in Fig. 7.7a.

7.9   A practical sampler has the pulse shape $p(t) = \cos 2\pi(4f_s)t$, $|t| \leq T_s/16$. If the sampling is instantaneous in the sense of Eq. (7.21), find the transfer function $H_{eq}(f)$ of the equalizing filter.

7.10   Two signals, $\cos 2\pi 10t$ and $\cos 2\pi 70t$, are sampled at $f_s = 80$. Find the sample values graphically, and show that both signals will have the same reconstructed waveform. Verify this from the spectra.

7.11   Use the method followed in deriving Eq. (7.5) to obtain the output spectrum of the bipolar chopper shown in Fig. 7.11. Note that the switching function is a square wave with no dc component.

7.12   Show that the chopper system of Fig. P 7.1 combines the operations of matrixing and subcarrier modulation and thereby produces the baseband spectrum required for FM stereo (Fig. 5.46). Note that there are effectively two switching functions having $\tau = T_s/2$ and differing by a time delay of $T_s/2$.

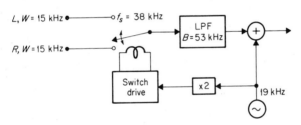

Fig. P 7.1

7.13   Explain why a single-channel PPM system requires a synchronizing signal, whereas PAM and PDM do not.

7.14   A PDM signal has $\tau \ll T_s$ and fixed leading edges at $t = mT_s$. By sketching typical waveforms, show that the device of Fig. P 7.2 will approximately reconstruct the modulating signal if $RC \gg T_s$ and the switch is closed momentarily whenever $t = mT_s$.

Fig. P 7.2

7.15   Derive an equation similar to (7.31) for the case of unipolar PAM wherein $A_{00} \neq 0$. Compare with Eq. (5.29).

7.16   Calculate the output signal-to-noise ratio, in terms of $z = P_T/\eta W$, for a PPM system with $f_s = 2.5W$, $W = 2$ kHz, $\tau = 20$ μsec, pulse displacement ± 50 μsec, and $B_T = 100$ kHz.

7.17 Write an expression equivalent to (7.32) for pulse-time modulation with RF pulses, interpreting P_T and B_T as properties of the RF wave. *Note:* This entails finding t_r and P_T in terms of the RF waveform and does not follow from substitution in (7.32).

7.18 The device shown in Fig. P 7.3 is inserted into a PPM receiver just before the PPM-to-PAM converter. Analyze the operation by sketching typical waveforms, including noise, and estimate the improvement in output signal-to-noise ratio made possible by the device.

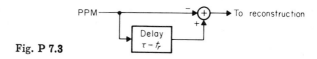

Fig. P 7.3

7.19 Consider a binary PCM signal transmitted as on-off RF pulses. If $f_s = 2.5W$, plot the output tone-to-noise ratio in decibels versus B_T/W for $q = 4, 16, 64$, and 256.

7.20 A voice signal bandlimited in 3 kHz is to be transmitted without carrier modulation over a channel of bandwidth 20 kHz. The output signal-to-noise ratio must be at least $18,000\overline{x^2}$. Design a PCM system to accomplish this task, specifying the values of ν, f_s, q, and μ.

7.21 Any M successive quantized samples having q quantum levels can be represented by *one* number having $Q = q^M$ possible values. This process is called *hyperquantization.* Discuss how PCM with hyperquantization can achieve bandwidth *compression*, that is, $B_T < W$.

7.22 Consider binary PCM transmitted as on-off pulses of amplitude A. The decision crossover level is set at $A/2$. If the noise is gaussian with $\sigma^2 = N$, find the average signal power S such that the decoding error probability is no greater than 10^{-3}. Explain why S can be less if the pulses are *bipolar*.

7.23 Twenty-five voice signals, each bandlimited in 3 kHz, are to be multiplexed in time and transmitted via PAM-AM. Allowing for a 2-kHz reconstruction guard band, determine the minimum transmission bandwidth required. Sketch block diagrams of the transmitter and receiver.

7.24 Four signals bandlimited in W, W, $2W$, and $4W$, respectively, are to be time-division-multiplexed. Devise a commutator configuration such that each signal is periodically sampled at its own Nyquist rate and the sample values are properly interlaced.

7.25 Ten signals, each bandlimited in 4 kHz, are to be transmitted via TDM-PPM. Taking $\tau = 1$ μsec and allowing a guard time of 2 μsec and reconstruction guard bands of 2 kHz, calculate to the maximum possible displacement per pulse. Estimate the resulting reduction in output signal-to-noise ratio per channel, compared to a single-channel PPM system having the same pulse duration, same guard band, one-tenth the average power, and one-tenth the transmission bandwidth.

7.26 Discuss the merits of TDM versus FDM when the transmission medium is subject to *selective fading;* i.e., a narrow band of frequencies suffers severe attenuation compared to the rest of the band. Consider two cases: (*a*) when the fading is slow and prolonged compared to message transmission time, and (*b*) when the fading is of relatively short duration.

SELECTED SUPPLEMENTARY READING

Though somewhat dated, Black (1953, chaps. 4, 5, and 15 to 20) has the best overall coverage of sampling and pulse modulation. The treatment of PCM is primarily qualitative but contains a wealth of practical information on equipment and allied problems. Advanced discussions of various topics can be found in: Linden (1959), who examines sampling theory; Rowe (1965, chap. 4), who gives spectral analysis of analog pulse modulation; Bennett (1948), who tackles the awesome problem of spectra for quantized signals; and Panter (1965, chap. 18), who gives cross-talk calculations in TDM.

In the professional literature two papers are of particular interest, Shannon (1949), and Oliver, Pierce, and Shannon (1948). The former was, among other things, the first exposition of sampling theory applied to electrical communication. The latter discusses the philosophy of PCM. Both papers are easy to read and strongly recommended.

8
Information Theory and Communication Systems

The past several chapters have dealt with electrical communication primarily in terms of signals, both desired and undesired. We have devised signal models, examined the effects of networks on signals, and analyzed modulation as a means of signal transmission. While many rewards and much insight have been gained by this approach, signal theory alone is not sufficient for a complete understanding of our subject matter, particularly when it comes to the design of new and improved systems. What is needed is a more encompassing view of the communication process, a broader perspective leading to basic principles for system design and comparison—in short, a general theory of communication.

Prior to the 1940s a few steps were taken toward such a theory, notably the investigations of telegraphy by Nyquist and Hartley. But in 1948, spurred by technological advances of World War II, two works of major impact appeared, Norbert Wiener's "Cybernetics," and A Mathematical Theory of Communication by Claude E. Shannon. Taken together, the ideas of Wiener and Shannon established the foundation of modern (statistical) *communication theory*. Both men were concerned with extracting information from a background of noise, and both applied statistical concepts to the problem. There were, however, differences in emphasis.

Wiener treated the case where the information-bearing signals are beyond the designer's control, in whole or part, all the processing being at the receiving end. (Radar is a good example of this situation, since the nature of the reflected signal depends largely on the shape and flight path of the target.) The problem then can be stated in this fashion: given the set of possible signals, not of our choosing, plus the inevitable noise, how do we make the best estimate of the present and future values of the signal being received? Optimum solutions to this and similar problems are sought in the disciplines known as *detection theory* and *statistical decision theory*.

Shannon's work is more nearly akin to what we think of as communication, where signal processing can take place at both transmitter and receiver. In a communication system the information to be transferred is the prerogative of the source, but the way in which it is actually transmitted—the type of signal used—is within the designer's control. Shannon therefore posed this problem: given the set of possible messages a source may produce, not of our choosing, how shall the messages be represented so as best to convey the information over a given system with its inherent physical limitations? To handle this problem in quite general terms it is necessary to concentrate more on the *information* per se than on the signals, and Shannon's approach was soon rechristened *information theory*.

Information theory is a mathematical subject dealing with three basic concepts: the measure of information, the capacity of a communication channel to transfer information, and coding as a means of utilizing channels at full capacity. These concepts are tied together in what can be called the *fundamental theorem* of information theory, as follows.

Theorem: Given an information source and a communication channel, there exists a coding technique such that the information can be transmitted over the channel at any rate less than the channel capacity and with arbitrarily small frequency of errors despite the presence of noise.

The surprising, almost astonishing aspect of this theorem is *error-free* transmission on a *noisy* channel, a condition which can be achieved through the use of coding. In essence, coding is used to match the source and channel for maximum reliable information transfer, roughly analogous to impedance matching for maximum power transfer.

But the study of coding is, by and large, tangential to our immediate aims.   Thus, with some reluctance, we limit this chapter primarily to the concepts of information measure and channel capacity, with emphasis on the latter.   By so doing we shall eventually arrive at answers to these significant questions:

1. Precisely how do the fundamental physical limitations (i.e., bandwidth and noise) restrict information transmission?
2. Is there such a thing as an *ideal* communication system, and, if so, what are its characteristics?
3. How well do existing communication systems measure up to the ideal, and how can their performance be improved?

Answers to these questions are certainly germane to electrical communication.   They will be explored in some detail at the close of the chapter. But we must begin with information theory.

8.1 INFORMATION MEASURE: ENTROPY

The crux of information theory is the measure of information.   Here we are using *information* as a technical term, not to be confused with its more conventional interpretations.   In particular, the information of information theory has little to do with knowledge or meaning, concepts which defy precise definition, to say nothing of quantitative measurement.

In the context of communication, information is simply that which is produced by the source for transfer to the user.   This implies that before transmission, the information was not available at the destination; otherwise the transfer would be zero.   Pursuing this line of reasoning, consider the following somewhat contrived situation.

A man is planning a trip to Chicago.   To determine what clothes he should pack, he telephones the Chicago weather bureau and receives one of the following forecasts:

The sun will rise.
It will rain.
There will be a tornado.

Clearly, the amount of information gained from these messages is quite different.   The first contains virtually no information, since we are reasonably sure in advance that the sun will rise; there is no uncertainty about this, and the call has been wasted.   But the forecast of rain does provide information not previously available to the traveler, for rain is not an everyday occurrence. The third forecast contains even more information, tornadoes being relatively rare and unexpected events.

Note that the messages have been listed in order of decreasing likelihood and increasing information. The less likely the message, the more information it conveys to the user. We are thus inclined to say that information measure is related to *uncertainty*, the uncertainty of the user as to what the message will be. Moreover, the amount of information depends only on the message uncertainty, rather than its actual content or possible interpretations. Had the Chicago weather forecast been "The sun will rain tornadoes," it would convey much information, being quite unlikely, but not much meaning.

Alternately, going to the transmitting end of a communication system, information measure is an indication of the *freedom of choice* exercised by the source in selecting a message. If the source can freely choose from many different messages, the user is highly uncertain as to which message will be selected. But if there is no choice at all, only one possible message, there is no uncertainty and hence no information.

Whether one prefers the uncertainty viewpoint or the freedom-of-choice interpretation, it is evident that the measure of information involves *probabilities*. Messages of high probability, indicating little uncertainty on the part of the user or little choice on the part of the source, convey a small amount of information, and vice versa. This notion is formalized by defining self-information in terms of probability.

Self-information

Consider a source which produces various messages. Let one of the messages be designated A, and let P_A be the probability that A is selected for transmission. Consistent with our discussion above, we write the self-information associated with A as

$$I_A = f(P_A)$$

where the function $f(\ )$ is to be determined.

As a step toward finding $f(\ )$, intuitive reasoning suggests that the following requirements be imposed: (1) the measure of self-information should be a positive real number—there is no reason for it to be otherwise; (2) if $P_A = 1$, the message is certain (like the rising of the sun) and carries no information, so $I_A = 0$; (3) if message A is less likely than message B, $P_A < P_B$, then we have $I_A > I_B$, since information should increase monotonically with uncertainty. Symbolically, and in terms of $f(\ )$, these conditions are

$$f(P_A) \geq 0 \qquad \text{where } 0 \leq P_A \leq 1 \tag{8.1}$$

$$\lim_{P_A \to 1} f(P_A) = 0 \tag{8.2}$$

$$f(P_A) > f(P_B) \qquad \text{for } P_A < P_B \tag{8.3}$$

Many functions satisfy (8.1) to (8.3).   The final and deciding factor comes from considering the transmission of *independent* messages.   When a message A is delivered, the user receives I_A units of information.   If a second message B is also delivered, the total information received should be the sum of the self-informations, $I_A + I_B$.   This summation rule is readily appreciated if we think of A and B as coming from different sources.

But suppose both messages come from the same source; we can then speak of the compound message $C = AB$.   If A and B are *statistically independent*, $P_C = P_A P_B$ and $I_C = f(P_A P_B)$.   But the received information is still $I_A + I_B$, so

$$I_C = I_A + I_B = f(P_A) + f(P_B)$$

and therefore

$$f(P_A P_B) = f(P_A) + f(P_B) \tag{8.4}$$

which is our final requirement for $f(\ )$.

There is one and only one function satisfying the conditions (8.1) to (8.4), namely, the *logarithmic function* $f(\ ) = -K \log_b (\ )$, where K is a positive constant and b is the logarithmic base.   Setting $K = 1$ for simplicity, self-information is defined as

$$I_A = -\log_b P_A = \log_b \frac{1}{P_A} \tag{8.5}$$

where b is unspecified for the moment.   The minus sign in $-\log_b P_A$ is perhaps disturbing at first glance.   But, since probabilities are bounded by $0 \leq P_A \leq 1$, the negative of the logarithm is positive, as desired. The alternate form $\log_b (1/P_A)$ helps avoid confusion on this score, and will be used throughout.

We have not proved that (8.5) is indeed the *only* function satisfying the imposed requirements, for the proof[1] is rather tedious.   However, the reader can verify that I_A, so defined, does satisfy (8.1) to (8.4) and agrees nicely with our notion of information and uncertainty.

Specifying the logarithmic base b is equivalent to selecting the *unit* of information.   While common or natural logarithms ($b = 10$ or $b = e$)

[1] See Ash (1965, chap. 1).

seem obvious candidates, the standard convention of information theory is to take $b = 2$. The corresponding unit of information is termed the *bit*, a contraction for *binary digit* suggested by J. W. Tukey. Thus

$$I_A = \log_2 \frac{1}{P_A} \qquad \text{bits}$$

The reasoning behind this rather strange convention goes like this. Information is a measure of choice exercised by the source; the simplest possible choice is that between two equiprobable messages, i.e., an unbiased binary choice. The information unit is therefore normalized to this lowest-order situation, and 1 bit of information is the amount required or conveyed by the choice between two equally likely possibilities; i.e., if $P_A = P_B = \frac{1}{2}$, then $I_A = I_B = \log_2 2 = 1$ bit.

Binary *digits* enter the picture simply because any two things can be represented by the two binary digits 0 and 1. Note, however, that 1 binary digit may convey more or less than 1 bit of information, depending on the probabilities. To prevent misinterpretation, binary digits as message elements are called *binits* in this chapter.

Since tables of base 2 logarithms are relatively uncommon, the following general conversion relationship is needed

$$\log_a u = \log_a b \, \log_b u = \frac{\log_b u}{\log_b a} \tag{8.6}$$

Hence, for numerical calculations,

$$\log_2 u = \log_2 10 \, \log_{10} u = 3.32 \log_{10} u \tag{8.7}$$

Thus, if $P_A = \frac{1}{10}$, $I_A = 3.32 \log_{10} 10 = 3.32$ bits.

In the remainder of this chapter, all logarithms will be base 2 unless otherwise indicated.

Example 8.1

The above discussion was couched in terms of messages and is therefore applicable to any type of information source whose possible messages can be enumerated. In the particular case of a discrete source, a source producing discrete symbols which go to make up the messages, it is possible to speak in terms of the self-information of the symbols themselves. To illustrate, suppose a source produces the symbols A, B, C, and D, with probabilities $\frac{1}{2}$, $\frac{1}{4}$, $\frac{1}{8}$, and $\frac{1}{8}$, respectively. (If we have listed all the possible symbols, the probabilities must of course sum to 1.) Since the probabilities are of the form $1/2^n$ in this example, self-information is readily calculated, namely, $\log 2^n = n$, and is tabulated below.

Symbol	P	$I = \log \dfrac{1}{P}$ bits
A	$\frac{1}{2}$	1
B	$\frac{1}{4}$	2
C	$\frac{1}{8}$	3
D	$\frac{1}{8}$	3

If the symbols are independent, the four-symbol message $X = BACA$ should then provide $2 + 1 + 3 + 1 = 7$ bits of information. Checking this result, $P_X = P_B P_A P_C P_A = 1/128 = 1/2^7$ and $I_X = \log 2^7 = 7$ bits.

Example 8.2 The information in a picture

It has often been said that one picture is worth a thousand words. With a little stretching, information measure supports this old saying.

For analysis we decompose the picture into a number of discrete dots, or elements, each element having a brightness level ranging in steps from black to white. The standard television image, for instance, has about $500 \times 600 = 3 \times 10^5$ elements and 8 distinguishable levels. Hence, there are $8 \times 8 \times \cdots = 8^{3 \times 10^5}$ possible pictures, each with probability $P = 8^{-(3 \times 10^5)}$ if selected at random. Therefore

$$I = \log 8^{3 \times 10^5} = 3 \times 10^5 \log 8 \approx 10^6 \text{ bits}$$

Alternately, assuming the levels to be equally likely, the information per element is $\log 8 = 3$ bits, for a total of $3 \times 10^5 \times 3 \approx 10^6$ bits, as before.

But what about the thousand words? Suppose, for the sake of argument, that a vocabulary consists of 100,000 equally likely words. The probability of any one word is then $P = 10^{-5}$, so the information contained in 1,000 words is

$$I = 1,000 \log 10^5 = 10^3 \times 3.32 \log_{10} 10^5 \approx 2 \times 10^4 \text{ bits}$$

or substantially less than the information in one picture.

The validity of the above assumptions is of course open to question; the point of this example is the method, not the results.

Entropy and information rate

Self-information is defined in terms of the individual messages or symbols a source may produce. It is not, however, a useful description of the source relative to communication. A communication system is not designed around a particular message but rather all possible messages, i.e., what the source could produce as distinguished from what it does produce on a given occasion. To describe the source we therefore must deal with the totality, or ensemble, of possible messages.

An interesting parallel comes to mind here. The instantaneous power of an electric signal is a function of time, continuously fluctuating. Yet we found it both useful and meaningful to speak of the average power. Similarly, though the instantaneous information flow from a source may be erratic, one might describe the source in terms of the *average information* produced. This average information is called the source *entropy*.

For a discrete source whose symbols are *statistically independent*, the entropy expression is easily formulated. Let m be the number of different symbols; the source is then said to have an *alphabet of size m*. When the jth symbol is transmitted, it conveys $I_j = \log (1/P_j)$ bits of information. In a long message of $N \gg 1$ symbols, the jth symbol occurs about NP_j times, and the total information in the message is approximately

$$NP_1I_1 + NP_2I_2 + \cdots + NP_mI_m = \sum_{j=1}^{m} NP_jI_j \qquad \text{bits}$$

Since there are N symbols all told, the average information per symbol is $\dfrac{1}{N} \sum_j NP_jI_j = \sum_j P_jI_j$. We therefore define the entropy of a discrete source as

$$H = \sum_{j=1}^{m} P_jI_j = \sum_{j=1}^{m} P_j \log \frac{1}{P_j} \qquad \text{bits/symbol} \tag{8.8}$$

It should be observed that (8.8) is a *statistical*, or ensemble, average. If the source is nonstationary, the symbol probabilities may change with time, and the entropy is not very meaningful. We shall henceforth assume that information sources are *ergodic*, so that time and ensemble averages are identical.

The name *entropy* and its symbol H are borrowed from a similar equation in statistical mechanics. Because of the mathematical similarity, various attempts have been made to relate communication entropy with thermodynamic entropy.[1] However, the attempted relationships

[1] See Brillouin (1956).

seem to cause more confusion than illumination, and it is perhaps wiser to treat the two entropies as different things with the same name. For this reason the alternate designation *comentropy* has been suggested for communication entropy.

But what is the meaning of communication entropy as written in Eq. (8.8)? Simply this: although one cannot say which symbol the source will produce next, on the average we expect to get H bits of information per symbol or NH bits in a message of N symbols, if N is large.

For a fixed alphabet size (fixed m) the entropy of a discrete source depends on the symbol probabilities but is bounded by

$$0 \leq H \leq \log m \tag{8.9}$$

These extreme limits are readily interpreted and warrant further discussion.

The lower limit, $H = 0$, implies that the source delivers no information (on the average), and hence there is no uncertainty about the message. We would expect this to correspond to a source that continually produces the same symbol; i.e., all symbol probabilities are zero, save for one symbol having $P = 1$. It is easily shown that

$$\lim_{P \to 0} P \log \frac{1}{P} = 0 \qquad \lim_{P \to 1} P \log \frac{1}{P} = 0$$

and hence if $P_1 = 1$ and $P_j = 0$ for $j \neq 1$, then

$$H = P_1 \log \frac{1}{P_1} + \sum_{j=2}^{m} P_j \log \frac{1}{P_j} = 0$$

At the other extreme, the maximum entropy must correspond to maximum uncertainty or maximum freedom of choice. This implies that all symbols are equally likely; there is no bias, no preferred symbol. A little further thought reveals that the symbol probabilities must be the same; i.e., all $P_j = 1/m$ since $\sum_{j=1}^{m} P_j = 1$. Thus, $H = \Sigma(1/m) \log m = \log m$. This result, $H_{\max} = \log m$, has particular significance for our later work.

The variation of H between the limits of (8.9) is best illustrated by considering a binary source ($m = 2$). The symbol probabilities are then related and can be written as p and $q = 1 - p$. Thus

$$H = p \log \frac{1}{p} + (1 - p) \log \frac{1}{1 - p}$$

which is plotted versus p in Fig. 8.1. Note the rather broad maximum centered at $p = 0.5$, the equally likely case, where $H = \log 2 = 1$ bit.

Bringing the time element into the picture, suppose two sources have equal entropies but one is "faster" than the other, producing more symbols per unit time. In a given period, more information must be transferred from the faster source than from the slower, which obviously places greater demands on the communication system. Thus, for our

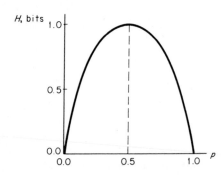

Fig. 8.1 *Entropy of a binary source,* $H = p \log (1/p) + q \log (1/q)$, $q = 1 - p$.

purposes, the description of a source is not its entropy alone, in bits per symbol, but its *entropy rate*, or information rate, in bits per second. The entropy rate of a discrete source is simply defined as

$$R = rH \qquad \text{bits/sec} \qquad (8.10)$$

where r is the symbol rate, i.e., the average number of symbols per second.

Having discussed discrete sources at some length, the next logical step would be the definition of entropy for *continuous* sources, sources whose messages are continuously varying functions of time. Such a definition is possible but will not be presented here. For one reason, the mathematics gets rather complicated and tends to obscure physical interpretation; for another, the entropy of a continuous source turns out to be a relative measure instead of an absolute measure of information.

Fortunately, the goals of this chapter can be achieved by sticking to the discrete formulation, and most of our conclusions will apply to continuous sources with but minor modification. To illustrate, a band-limited time function produced by a continuous source can be described to any reasonable degree of accuracy by quantized sample values. And sampling plus quantizing converts the output from continuous to discrete, at least for purposes of analysis.

But more important, we shall find that because of the fundamental physical limitations, communication is inherently a *discrete process* regardless of the source. This striking conclusion is one of Shannon's principal contributions to the theory of communication, but it was noted by Hartley as far back as 1928.

Example 8.3

Applying Eq. (8.8) to the source in Example 8.1, the entropy is

$$H = \tfrac{1}{2} \log 2 + \tfrac{1}{4} \log 4 + \tfrac{1}{8} \log 8 + \tfrac{1}{8} \log 8$$
$$= 1.75 \text{ bits/symbol}$$

Note that $I_A = 1$ bit is less than H, while $I_C = 3$ bits is greater than H.

Had the symbols been equally likely, we would have had the maximum entropy for a four-symbol source, namely, $H_{\max} = \log 4 = 2$ bits/symbol. Under this condition, all symbols have the same amount of self-information, and $I_j = H_{\max}$.

Example 8.4

As an example of entropy applied to our earlier studies, consider a PCM system whose input is the continuous signal $x(t)$ bandlimited in $W = 50$ Hz. Suppose $x(t)$ is sampled at the Nyquist rate $f_s = 2W$, and let there be 4 quantum levels such that the quantized values have probabilities $1/2, 1/4, 1/8$, and $1/8$. Identifying each possible quantized value as a "symbol," the output of the quantizer then looks like a discrete source with $m = 4$, entropy $H = 1.75$ bits/symbol, symbol rate $r = 2W = 100$ symbols/sec, and information rate $R = rH = 175$ bits/sec. Thus, it should be possible to represent this same information by equiprobable binary digits (binits) generated at a rate of 175 binits/sec.

To check this idea, suppose that the system is in fact *binary* PCM with the quantized samples transmitted as coded binary pulses. Labeling the quantum levels A, B, C, and D, let us assume the following coding:

Quantized value	Probability	Binary code
A	$\tfrac{1}{2}$	00
B	$\tfrac{1}{4}$	01
C	$\tfrac{1}{8}$	10
D	$\tfrac{1}{8}$	11

Since there are 2 binary digits for each sample, the PCM pulse rate is 200 pulses per second. Now we argued that 175 binary digits per second should be sufficient; yet 200 per second is required with this code.

The anomaly is quickly resolved by noting that the binary digits are not equally likely; in fact, $P_0 = {}^{11}\!/_{16}$ and $P_1 = {}^{5}\!/_{16}$, as the reader can verify. Clearly, the suggested code is not optimum. On the other hand, it is simple and reasonably efficient.

Pressing onward, the binit rate and probabilities given above suggest that the information rate at the encoder output is

$$R' = 200 \; (^{11}\!/_{16} \log {}^{16}\!/_{11} + {}^{5}\!/_{16} \log {}^{16}\!/_{5})$$
$$= 200 \times 0.897 = 179 \text{ bits/sec}$$

Again something is wrong, for the information rate into the encoder is only 175 bits/sec. Surely direct encoding does not *add* information!

To explain the discrepancy we must recall that the entropy equation (8.8) is based on statistically *independent* symbols. And, while it may be true that successive quantized samples are independent, the successive binary pulses are not, because we have encoded in groups of two. For the case of dependent symbols, we must modify the measure of information and introduce conditional entropy.

Conditional entropy and redundancy

Discrete sources are often constrained by certain rules which limit the choice in selecting successive symbols. The resulting *intersymbol influence* reduces uncertainty and thereby reduces the amount of information produced. We account for this effect by using conditional probabilities and conditional entropy.

Written text, being governed by rules of spelling and grammar, is a good example of intersymbol influence. On a relative-frequency basis, the probability[1] of U in printed English is $P_U = 0.02$; but if the previous letter is Q, the conditional probability of U given Q is $P(U|Q) \approx 1$, whereas in contrast $P(U|W) < 0.001$. The influence may well extend over several symbols, phrases, or even complete sentences, as illustrated by the fact that in most textbooks, including this one, $P(\text{THAT}|\text{IT CAN BE SHOWN}) \approx 1$.

The expression for conditional entropy therefore is formulated by considering the entire past history of the source—more precisely, all possible past histories. Thus, if j represents the next symbol (or group of symbols) and i represents the preceding sequence, the information conveyed by j given i is $\log [1/P(j|i)]$. Averaging over all j's and i's gives the conditional entropy

$$H_c = \sum_i \sum_j P_i P(j|i) \log \frac{1}{P(j|i)} \tag{8.11}$$

---

[1] Letter frequencies have long been known to cryptographers; see Pratt (1942), the standard reference.

In general $H_c \leq H$; the equality applies only when the symbols are independent and $P(j|i) = P_j$.

A source producing dependent symbols is said to be *redundant*, meaning that symbols are generated which are not absolutely essential to convey the information.   (Is it really necessary to explicitly write the U following every Q in English?)   Redundancy, usually expressed as a percentage, is defined by

$$E = 1 - \frac{H_c}{H_{\max}} \tag{8.12}$$

and $H_c/H_{\max}$ is known as the *relative entropy*.

From the viewpoint of efficient communication, redundancy in a message is undesirable; the same information could be sent with fewer nonredundant (independent) symbols.   Thus, coding to reduce inter-symbol influence is a method of improving efficiency.   On the other hand, redundancy is a definite aid in resolving ambiguities if the message is received with *errors*, a not uncommon phenomenon in telegraphy, for example.   Indeed, coding for error protection is based on the insertion of redundant symbols.

Optimum transmission therefore entails coding to reduce the inefficient redundancy of the message, plus coding to add "efficient" redundancy for error control.   Much has been done in the area of error-detecting and error-correcting codes, but coding to reduce message redundancy is far more difficult, and relatively little has been accomplished in this direction.   Oliver (1952) gives an excellent discussion of the practical problems and some implementations.

Example 8.5   The entropy of English text

The study of written language is very important in information theory and sheds considerable light on human communication.[1]   For this purpose it is convenient to deal with *telegraph language*, where the only symbols are the letters of the alphabet plus space.

In the case of English, $m = 27$ and

$$H_{\max} = \log 27 = 4.8 \text{ bits/symbol}$$

Of course some symbols are known to be more frequent than others. The probabilities range from $P_{\text{space}} = 0.19$ and $P_E = 0.10$ down to $P_Z = 0.0005$ in representative passages.   Based on these unconditioned

[1] See Shannon (1951).

probabilities, the entropy of typical English text is

$$H = 4.1 \text{ bits/symbol}$$

a value not too different from H_{max}.

However, accounting for intersymbol influence extending over 8-symbol groups, the conditional entropy is found to be about 2 bits/symbol. Thus the redundancy of English is $E \approx 1 - (2/4.8)$, or roughly 50 percent. This implies that in the long run, half the symbols are unnecessary: yu shld babl t read ths evntho sevrl ltrs r msng. The reader may wish to ponder the observation that without redundancy, abbreviation would be impossible, and any two-dimensional array of letters would form a valid crossword puzzle.

For very long passages of printed English, the conditional entropy may be as low as 0.5 to 1.0 bits/symbol because of contextual inferences. Thus, with suitable coding, printed English theoretically could be transmitted in binary form with an average of 1 binary digit per symbol. Contrast this with existing teletype systems that use 5 binary digits per character. (However, there are 32 characters on the teletype keyboard.)

8.2 INFORMATION TRANSMISSION: CHANNEL CAPACITY

We mentioned in Chap. 1 that it is often convenient to treat the terminal equipment of a communication system as being perfect (noise-free, distortionless, etc.) and think of all undesired effects as taking place in the transmission process from transmitter to receiver. With this interpretation, the term *channel* used in information theory is synonymous with what we have previously called the transmission medium. The communication channel is therefore an abstraction, a model representing the vehicle of transmission plus all phenomena which tend to restrict transmission. The fact that there are fundamental physical limitations to information transfer by electrical means leads to the notion of *channel capacity*.

Just as entropy rate measures the amount of information produced by a source in a given time, capacity is a measure of the amount of information a channel can transfer per unit time. Channel capacity is symbolized by C, and its units are bits per second. Restating the fundamental theorem in terms of R and C, we have:

Theorem: Given a channel of capacity C and a source having entropy rate R, then if $R \leq C$, there exists a coding technique such that the output of the source can be transmitted over the channel with an arbitrarily small frequency of errors, despite the presence of noise. If $R > C$, it is not possible to transmit without errors.

Although we shall attempt to make the theorem plausible, its proof involves a great deal of coding theory and is omitted here.[1]   Instead, we shall concentrate on aspects which are more pertinent to electrical communication, particularly the relationship of channel capacity and system parameters.

Channel capacity

The fundamental theorem implicitly defines channel capacity as the maximum rate at which the channel supplies reliable information to the destination.   With this interpretation in mind we can formulate a general expression for capacity by means of the following argument.

Consider all the different messages, T sec in length, a source might produce.   If the channel is noisy, it will be difficult to decide at the receiver which particular message was intended, and the goal of information transfer, i.e., reduction of uncertainty, is partially defeated.   But suppose we restrict the messages to only those which are "very different" from each other, such that the received message can be correctly identified with sufficiently small probability of error.   Let $M(T)$ be the number of these very different messages of length T.

Now, insofar as the destination or user is concerned, the source-plus-channel combination may be regarded as a new source generating messages at the receiving end.   With the above message restriction, this equivalent source is discrete and has an alphabet of size $M(T)$.   Correspondingly, the maximum entropy produced by the equivalent source is log $M(T)$, and the maximum entropy rate at the destination is $1/T$ log $M(T)$.   Hence, letting $T \to \infty$ to ensure generality,

$$C = \lim_{T \to \infty} \frac{1}{T} \log M(T) \qquad \text{bits/sec} \tag{8.13}$$

becomes an alternate definition for channel capacity.   The following discussion of discrete channels shows that (8.13) is an intuitively meaningful definition.

Discrete channels

A discrete channel is one which transmits information by successively assuming various disjoint electrical states—voltage levels, instantaneous frequency, etc.   Each state or channel symbol has associated with it a minimum time duration that may differ from symbol to symbol.   However, if all durations are equal, as is often true, then the channel has a uniform signaling speed[2] given by the reciprocal of the symbol duration.

[1] See Shannon (1948), Abramson (1963), Harman (1963), etc., for the details.
[2] Signaling speed is measured in *bauds,* the speed in bauds equaling the number of channel symbols per second.

Consider now a discrete channel having μ states and uniform signaling speed $s = 1/\tau$, τ being the duration per state. If the signal-to-noise ratio is sufficiently large, the error probability at the receiver can be extremely small, so small that to all intents and purposes, the channel is deemed to be noiseless. Under this assumption, any sequence of symbols will be correctly identified, and the capacity calculation is straightforward, as follows.

A received message of length T will consist of $T/\tau = sT$ symbols, each symbol being one of the μ possible states. The number of different messages is thus $M(T) = \mu^{sT}$, and hence

$$
\begin{aligned}
C &= \lim_{T \to \infty} \frac{1}{T} \log \mu^{sT} = \lim_{T \to \infty} \frac{sT}{T} \log \mu \\
&= s \log \mu \qquad \text{bits/sec}
\end{aligned}
\tag{8.14}
$$

The capacity of a noiseless discrete channel is therefore proportional to the signaling speed and the logarithm of the number of states. For a binary channel ($\mu = 2$) the capacity is numerically equal to the signaling speed, that is, $C = s$.

According to (8.14), one can double channel capacity by doubling the signaling speed (which is certainly reasonable) or by *squaring* the number of states (somewhat more subtle to appreciate). In regard to the latter, suppose two identical but independent channels are operated in parallel so their combined capacity is clearly $2(s \log \mu) = 2s \log \mu = s \log \mu^2$. At the output we can say that we are receiving $2s$ symbols/sec, each symbol being drawn from an alphabet of size μ; or the output can be viewed as s *compound* symbols per second, each compound symbol being drawn from an alphabet of size $\mu \times \mu = \mu^2$.

When channel noise cannot be ignored, the capacity is less than $s \log \mu$ because of the errors. We calculate the capacity reduction by thinking of a fictitious *error compensator* (Fig. 8.2), which examines the

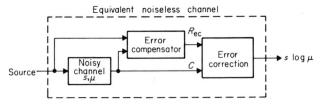

Fig. 8.2

channel input and output and tells us what corrections should be made. Let the channel be operating at its maximum of C bits/sec, and let the information rate supplied by the compensator be R_{ec} bits/sec. Then, since the noisy channel plus compensator is equivalent to the same channel without noise, the net information rate over the noisy channel is $s \log \mu$ minus the information rate from the compensator, that is,

$$C = s \log \mu - R_{ec} \tag{8.15}$$

If $R_{ec} < s \log \mu$, the fundamental theorem asserts that it is possible to get a nonzero rate of virtually errorless information at the channel output.

Examples 8.6 and 8.7 below demonstrate the validity of the fundamental theorem for two particular (and carefully selected) cases. Example 8.7 also shows that (8.15) agrees with the $M(T)$ capacity expression, Eq. (8.13).

Example 8.6

Again consider the source in Example 8.1, taking the source rate to be $r = 100$ symbols/sec so $R = rH = 175$ bits/sec. In theory a noiseless binary channel will suffice to convey the information if the signaling speed is $s \geq 175$ binits/sec.

If minimum signaling speed is desired, we cannot use the code suggested in Example 8.4, because it requires 2 binary digits per source symbol and $s = 200$. A more efficient code is as follows.

Source symbol	Probability	Code
A	$\frac{1}{2}$	0
B	$\frac{1}{4}$	10
C	$\frac{1}{8}$	110
D	$\frac{1}{8}$	111

In a long message of N symbols there will be approximately $N/2$ A's, $N/4$ B's, etc. Hence, using the above code, N successive source symbols require the transmission of $N/2 + 2(N/4) + 3(N/8) + 3(N/8) = 1.75N$ channel symbols, that is, 1.75 binary digits per source symbol. The required signaling speed is then $s = 1.75r = 175$, and we have achieved transmission at $R = C$; the encoding has produced a perfect match between source and channel.

Although this example is admittedly a special case, there are some general hints about efficient encoding to be gained from it. First, the code is such that the channel symbols 0 and 1 are equally likely and statistically independent. (The skeptical reader should verify this.) Second, the source symbol with the highest probability is assigned the shortest code, and so forth, down the line to the least probable symbol, which gets the longest code. Third, the code is uniquely decipherable without spaces or commas separating the code groups; that is, 0101100111 can only mean *ABCAD*. Systematic procedures for devising such codes are described in the literature, e.g., Abramson (1963, chap. 4).

Somewhat parenthetically we might note that assigning shorter code groups to the more probable symbols is just common sense. Over a century ago, long before Shannon, Samuel Morse constructed his telegraph code using this very principle, representing the letter E by a single dot, etc. Lacking the necessary data, Morse estimated letter frequencies by counting the distribution of type in a printer's font.

Example 8.7

Consider a binary channel with noise that affects the channel symbols in blocks of three such that a block is either received without error or there is exactly one error in the first, second, or third symbol. These four possibilities are equally likely.

If an error compensator were available, it would merely have to indicate which of the four situations applied to a given block. (If we know which binary digit is wrong, then we know what it should be; this is the unique advantage of binary codes.) Since the four possible "messages" from the compensator are equally likely, the information provided by it is log 4 = 2 bits/block, and for a channel signaling speed of s binits/sec, there are $s/3$ blocks/sec, so $R_{ec} = s/3 \times 2 = \frac{2}{3}s$ bits/sec. Thus the channel capacity is $C = s - \frac{2}{3}s = s/3$ bits/sec, or one-third of the capacity without noise.

Can we in fact achieve error-free transmission at $R = s/3$ bits/sec? Absolutely, if messages are represented by blocks of 3 identical binary digits, namely, 000 and 111. Instead of transmitting 01011, for example, we must transmit 000111000111111. A single error in any block can be immediately identified and corrected using a majority-rule decision: 010 received must be 000 with one error, 110 received is 111 with one error, etc. (Note the use of redundancy for *error correction*.) Since each block can convey 1 bit of information and there are $s/3$ blocks/sec, $R = s/3 \times 1 = s/3$ bits/sec $= C$, as desired.

In essence we are saying that in time $T = 3\tau = 3/s$ there are only $M(T) = 2$ "very different" messages instead of the $2^{sT} = 2^3 = 8$ which would be allowed in the absence of noise. Thus $C = 1/T \log M(T) = s/3$ log 2 = $s/3$, as calculated above.

Like the previous example, this one was artificially contrived for purposes of illustration. It would be more realistic to allow for independent errors in any or all of the transmitted symbols or errors occurring in bursts. Error-detection and error-correction codes of more general use are explored in Chap. 9.

Channel capacity and channel parameters

The capacity of a discrete channel depends on signaling speed and the number of states, yet the physical limitations to transmission in electrical form were said to be bandwidth and noise. Here we attempt to link channel capacity and channel parameters. It will be shown that signaling speed s is related to bandwidth B; attempting to signal at $s > 2B$ generally results in overlapping waveforms and hence *intersymbol interference*, not to be confused with the *intersymbol influence* of a redundant source. Similarly the number of states μ is dictated by the signal-to-noise ratio; attempting to increase μ without a suitable S/N increase generally results in excessive *errors*. For simplicity we deal with the *baseband* channel, a channel with a lowpass frequency response and whose states are simply voltage levels. The signaling wave thus consists of pulses having quantized amplitudes. Our baseband results can be extended to other channel types by modulation analysis.

Signaling speed has been studied since the earliest days of telegraphy, but it was Harry Nyquist (1924, 1928) who first stated the bandwidth–signaling speed relationship:

> Given an ideal lowpass channel of bandwidth B, it is possible to send independent symbols at a rate $s \leq 2B$ without intersymbol interference. It is not possible to send independent symbols at $s > 2B$.

This relationship may have a familiar ring, and well it might since it is essentially the sampling theorem turned inside out. The Nyquist sampling rate ($f_s = 2W$) is analogous to the Nyquist signaling rate[1] ($s = 2B$). Note also that since $s = 1/\tau$, $s \leq 2B$ agrees with the pulse-resolution requirement $B \geq 1/2\tau_{\min}$ of Sec. 3.4.

It is an easy matter to prove the second part of the relationship, for suppose we try to signal at $2(B + \epsilon)$ symbols/sec, ϵ being positive but arbitrarily small. One possible message sequence we may be required to send consists of two symbols alternating indefinitely, $01010101 \cdots$, for example. The resulting channel waveform is *periodic* with period $1/(B + \epsilon)$ and contains only the fundamental frequency $f_0 = B + \epsilon$ plus its harmonics. Since no frequency greater than B is passed by the

---

[1] Certain methods for signaling at greater than the Nyquist rate have been proposed, but closer inspection shows they do not violate the above theorem. See Bennett and Davey (1965, chap. 7) or Black (1953, chap. 7) for further details.

channel, the channel output will be zero aside from a possible but useless dc component.

That we can signal at $s = 2B$ was implied by our study of sampling reconstruction and Fig. 7.5. Specifically, if the channel has an ideal response over $|f| \leq B$ and impulses of spacing $1/2B$ are applied at the input, they appear at the output as *sinc pulses* whose zero crossings fall at the peaks of all other pulses; hence $s = 2B$, and there is no intersymbol interference. However, if the impulse spacing is less than $1/2B$, the zero-crossing property is destroyed, the pulses overlap, and there is intersymbol interference. For a practical channel whose frequency response is not ideal, the maximum signaling speed must be reduced somewhat. Values of s between B and $1.5B$ are attained in practice, B being the nominal bandwidth.

Turning to the number of channel states (voltage levels or pulse amplitudes), there is no inherent limit to μ on a truly noiseless channel; voltage levels having arbitrarily small spacing still could be distinguished at the output. The value of μ on a noisy channel is estimated in terms of the signal-to-noise ratio in the following manner.

Let the average signal power and noise power at the channel output be S and N, respectively, so the total received power is $S + N$, and the rms output voltage is $\sqrt{S + N}$. Because of noise corruption one can never exactly identify the intended signal voltage. But, if the voltage levels are separated by an amount equal to or exceeding the rms noise voltage σ, they can be identified with reasonably low probability of error. Thus, at the receiving end we have voltage levels *spaced* by $\Delta V \geq \sigma = \sqrt{N}$ and an rms voltage *range* of $\sqrt{S + N}$. The maximum number of channel states is therefore approximately

$$\mu = \frac{\sqrt{S + N}}{\sqrt{N}} = \left(1 + \frac{S}{N}\right)^{\frac{1}{2}}$$

For example, a binary channel would require $(1 + S/N)^{\frac{1}{2}} = 2$, or $S/N = 3$; with gaussian noise, the corresponding error probability is about 0.05.

Combining these values for s and μ with the capacity equation (8.14) yields

$$C = s \log \mu = 2B \log \left(1 + \frac{S}{N}\right)^{\frac{1}{2}}$$

Hence,

$$C = B \log \left(1 + \frac{S}{N}\right) \qquad \text{bits/sec} \qquad (8.16)$$

where B is the channel bandwidth and S/N is the signal-to-noise ratio.

This famous equation is called the *Hartley-Shannon law*. (Hartley did the preliminary spadework and Shannon derived it with rigor, far more rigor than was attempted here.) Examining (8.16), we see it is written in terms of parameters that apply equally well to discrete or *continuous* channels, suggesting that the capacity of a continuous channel is $B \log (1 + S/N)$. In fact, Shannon's derivation was based on the continuous channel and will be discussed shortly.

For the moment it should be observed that the Hartley-Shannon law, coupled with the fundamental theorem, has two important implications for communication engineers. First, it tells us the absolute best that can be done in the way of reliable information transmission, given the channel parameters. Second, for a given information rate, it says we can reduce signal power providing we increase the bandwidth an appropriate amount, and vice versa.

The exchange of bandwidth for power or signal-to-noise ratio is of course not new to us, having noted the effect in wideband noise-reduction systems such as FM and PCM. But the Hartley-Shannon law specifies the *optimum* possible exchange and further implies that bandwidth *compression* is possible.

To illustrate, suppose it is desired to transmit digital data at a rate of 30,000 bits/sec. According to (8.16), we could in theory use a channel having $B = 30$ kHz and $S/N = 1$, since

$$C = 30 \times 10^3 \log (1 + 1) = 3 \times 10^4 \text{ bits/sec}$$

Alternately, the bandwidth can be reduced to $B = 3$ kHz if the power is increased by a factor of 1,000, that is, $S/N = 10^3 = 30$ db. (Note the handy approximation $10^3 \approx 2^{10}$.) Incidentally, the latter parameters are typical of standard voice telephone circuits; but when used for digital signals, the data rate on such channels is normally 4,800 bits/sec or less, indicating considerable room for improvement.

Some of the above matters are further pursued in Sec. 8.4 after we examine the continuous channel more closely. Since practically all communication systems are capable of handling continuous signals and all systems have noise, the noisy continuous channel merits detailed investigation.

8.3 SIGNAL SPACE AND CONTINUOUS CHANNELS ★

Not so long ago imaginary numbers were playthings of pure mathematics, deemed to have no practical value. But physicists and electrical engineers have since assigned a useful interpretation to $\sqrt{-1}$, and it is now almost unthinkable to discuss signal analysis, electromagnetic waves, or system theory without the aid of this tool.   Similarly, the once esoteric geometry of multidimensional spaces (*hyperspace*) was given new meaning by Shannon for the study of communication on continuous channels, reducing an otherwise intractable problem to more familiar terms.   This section outlines his derivation of the Hartley-Shannon law using concepts of signal space; the material closely follows Shannon (1949), to which the reader is referred for complete details.

Signal space and signal vectors

Consider a continuous channel of bandwidth B, assumed for simplicity to be a baseband channel.   Of necessity, all signals transmitted over the channel are bandlimited in B.   Thus, if $x(t)$ is one of the channel signals, we can apply sampling theory and write it as

$$x(t) = \sum_m x_m \text{ sinc } (2Bt - m) \tag{8.17}$$

where $x_m = x(mT_s)$ and $T_s = 1/2B$.   Equation (8.17) follows from (7.11) with sampling at the Nyquist rate.

Now suppose that $x(t)$ is essentially zero outside an interval of time T sec long; then it is completely described by $D = 2BT$ sample values, which we label $x_1, x_2, \ldots, x_D$.   [True, $x(t)$ cannot be identically zero outside the interval T, for the channel signals are bandlimited, and simultaneous bandlimiting and timelimiting is impossible.   Eventually we shall let $T \to \infty$ to compensate for this.]   The fact that $x(t)$ is uniquely identified by D sample values leads to the notion of *signal space*, a D-dimensional space in which $x(t)$ is represented by the *signal vector*

$$\mathbf{x} = (x_1, x_2, \ldots, x_D) \qquad D = 2BT$$

The vector $\mathbf{x}$ starts at the origin and terminates at a point whose coordinates are the sample values $x_1, x_2, \ldots, x_D$.

To enhance the meaning of signal space we now pause to consider geometric interpretations for various types of signal processing.   *Linear operations* on $x(t)$, for instance, become *linear vector transformations* in signal space; on the other hand, *nonlinear distortion* alters the coordinate values of $\mathbf{x}$ in a nonlinear manner and is thus equivalent to *warping* the signal space.   Reducing the dimensionality $D = 2BT$ of a signal, either by *filtering* or *gating* in time, *projects* the signal vector onto a space of lower dimensionality, i.e., a line, a plane, or a hyperplane.   *Modulation* is a *one-to-one mapping* of the message space into the (modulated) signal space; if the modulation is wideband, the mapping is between spaces of different dimensionality.

Relative to the last remark, a well-known theorem in topology says that any one-to-one mapping between spaces of different dimensionality must be *discontinuous,* in the sense that a continuous path in one space maps into a broken path in the other. Hence, nearly identical vectors in the message space may be widely separated in the signal space, and vice versa. Abstract as this seems, it later helps explain the threshold effect of wideband modulation! At present, however, we need more information about the properties of multidimensional spaces.

D-dimensional space is like ordinary space save that it has D mutually perpendicular axes. And though we cannot construct more than three such axes in our three-dimensional world,[1] we can deal logically and mathematically with spaces of higher dimensionality. In particular, signal space is *euclidean* in that the square of the distance from the origin to any point is the sum of the squares of the coordinates. The magnitude squared of a signal vector is therefore

$$|\mathbf{x}|^2 = x_1{}^2 + x_2{}^2 + \cdots + x_D{}^2$$

Moreover, vector operations such as addition and dot-product multiplication take the usual form

$$\mathbf{x} + \mathbf{y} = (x_1 + y_1, x_2 + y_2, \ldots, x_D + y_D) \qquad \text{a vector}$$
$$\mathbf{x} \cdot \mathbf{y} = x_1 y_1 + x_2 y_2 + \cdots + x_D y_D \qquad \text{a scalor}$$

We note in passing that

$$|\mathbf{x}|^2 = \mathbf{x} \cdot \mathbf{x} = \sum_{m=1}^{D} x_m{}^2 \tag{8.18}$$

If the tip of a signal vector $\mathbf{x}$ is swept through all possible positions, the surface generated is a *hypersphere* of radius $r = |\mathbf{x}|$. The "volume" enclosed by such a sphere is[2]

$$V_D = K_D r^D \tag{8.19}$$

where the constant K_D does not particularly concern us here. A curious consequence of Eq. (8.19) is that most of the volume of a hypersphere of high dimensionality ($D \gg 1$) is concentrated at the surface. To illustrate, the relative volume between $r/2$ and r is $1 - 2^{-D}$, so if $D = 3$ (a conventional sphere), 87.5 percent of the volume is in the outer "half"; if $D = 100$, the relative volume of the outer portion is approximately $1 - 10^{-30}$.

This *volume-concentration* effect proves to be useful in our development, for the dimensionality of typical signal spaces is indeed large.

---

[1] Abbott (1950) is an amusing discourse on the mysteries of three-dimensional space as viewed by a native of Flatland, a two-dimensional space.

[2] Sommerville (1929).

For example, a voice signal bandlimited to 4 kHz and lasting 3 min requires $D = 2 \times 4 \times 10^3 \times 180 = 10^6$ sample points and coordinate axes.

Returning to the signal $x(t)$, its total energy is given by

$$E = \int_{-\infty}^{\infty} x^2(t) \, dt$$

or, using (8.17),

$$E = \int_{-\infty}^{\infty} \left[\sum_m \sum_n x_m x_n \text{ sinc } (2Bt - m) \text{ sinc } (2Bt - n) \right] dt$$

Interchanging summation and integration and noting, from (7.16), that the sinc functions are orthogonal yields

$$E = \frac{1}{2B} \sum_m x_m{}^2 = \frac{1}{2B} (\mathbf{x} \cdot \mathbf{x}) = \frac{|\mathbf{x}|^2}{2B} \qquad (8.20)$$

Since eventually we shall let $T \to \infty$, the average signal power is

$$S = \frac{E}{T} = \frac{|\mathbf{x}|^2}{2BT}$$

Hence, the length of the signal vector is related to the average power by

$$|\mathbf{x}| = \sqrt{2BTS} = \sqrt{DS} \qquad (8.21)$$

Generalizing this result, if a channel has an average power constraint and the source is ergodic, so that, as $T \to \infty$, all possible channel signals have the same average power S, then all signal vectors terminate at the surface of a hypersphere of radius $r = \sqrt{DS}$.

These arguments also apply to the channel *noise* providing it is gaussian white noise from an ergodic source, bandlimited in B. With this condition, sample values spaced by $1/2B$ are uncorrelated and statistically independent. The noise energy in time T is then very nearly NT, N being the average noise power. Hence, the noise is represented in signal space by a vector of length $\sqrt{DN}$, and all possible noise signals are contained within a sphere of that radius. Because the noise is *random*, it might seem that the noise sphere should be "fuzzy"; i.e., a particular sample function of length T may have an energy quite different than NT. But if the dimensionality is high, volume concentration indicates that the noise sphere is quite sharply defined, like a Ping-Pong ball rather than a cloud of gas.

Communication on a noisy channel

Consider the state of affairs at the channel output, where we have the desired signal contaminated by noise. Under the usual assumption that

signal and noise are independent, their average powers add, and the
received signal plus noise is a vector of length $\sqrt{D(S + N)}$. Figure
8.3a shows the geometric interpretation of the transmitted signal, added

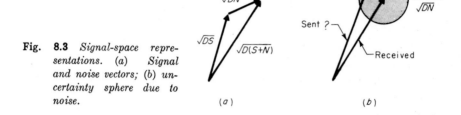

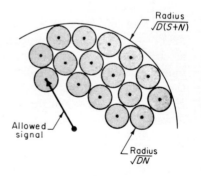

Fig. 8.3 *Signal-space representations. (a) Signal
and noise vectors; (b) uncertainty sphere due to
noise.*

noise, and signal plus noise.   The transmitted signal is seen to lie within
a sphere of radius $\sqrt{DN}$ at the tip of the signal-plus-noise vector (Fig.
8.3b), and this noise sphere indicates the uncertainty of the receiver as to
which signal was intended.

If the possible transmitted signals are known in advance at the
receiver, and if the sphere of uncertainty contains the tip of one and only
one of the possible signal vectors, then the intended signal can be *exactly
determined* despite the noise.   Thus, suppose we put into the hypersphere
of radius $\sqrt{D(S + N)}$ a large number of nonoverlapping noise spheres
of radius $\sqrt{DN}$ and then send only those signals corresponding to the
center points of the noise spheres (Fig. 8.4).   When transmitted signals

Fig. 8.4 *Signal vectors for virtually
errorless transmission.*

are selected in this fashion, it is possible to convey information over a
noisy continuous channel with vanishingly small error probability.

How many little noise spheres can be packed into the big signal-
plus-noise sphere without overlapping?   The calculation is important,
for it tells us $M(T)$, the number of "very different" signals (messages)

of length T that can be correctly identified at the channel output, from which we then can find the channel capacity. Clearly, $M(T)$ does not exceed the volume of the big sphere divided by the volume of one of the little spheres, i.e., using (8.19),

$$M(T) \leq \frac{K_D\,[\sqrt{D(S+N)}]^D}{K_D\,[\sqrt{DN}]^D} = \left(1 + \frac{S}{N}\right)^{D/2}$$

Note that $M(T)$ is finite for all but truly noiseless channels. Moreover, because a real channel has noise and $M(T)$ is *finite*, communication over a continuous channel is inherently a *discrete* process.

Setting $D = 2BT$ and inserting $M(T)$ into (8.13) gives

$$C \leq \lim_{T \to \infty} \frac{1}{T} \log \left(1 + \frac{S}{N}\right)^{BT} = B \log \left(1 + \frac{S}{N}\right)$$

This is an *upper* bound on the capacity of a continuous channel; Shannon then goes on to show that it is also the *lower* bound in the limit as $T \to \infty$. Therefore

$$C = B \log \left(1 + \frac{S}{N}\right) \tag{8.16}$$

which is the Hartley-Shannon law as previously obtained by cruder arguments. Summarizing the conditions on (8.16), the channel has bandwidth B and an average signal-power constraint S; the noise is additive gaussian white noise of average power N; both signal and noise come from ergodic sources.

It is important to observe that attaining information transmission at a rate of $B \log (1 + S/N)$ bits/sec requires $D = 2BT \gg 1$ and $M(T) = (1 + S/N)^{BT}$; in other words, the noise spheres must be packed as closely as possible without overlapping. Now suppose the signal-to-noise ratio drops slightly below the design value. The noise spheres will then overlap, and the receiver will make frequent decoding errors. Hence there is a sharp *threshold effect* in that a small increase of noise power (or a small decrease of signal power) produces a large increase in the probability of error. As a result of these errors the information is lost.

Recalling the discontinuous mapping property of wideband modulation, the same explanation holds for threshold effects in analog message transmission. Specifically, since adjacent vectors in the modulated signal space do not necessarily represent adjacent vectors in the message signal space, a slight overlapping of the noise spheres may cause the receiver to "demodulate" a message totally unlike the one intended; again the message is lost. This further implies that threshold effect is inevitable in all types of wideband modulation and is not an exclusive property of FM or PCM.

Ideal communication systems

A communication system capable of transmitting without errors at a rate of $B \log (1 + S/N)$ bits/sec, where B and S/N are the channel parameters, is called an *ideal* system. While no practical system is or can be ideal, it is possible to visualize systems whose performance approaches that of the ideal. As an aid to the design of such systems, let us examine the characteristics of a nearly ideal system.

To begin with, the number of different signals (or messages or symbols) of length T is $M = (1 + S/N)^{BT}$, and the maximum information rate is $R = 1/T \log M$. The channel signals are chosen such that they can be identified at the receiver with very small probability of error; i.e., the signals are those represented in signal space by the center points of nonoverlapping noise spheres. Shannon has shown that if the signals themselves are randomly selected sample functions of gaussian white noise and if $2BT \gg 1$, then the above condition is closely approximated.

The information produced by a source is conveyed over the system in the following fashion. The source output is observed for T sec, and the observed message is represented (encoded) as one of the noiselike channel signals which is then transmitted; thus, the information is encoded in *blocks* of length T. (This implies that the number of possible source messages of length T is not greater than M.) At the output, the received signal plus noise is compared with stored copies of the channel signals. The one that best matches the signal plus noise is presumed to be the signal actually transmitted, and the corresponding message is decoded. A total time delay of $2T$ is therefore required for the encoding and decoding operations.

One method for determining the best match is to evaluate mean-square errors, as indicated in Fig. 8.5. Let the set of channel signals be

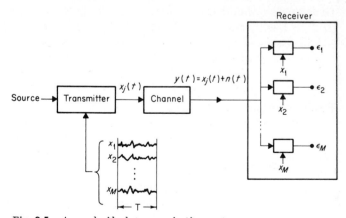

Fig. 8.5 *A nearly ideal communication system.*

$x_i(t)$, from which $x_j(t)$ is chosen for transmission. The received signal plus noise is $y(t) = x_j(t) + n(t)$. Using stored copies of $x_i(t)$, the receiver calculates the mean-square errors

$$
\epsilon_i = \frac{1}{T} \int_T [y(t) - x_i(t)]^2 \, dt
$$
$$
\approx \begin{cases} 2S + N & i \neq j \\ N & i = j \end{cases}
$$

where the approximate values require $2BT \gg 1$. The error meter with the lowest output, namely, $\epsilon_j = N$, then indicates which signal was intended.

These receiver operations are equivalent to crosscorrelating $y(t)$ with $x_i(t)$, and are called *correlation detection*. Essentially the same results can be achieved with a bank of *matched filters* having impulse responses $h_i(t) = x_i(-t)$, thereby eliminating the need for time-domain copies of the channel signals. Harman (1963, chap. 6) gives further details.

Throughout the above description it has been tacitly assumed that $T \to \infty$, for only in this limit are all the conditions satisfied so that $R = B \log (1 + S/N)$. Thus, the characteristics of an ideal system are as follows:

1. The information rate approaches $B \log (1 + S/N)$.
2. The frequency of errors approaches zero.
3. The statistical properties of the transmitted signal approach those of bandlimited gaussian white noise.
4. The coding time delay increases indefinitely.
5. There is a sharp threshold effect.

Because of the infinite coding delay, an ideal system is impractical. Of course, if the channel bandwidth is large enough, we can still have $2BT \gg 1$ with reasonable values of T. However, the design of nearly ideal systems is no trivial matter, for one must balance off coding delay and signal selection against reliability.

Accurate signal identification, say by means of correlation detection, argues for large T; the longer the integration time, the better the accuracy. But as T is increased, the delay increases, and M must increase exponentially to maintain constant information rate $R = 1/T \log M$. Rice (1950) has shown that to achieve $R = 0.96C$ with $S/N = 10$ and an error probability of 10^{-5}, the number of channel signals required is $M = 2^{10,000}$!

Clearly, efficient *block coding* with its extravagant number of channel signals is just as impractical as the infinite coding delay of an ideal system.

Alternate schemes, using *sequential coding*,[1] are nearly as efficient and require relatively simple equipment.

8.4 SYSTEM COMPARISONS

The Hartley-Shannon law applies to a restricted class of channels, namely, continuous channels having an average power limitation and additive gaussian white noise. But this description fits many practical communication systems to a reasonable degree, so the hypothetical ideal system that delivers information at a rate given by (8.16) is the generally accepted standard for system comparisons. In this section we reexamine various existing systems in the light of information theory and see how they measure up against the ideal.

One aspect of particular interest is wideband noise reduction. Hence, as a preliminary, we shall investigate the exchange of bandwidth for signal-to-noise ratio (or transmitted power) implied by the Hartley-Shannon law, for this is the optimum bandwidth-power exchange.

Optimum bandwidth-power exchange

Suppose it is desired to transmit a signal, bandlimited in W, such that the output signal-to-noise ratio at the destination is $(S/N)_D$. No matter how the transmission is accomplished, the information rate at the output can be no greater than $R_{\max} = W \log [1 + (S/N)_D]$. Further suppose that an ideal system is available for this purpose and that the channel (transmission) bandwidth is B_T, the noise power density is η, and the average transmitted power at the receiver is P_T. In other words, the channel capacity is $C = B_T \log [1 + (S/N)_T]$, where $(S/N)_T = P_T/\eta B_T$. These factors are summarized in Fig. 8.6.

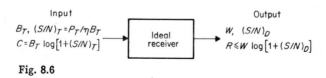

Fig. 8.6

If information is neither destroyed nor accumulated in the receiver, the output rate must equal the information rate on the channel. Assuming the system is operating at capacity and the output rate is maximum,

[1] Wozencraft and Reiffen (1961).

then $R_{max} = C$, so

$$B_T \log \left[1 + \left(\frac{S}{N} \right)_T \right] = W \log \left[1 + \left(\frac{S}{N} \right)_D \right]$$

Solving for $\left(\frac{S}{N} \right)_D$ yields

$$\left(\frac{S}{N} \right)_D = \left[1 + \left(\frac{S}{N} \right)_T \right]^{B_T/W} - 1 \tag{8.22}$$

which shows that the optimum exchange of bandwidth for power is *exponential*. To emphasize this relation, note that $(S/N)_D \approx (S/N)_T{}^{B_T/W}$ if the signal-to-noise ratios are large. The exponential trade-off is realized by an ideal system operating to its fullest capacity.

However, with fixed channel *noise density*, as distinguished from fixed noise power, (8.22) does not tell the exact story. For as channel bandwidth is increased, the noise power $N = \eta B_T$ is likewise increased, and $(S/N)_T$ decreases. A more equitable basis for comparison is obtained by rewriting (8.22) in terms of the normalized parameters

$$z = \frac{P_T}{\eta W} \quad \text{and} \quad b = \frac{B_T}{W}$$

used in previous chapters.

The signal-to-noise ratio at the receiver input is then $(S/N)_T = (W/B_T)(P_T/\eta W) = z/b$, and (8.22) becomes

$$\left(\frac{S}{N} \right)_D = \left(1 + \frac{z}{b} \right)^b - 1 \tag{8.23}$$

$$\approx \left(\frac{z}{b} \right)^b \quad \frac{z}{b} \gg 1$$

Thus, while the exchange is not strictly exponential, it is very nearly so for large signal-to-noise ratios. This means that doubling the transmission bandwidth of an ideal system squares (approximately) the output signal-to-noise ratio. Alternately, since z is proportional to P_T, the transmitted power can be reduced to about the square root of its original value, without reducing $(S/N)_D$, if bandwidth is increased by a factor of 2. As demonstrated shortly, this exchange is considerably better than that of most existing systems.

Equation (8.23) also shows what is involved in bandwidth *compression*—transmitting a signal of bandwidth W over a channel of bandwidth $B_T < W$, so that $b < 1$. Inserting typical values, one finds that such compression is exceedingly costly in terms of transmitted power. To illustrate, suppose we want $(S/N)_D = 10^4$ and suppose that $\eta W = 10^{-3}$. Transmission at baseband ($b = 1$) requires $P_T = \eta W (S/N)_D = 10$ watts.

But to compress bandwidth by a factor of $\frac{1}{2}$ we need $z = b[1 + (S/N)_D]^{1/b} - b \approx \frac{1}{2}(10^4)^2 = 5 \times 10^7$, or $P_T = 50$ kw, which is 5,000 times the baseband power. Similarly, for $b = \frac{1}{10}$, the power requirement is a colossal 10^{36} watts!

As a general conclusion we can say that even the optimum bandwidth-power exchange is practical in one direction only, the direction of increasing bandwidth and decreasing power.

System comparisons—analog signal transmission

It is a difficult matter to assess the information rate of analog signals: voice and music waveforms, television video, etc. Furthermore, communication systems designed for such signals have as their goal reasonably faithful reproduction of the signals themselves, with a minimum of noise and distortion. This goal is not quite the same thing as reliable information transfer in the sense of information theory; i.e., the communication engineer may be more concerned with transmission bandwidth, threshold power requirements, and signal-to-noise ratios than he is with channel capacity and its utilization.

Nonetheless, information theory does have something to say in regard to analog signal transmission. Specifically, by Eq. (8.23), it tells us the best signal-to-noise ratio that can be obtained with given channel parameters; it tells us the minimum power required to achieve a specified signal-to-noise ratio, as a function of bandwidth; and it indicates the optimum possible exchange of bandwidth for power. Therefore, let us compare the performance of existing systems with that of an ideal system as described by (8.23).

Table 8.1 summarizes many of our earlier results in the notation of

Table 8.1   Comparison of analog message transmission systems

System	B_T	$\left(\dfrac{S}{N}\right)_D$	Comments
AM	$2W$	$\frac{1}{3}z$	$m = 1$
DSB	$2W$	z	Synchronous detection
SSB	W	z	Synchronous detection
PM	$2(\phi_d + 1)W$	$\frac{1}{8}(b - 2)^2 z$	$\phi_d \leq \pi$
FM	$2(D + 1)W$	$\frac{3}{8}(b - 2)^2 z$	Deemphasis not included
PDM	$1/2t_r$	$\frac{1}{4}bz$	See (7.34)
PPM	$1/2t_r$	$\frac{1}{16}b^2 z$	See (7.33)
PCM	νW	$\frac{3}{2}\mu^{2b}$	$\nu = \log_\mu q$
Ideal	bW	$(z/b)^b$	Unrealizable

this section. Specifically, the postdetection signal-to-noise ratios are written as functions of $z = P_T/\eta W$ and $b = B_T/W$. It is assumed that the signal-to-noise ratios are large, all systems are above threshold, and the message is normalized so that $\overline{x^2} = \frac{1}{2}$. The values for the pulsed systems assume $f_s = 2W$, etc., and would not be achieved in practice.

Clearly, none of the practical systems exhibit the output improvement that can be had in an ideal system by increasing *either* z (power) or b (bandwidth). PCM does have an exponential bandwidth dependence, but, once above threshold, increasing transmitted power yields no further improvement of $(S/N)_D$; its value is determined by the quantization. It also might be noticed that SSB is just as good as an ideal system having $b = 1$. Of course the SSB bandwidth ratio is fixed at $b = 1$, so there is no possibility of wideband noise reduction.

Since, in general, the output signal-to-noise ratio depends on both z and b, it is difficult to give a complete graphical display of the relations in Table 8.1. As an alternate we can plot $(S/N)_D$ as a function of z for typical bandwidth ratios (Fig. 8.7) or plot the value of z required for a specified $(S/N)_D$ as a function of b (Fig. 8.8).

Figure 8.7 repeats some of the curves from Chaps. 6 and 7, with the addition of a curve for an ideal system having $b = 6$. Also shown are the threshold points of the practical systems, which is precisely the reason for the figure. It can be seen that practical wideband noise-reduction systems, e.g., FM, PPM, and PCM, fall short of ideal performance primarily because of threshold limitations. For example, FM and PCM

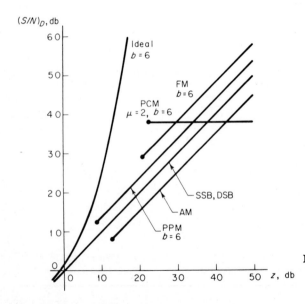

Fig. 8.7 *Postdetection signal-to-noise ratios versus $z = P_T/\eta W$.*

with $b = 6$ have threshold points offset horizontally by about 10 db from an ideal system with the same bandwidth ratio. The PPM threshold is much closer but occurs at too low a value of $(S/N)_D$ to be useful for analog signals.

As to the exchange of bandwidth for power, Fig. 8.8 shows the minimum value of z needed for $(S/N)_D = 10^5$ as a function of bandwidth

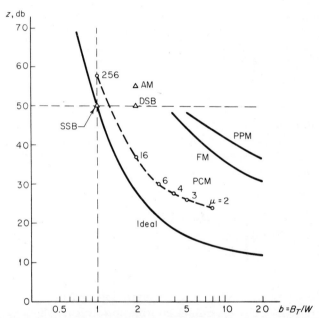

Fig. 8.8 *Transmitted power $z = P_T/\eta W$ required for $(S/N)_D$ $= 10^5$ as a function of bandwidth ratio $b = B_T/W$.*

ratio. For this relatively high output S/N we see that PCM does considerably better than FM or PPM but requires 8 to 10 db more power than an ideal system. (Just how the PCM curve was obtained is discussed under the next heading.) We might also point out the sharp increase in z for the ideal system when $b < 1$, echoing our earlier observation about bandwidth compression.

In summary, at high signal-to-noise ratios FM and PCM give the best wideband performance, PCM being somewhat better. From the power-bandwidth viewpoint, all practical wideband systems are an order of magnitude below the ideal. At low signal-to-noise ratios, only SSB and DSB are useful, having no threshold effect.

The channel capacity of PCM

Of the various systems we have discussed, PCM is the most amenable to direct analysis in terms of information theory. This is because the transmitted signal is discrete, even though it represents an analog signal, and the information rate can be readily calculated. Therefore, let us find the channel capacity of PCM and make a direct comparison with the Hartley-Shannon law.

Consider a baseband PCM system having transmission bandwidth B_T, μ equally spaced coded pulse amplitudes, and channel signal-to-noise ratio $(S/N)_T$. Since the entropy of the pulsed signal is $H \leq \log \mu$ and the pulse rate is $s \leq 2B_T$, the information rate on the channel is $R \leq 2B_T \log \mu$. Hence

$$C = R_{\max} = 2B_T \log \mu = B_T \log \mu^2 \qquad (8.24)$$

providing decoding errors can be ignored. The next step is to determine the signal-to-noise ratio which ensures that decoding errors are negligible.

We have previously shown that the error probability P_e is small if the voltage spacing between pulse amplitudes is $\Delta V = K\sigma$, where σ is the rms noise voltage and $K \geq 1$. Taking the pulses to be *bipolar*, the pulse amplitudes are

$$A_j = \begin{cases} 0, \ \pm K\sigma, \ \pm 2K\sigma, \ \ldots, \ \pm \dfrac{\mu - 1}{2} K\sigma & \mu \text{ odd} \\[2ex] \pm \tfrac{1}{2}K\sigma, \ \pm \tfrac{3}{2}K\sigma, \ \ldots, \ \pm \dfrac{\mu - 1}{2} K\sigma & \mu \text{ even} \end{cases}$$

(A bipolar waveform is assumed since otherwise the signal contains a dc component which cannot convey information and would be a waste of power.) For maximum information transfer the μ amplitudes should be equally likely, and there should be no "spaces" between pulses. With these conditions the average signal power is

$$S = \sum_{j=1}^{\mu} P_j A_j^2 = \frac{1}{\mu} \sum_{j=1}^{\mu} A_j^2 = \frac{\mu^2 - 1}{12} (K\sigma)^2 \qquad (8.25)$$

which holds for any value of μ, even or odd, as the reader can check.

In effect, we have calculated the *threshold power requirement* (or minimum channel signal-to-noise ratio) as a function of μ. To clarify this, note that $\sigma^2 = N_T$, the channel noise power, so (8.25) becomes

$$\left(\frac{S}{N}\right)_T = \frac{K^2}{12} (\mu^2 - 1)$$

and hence

$$\left(\frac{S}{N}\right)_{T\text{th}} = \frac{K_{\min}^2}{12} (\mu^2 - 1) = \gamma(\mu^2 - 1) \qquad (8.26)$$

where $\gamma = K_{\min}^2/12$ and $K_{\min}^2$ is determined by the minimum permitted spacing between pulse amplitudes for the specified decoding error probability. Equation (8.26), with $\gamma = 10$, previously appeared without derivation in Chap. 7, and was used to calculate the PCM threshold powers for Figs. 7.28, 8.7, and 8.8.

Since a PCM system is most efficient just above threshold, we take $(S/N)_T = (S/N)_{T_{\text{th}}}$ and solve (8.26) for μ^2 as

$$\mu^2 = 1 + \frac{1}{\gamma}\left(\frac{S}{N}\right)_T$$

Substituting in (8.24) then gives

$$C = B_T \log\left[1 + \frac{1}{\gamma}\left(\frac{S}{N}\right)_T\right] \qquad \text{bits/sec} \qquad (8.27)$$

or, if $(S/N)_T \gg \gamma$,

$$C \approx B_T \log\left[\left(\frac{S}{N}\right)_T \gamma^{-1}\right] = C_{\text{ideal}} - B_T \log \gamma$$

To complete the comparison one must evaluate γ, an evaluation that depends on two factors: (1) what is considered to be sufficiently small error probability, and (2) the nature of the noise. There is a further complication that for a specified value of P_e, $K_{\min}$ has a slight dependence on the number of pulse levels μ.

Taking gaussian noise and $P_e = 10^{-5}$, Shannon (1949) has carried out the calculations, with results as shown in Fig. 8.9, where C/B_T is

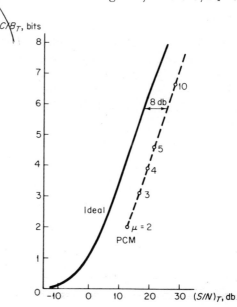

Fig. 8.9 *Channel capacity of PCM (with error probability $P_e = 10^{-5}$) compared with an ideal system.*

plotted versus $(S/N)_T$. The corresponding curve for an ideal system is also given by way of comparison. Viewed in this light, PCM is seen to require about 8 db more power than an ideal system with its prohibitively complex encoding requirements, long time delays, etc. However, it is well to bear in mind that an ideal system would have vanishingly small error probability, whereas the PCM curve is for $P_e = 10^{-5}$.

The reason why PCM compares as favorably as it does to an ideal system stems from the earlier conclusion that because of channel noise, electrical communication is inherently a discrete process. The PCM design recognizes and accepts this fact; the transmitted PCM signal, being discrete, is better suited to the noisy channel than uncoded continuous signals.

Communication efficiency

To conclude our comparisons, let us examine transmission power requirements in terms of information rate. This viewpoint is particularly relevant to long-range minimum-power systems—e.g., satellite relay, space probes—and leads to the so-called *communication efficiency*.

Suppose a system requires P_T watts of power to deliver R bits/sec with low error probability in the presence of channel noise having density η. The ratio P_T/R is the transmitted power per bit per second, or simply the signal *energy* in joules per bit. Normalizing by the noise density gives a system figure of merit

$$\beta = \frac{P_T}{\eta R} \tag{8.28}$$

which is the signal energy per bit per unit background noise. Note carefully that efficient systems should have *small* values of β, since the required power is $P_T \geq \beta \eta R$.

Now consider an ideal system whose maximum information rate is $R = B_T \log [1 + (S/N)_T]$. Since $(S/N)_T = P_T/\eta B_T = \beta(R/B_T)$, we have

$$\frac{R}{B_T} = \log_2 \left(1 + \frac{R}{B_T} \beta \right)$$

so that

$$\beta = \frac{B_T}{R} (2^{R/B_T} - 1) \tag{8.29a}$$

or, in terms of $(S/N)_T$,

$$\beta = \frac{(S/N)_T}{\log [1 + (S/N)_T]} \tag{8.29b}$$

Referring to (8.29a), we ask a significant question: For a given information rate, what is the *minimum* value of β? This minimum figure of merit (minimum power requirement) for the ideal system will serve as a reference point and is designated β_0. Using the series expansion

$$a^u = 1 + u \ln a + \frac{(u \ln a)^2}{2!} + \cdots$$

where $\ln a = \log_e a$, it then follows that

$$\beta_0 = \lim_{B_T/R \to \infty} \left[\frac{B_T}{R} \left(2^{B_T/R} - 1 \right) \right]$$

$$= \lim_{u \to 0} \frac{1}{u} \left[u \ln 2 + \frac{(u \ln 2)^2}{2!} + \cdots \right]$$

$$= \ln 2 = 0.693 \tag{8.30}$$

But we have specified constant information rate R, so β_0 is attained in the limit as $B_T \to \infty$. In other words, an ideal system achieves minimum power only when the coding is such that the transmission bandwidth approaches infinity.

Alternately, drawing upon Eq. (8.29b), we can plot β as a function of $(S/N)_T$ (Fig. 8.10). Maximum efficiency ($\beta = \beta_0$) is then seen to

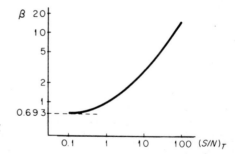

Fig. 8.10 β versus $(S/N)_T$ for an ideal system.

correspond to $(S/N)_T \to 0$, which agrees with the above limiting, since, for fixed η, $N_T = \eta B_T \to \infty$ as $B_T \to \infty$. Nonetheless, it is rather disconcerting to realize that *increasing* the signal-to-noise ratio actually *decreases* the efficiency!

Taking $\beta_0 = \ln 2$ as reference, the *communication-efficiency factor* of any system is measured by β/β_0, where β is the figure of merit for the system in question. As illustration, an ideal system operating at $(S/N)_T = 10^3$ has $\beta/\beta_0 = 100/(\ln 2) = 144$, or an efficiency of about 7 percent in the sense that an ideal system operating at $(S/N)_T \to 0$ requires less power by a factor of $\frac{1}{144} = 0.07$. Note, however, that β/β_0 is a highly specialized measure, concerned only with power requirements relative to information rate and oblivious of bandwidth considerations. In fact, β/β_0 is referenced to an ideal system having *infinite* bandwidth and therefore is meaningful only when power is the dominant consideration and bandwidth conservation is unimportant.

To emphasize the meaning of communication efficiency, let us draw upon the fact that channel noise density can be expressed as

$$\eta = kT_N \qquad \text{watts/Hz}$$

where $k = 1.4 \times 10^{-23}$ joule/deg is Boltzmann's constant and T_N is the system *noise temperature* in degrees Kelvin. (High-quality low-noise communication systems have noise temperatures in the vicinity of 30 to 100°K.) For a specified information rate the minimum power required is

$$P_T = \beta\eta R = \frac{\beta}{\beta_0}(\ln 2)kT_N R$$

$$\approx 10^{-23}\frac{\beta}{\beta_0}T_N R \qquad \text{watts} \tag{8.31}$$

Inserting representative values in the above shows that P_T can be exceedingly small. But then we must remember that P_T is measured at the *receiving* end, and long-range systems have tremendous amounts of attenuation causing the received power to be many orders less than that transmitted.

Illustrating the implications of (8.31), suppose it is desired to transmit a still photograph from Mars to earth with a signal power of 10 watts. Accounting for propagation loss and antenna gain, the net attenuation might be about 200 db, so at the receiver $P_T = 10^{-20} \times 10 = 10^{-19}$ watt. If the noise temperature is $T_N = 50°$K and the system has 100 percent efficiency ($\beta/\beta_0 = 1$), the maximum information rate permitted is

$$R = \frac{P_T}{10^{-23}T_N} = 200 \text{ bits/sec}$$

Assuming that the photograph is quantized into $200 \times 200 = 4 \times 10^4$ elements, each element having one of 64 possible brightness levels, the total information to be transferred is $I = 24 \times 10^4$ bits (see Example 8.2). Therefore, the total transmission time T must be

$$T = \frac{I}{R} = \frac{24 \times 10^4}{200} = 1{,}200 \text{ sec} = 20 \text{ min}$$

The Mariner IV mission, using a less efficient but practical system, required 8 hr to transmit each picture.

The above discussion of an ideal system and its efficiency is far more than idle speculation; the conclusions arrived at have significant implications for the design of efficient practical systems. Specifically, if conservation of power is the prime objective, the system should have both large transmission bandwidth *and* small channel signal-to-noise ratio. We can get large B_T with wideband modulation techniques, but the inevitable threshold effect prohibits very small $(S/N)_T$ and thereby precludes operation at highest efficiency. This point is further demonstrated by examining the figure of merit for practical systems, beginning with PCM.

We have previously estimated the maximum information rate of PCM to be $R = B_T \log\left[1 + \dfrac{1}{\gamma}\left(\dfrac{S}{N}\right)_T\right]$, and hence

$$\beta = \frac{P_T}{\eta B_T \log\left[1 + (1/\gamma)(S/N)_T\right]} = \frac{(S/N)_T}{\log\left[1 + (1/\gamma)(S/N)_T\right]}$$

Setting $u = (S/N)_T$ and using the expansion $\ln(1 + a) = a - a^2/2 + \cdots$ for $|a| < 1$, the minimum value of β is

$$\beta_{\min} = \lim_{u \to 0} \frac{u}{\log(1 + u/\gamma)}$$

$$= \lim_{u \to 0} \frac{\gamma u \ln 2}{u - u^2/2 + \cdots} = \gamma\beta_0$$

corresponding to $(S/N)_T \to 0$, as might be expected.

However, from (8.26), the threshold condition is $(S/N)_{T_{\text{th}}} = \gamma(\mu^2 - 1)$, so we cannot take $(S/N)_T$ arbitrarily small. The best we can do is $\mu = 2$ (*binary* PCM), which gives the largest transmission bandwidth, the smallest channel signal-to-noise ratio, and the highest efficiency. Several times in the past we have suspected that binary PCM is superior to PCM with $\mu > 2$; that suspicion is now confirmed from the power viewpoint. For $\mu = 2$, $(S/N)_{T_{\text{th}}} = 3\gamma$, and

$$\beta = \tfrac{3}{2}\gamma = 2.2\gamma\beta_0 \qquad \text{binary PCM}$$

Therefore, the efficiency is in the neighborhood of 5 percent ($\beta/\beta_0 = 20$), depending on the assumed value of γ. Such a low efficiency may seem discouraging at first. Putting the matter in proper perspective, it is better to say that binary PCM requires about 13 db more power than an ideal system with *infinite* bandwidth.

Turning to analog modulation, we are faced with the problem of estimating the information rate R of an analog signal.[1] A crude but simple expedient is to take the upper bound $R \leq W \log[1 + (S/N)_D]$, where $(S/N)_D$ and W are the signal parameters after demodulation.

---

[1] Sanders (1960) gives a more complete discussion.

This gives a *lower bound* for β, namely,

$$\beta \geq \frac{P_T}{\eta W \log [1 + (S/N)_D]} = \frac{z}{\log [1 + (S/N)_D]}$$

where $z = P_T/\eta W$ and $(S/N)_D$ is a function of z.

For suppressed-carrier linear modulation (SSB and DSB) we have $(S/N)_D = z$. Such systems are normally operated with postdetection signal-to-noise ratios of 30 to 40 db, so $\beta/\beta_0 \geq 100$ to $1,000$. The corresponding efficiency is substantially less than 1 percent, not unexpected since the transmission bandwidth is relatively small. On the other hand, wideband analog modulation should prove to be better, and in fact it is, as the following calculations indicate.

Consider FM where $(S/N)_D = \frac{3}{2}D^2z$ if $\overline{x^2} = \frac{1}{2}$. Then

$$\beta \geq \frac{z}{\log (1 + \frac{3}{2}D^2z)}$$

and maximum efficiency suggests letting $z \rightarrow 0$. However, one cannot violate the threshold condition of Eq. (6.43), namely,

$$z \geq z_{\text{th}} \approx 20 \frac{B_T}{W} = 40M(D)$$

Taking $z = z_{\text{th}}$ gives

$$\beta \geq \frac{40M(D)}{\log [1 + 60D^2M(D)]}$$

and hence β is a function of the deviation ratio $D = f_d/W$. A plot of β versus D reveals a rather broad minimum for deviation ratios between 1 and 4, corresponding to $\beta_{\min} \geq 20\beta_0$ and a 5 percent efficiency. (A direct comparison with PCM is, however, not meaningful because of the inequality.) In effect, there is an *optimum deviation ratio* $D_{\text{opt}} \approx 2$; for $D < 2$ the loss of wideband noise reduction decreases efficiency, while for $D > 2$ the higher threshold level counterbalances the S/N improvement and again decreases efficiency.

Clearly, the *threshold-extension* techniques alluded to in Chap. 6 would be of benefit here since they lower the value of z_{th}. In theory, FM with frequency-compressive feedback in the receiver (FMFB) is capable of achieving $\beta \geq 2\beta_0$, for a 50 percent maximum efficiency; in practice, efficiencies of 25 to 30 percent are possible. As an example, the FMFB receiver used in the Telstar satellite system[1] has $z_{\text{th}} \approx 50$ with $D = 5$, giving $\beta \geq 6\beta_0$.

[1] Giger and Chaffee (1963).

Because of its low threshold level, PPM appears to have efficiencies comparable to FMFB.

PROBLEMS

8.1 A card is selected at random from a deck of 52. You are told it is a heart. How much information (in bits) have you received? How much more information is needed to completely specify the card?

8.2 Calculate the amount of information needed to open a lock whose combination consists of three numbers, each ranging from 00 to 99.

8.3 In addition to the bit there are two other units of information: the dit, or Hartley ($b = 10$), and the nat, or nit ($b = e$). Evaluate each of these in terms of bits.

8.4 A source produces 6 symbols with probabilities $\frac{1}{2}$, $\frac{1}{4}$, $\frac{1}{8}$, $\frac{1}{16}$, $\frac{1}{32}$, and $\frac{1}{32}$. Find the entropy H.

8.5 A source has an alphabet of size $m \gg 1$. One symbol has probability $\epsilon \ll 1/m$, while the other symbols are equally likely. Show that the source entropy is approximately

$$H = \log_2 m - \epsilon \log_2 \epsilon - (m \ln 2)^{-1}$$

8.6 Suppose you are given nine pennies, eight of which are good. The remaining coin is counterfeit and weighs more or less than a good one. Given an uncalibrated balancing scale, how many weighings are necessary in theory to locate the bad coin and determine whether it is light or heavy? *Hint:* The scale has three possible positions, balanced and unbalanced on one side or the other. The average information per weighing, i.e., the entropy, is maximized if each of these positions is equally likely. Although it is not required, you may wish to devise the actual weighing procedure.

8.7 Conventional telegraphy uses two symbols, the dot and dash. Assuming the dash is twice as long as the dot and half as probable, find the average symbol rate and the entropy rate.

8.8 A certain data source has 8 symbols that are produced in blocks of 3 at a rate of 1,000 blocks per second. The first symbol in each block is always the same (presumably for synchronization); the remaining two places are filled by any of the 8 symbols with equal probability. What is the entropy rate R?

8.9 Analogous to Eq. (8.8), the entropy of a continuous signal $x(t)$ can be defined as

$$H(x) = -\int_{-\infty}^{\infty} p(x) \log p(x) \, dx$$

where $p(x)$ is the probability density function. Show that this is a relative rather than absolute measure of information by considering $H(x)$ and $H(y)$ when $y(t) = Kx(t)$ and $x(t)$ is uniformly distributed.

8.10 Referring to Example 8.4, suppose the sampling frequency is greater than the Nyquist rate. Does this increase the entropy rate of the sampled and quantized signal? Explain.

8.11 A binary data source has $P_0 = \frac{3}{8}$, $P_1 = \frac{5}{8}$, and intersymbol influence extending over groups of two successive symbols such that $P(1|0) = \frac{3}{4}$ and $P(0|1) = \frac{1}{16}$. Calculate the conditional entropy H_c and the redundancy $E = 1 - H_c/H_{\max}$.

8.12 Is the binary encoding shown below optimum and uniquely decipherable?

Source symbol	Probability	Binary code
A	$\frac{1}{4}$	00
B	$\frac{1}{4}$	01
C	$\frac{1}{4}$	10
D	$\frac{1}{8}$	101
E	$\frac{1}{16}$	1110
F	$\frac{1}{16}$	1111

8.13 A discrete source produces the symbols A and B with $P_A = \frac{3}{4}$ and $P_B = \frac{1}{4}$ at a rate of 100 symbols/sec. In an attempt to match the source to a binary channel, the symbols are grouped in blocks of 2 and encoded as follows:

Grouped symbol	Binary code
AA	1
AB	01
BA	001
BB	000

By calculating the source entropy rate and the channel symbol rate, show that this code is not optimum but is reasonably efficient.

8.14 The so-called *binary symmetric channel* (BSC) is a binary channel with noise such that the error probability is the same for both channel symbols, that is, $P_e(0) = P_e(1) = p$, and errors are statistically independent. Using Eq. (8.15), show that the channel capacity is

$$C = s[1 + p \log p + (1 - p) \log (1 - p)]$$

8.15 Referring to Prob. 8.14, find the capacity of a BSC when $p = 0, \frac{1}{4}, \frac{1}{2}$, and 1. Interpret the meaning of your results.

8.16 A discrete channel is to convey information at a rate of 9,000 bits/sec. Determine the minimum number of channel states required if the bandwidth is 1 kHz or 5 kHz.

8.17 Consider the set of channel signals

$$x_i(t) = \text{sinc } (2Bt - i) \qquad i = 0, \pm 1, \pm 2, \ldots$$

Discuss the signal-space interpretation and determine the upper limit on $|i|$ for a space of D dimensions.

8.18 Referring to the correlation detection system of Fig. 8.5 in the limit as $T \to \infty$, write the mean-square errors ϵ_i in terms of the various autocorrelation and crosscorrelation functions and justify the approximation

$$\epsilon_i \approx \begin{cases} 2S + N & i \neq j \\ N & i = j \end{cases}$$

8.19 Verify Eq. (8.25) for μ even and odd. *Hint:*

$$\sum_{k=1}^{n} k^2 = \frac{n(n+1)(2n+1)}{6}$$

8.20 With the aid of Eq. (8.29a) plot β/β_0 versus B_T/R for an ideal communication system. What does this curve imply?

8.21 Suppose it is desired to have real-time voice transmission from Mars to earth, save of course for the propagation delay. Assume a binary PCM system with $f_s = 8$ kHz and $q = 64$. By inserting typical values into Eq. (8.31) estimate the power requirements at the Mars transmitter.

SELECTED SUPPLEMENTARY READING

The literature of information theory applied to communications is indeed rich and abundant. The classic papers, Nyquist (1924, 1928), Hartley (1928), and Shannon (1948, 1949), deserve at least a casual perusal. Shannon (1949) is relatively nonmathematical and highly recommended.

Other useful texts at about the same level as this chapter are Woodward (1953), Goldman (1953), Hancock (1961), Harman (1963), and Beckmann (1967). Goldman further develops the concepts and interpretations of signal space, while Harman treats a number of related topics including statistical decision theory.

Abramson (1963) is a good introduction to the coding aspects of information theory. For the more mathematically inclined, Wozencraft and Jacobs (1965) is definitive, demanding, and rewarding.

Finally, the reader's attention is directed to Pierce (1961). Written for the layman, this fascinating book covers (and quite lucidly) not only the basics of information theory but also its implications for art, music, psychology, etc. Anyone with an interest in these subjects will profit from Pierce's fresh interpretations.

9
Digital Data Systems

Until about 1950, analog signal transmission was the primary stock-in-trade of the communications industry, save, of course, for teletype and telegraph. But that was before the advent of automation and the computer revolution. Today, enormous quantities of digital data are being generated in government, commerce, and science. Furthermore, the need to eliminate manual handling and human error has led to the concept of integrated data processing systems. These developments, coupled with a growing trend toward decentralization, have made the transmission and distribution of digital signals a major task of electrical communication. This chapter is a survey of digital data systems, their design principles, and problems.

While, according to information theory, highly efficient and virtually errorless data transmission is possible, the vast majority of applications simply do not warrant the cost and complexity demanded by a near-ideal communication system. This is not to say that information theory has no place in the study of data transmission but rather that it may be better to do the job now, in less than optimum fashion, than wait until technological breakthroughs permit a more sophisticated solution.

Consequently, the design of a practical data system is usually quite pragmatic, based on two elementary considerations: (1) the gross source rate in, say, binary digits per second, as distinguished from the entropy rate in bits per second, and (2) the desired transmission reliability, i.e., the tolerated error rate. These factors in turn dictate signaling speed and error-probability requirements; indeed, signaling speed and error probability are the defining parameters of digital systems, equivalent to bandwidth and signal-to-noise ratio for analog systems. Thus, before going on, it is perhaps helpful to get a feeling for typical values of these parameters.

Just as analog signals have a wide range of bandwidths, so digital signals have a wide range of *bit rates*,[1] depending on the source and application. The output of manual keyboard devices, such as the teletypewriter, is seldom more than 100 bits/sec. Telemetry rates may be even lower. At the other extreme, internal operations in a modern computer can exceed 10^9 bits/sec. Intermediate rates of 1 to 100 kilobits/sec are attained by electromechanical and electromagnetic data readers. Currently, the best wideband systems have speed capabilities of order 10^7 bits/sec, accommodating all but direct computer-to-computer links. But most digital transmission is via standard telephone voice circuits with top speeds of 1,000 to 5,000 bits/sec. Thus, for the higher-rate sources, buffering is required.

Unlike signaling speed, the question of acceptable error rate is rather nebulous. Obviously, errors are of grave concern in defense surveillance systems, whereas requirements can be quite relaxed if one is telemetering data from measuring instruments having 10 percent accuracy. For the former an error probability of 10^{-5} is probably intolerable; for the latter a probability of 10^{-3} would be an extravagant waste. Nonetheless, error probabilities of order 10^{-4} are representative and suitable to many applications. We shall take this value as a guideline for comparison purposes, bearing in mind its arbitrary nature.

Our earlier studies have shown a direct correspondence between the parameters of analog and digital transmission, specifically, bandwidth versus signaling speed and signal-to-noise ratio versus error probability. Here these relationships are developed in more detail, and we shall frequently rate digital system performance in terms of the usual analog parameters. However, one fundamental distinction must be noted at the start: in analog transmission the goal is to reproduce the actual message *waveform* as closely as possible; in digital transmission

---

[1] Bowing to convention, we use the terms *bit* and *binary digit* interchangeably in this chapter.

the goal is to deliver the message *information* as accurately as possible. Since a digital message consists of a finite number of different symbols and those symbols can be represented in a variety of ways, the digital system designer is not bound to any particular waveform and therefore enjoys considerable latitude.

For example, because a digital signal is *discrete* in nature, system refinements such as *regeneration* and *error-control coding* can be employed. Moreover, it is possible to provide the receiver with a "dictionary" listing what the message elements should look like in the absence of noise and other contamination. Then, given the contaminated signal at the receiver, transmission reliability can be further improved by a matching procedure wherein the received waveform is compared with the possible transmitted waveforms. This is called *coherent detection* and leads to system *optimization* through the use of statistical decision theory.

These special techniques are also examined in this chapter, albeit briefly, after we discuss the fundamentals of data transmission in the context of baseband systems.

9.1 BASEBAND DATA TRANSMISSION

One normally thinks of a digital signal as being a string of discrete-amplitude rectangular pulses. And, in fact, that is often the way it comes from the data source. By way of illustration, Fig. 9.1a shows the binary message 10110100 as it might appear at the output of a digital computer. This waveform, a simple on-off sequence, is said to be *unipolar*, because it has only one polarity, and *synchronous*, because all pulses have equal duration and there is no separation between them.

Generally speaking, unipolar signals contain a nonzero dc component that is difficult to transmit, carries no information, and is a waste of power. Similarly, synchronous signals require timing coordination at transmitter and receiver (like synchronous demodulation), which means design complications. The *polar*[1] (two-polarity) *return-to-zero* signal of Fig. 9.1b gets around both of these problems, but the "spaces" making the signal self-clocking are a waste of transmission time. If efficiency is a dominant consideration, the *polar synchronous* signal of Fig. 9.1c would be preferable. Again, as in analog systems, we have a trade-off between efficient transmission and equipment simplification.

Other representations of the same binary message are shown in Fig. 9.1d and 9.1e. The *dicode* wave consists of polar pulses denoting

---

[1] In this chapter *polar* has the same meaning as *bipolar* did in Chap. 7. The designation bipolar is reserved for a specific type of binary signal in which 1s are represented by pulses of alternating polarity and 0s by no pulse.

only the symbol transitions, while the *quaternary* (four-level) signal is derived by grouping the binary digits in blocks of two. The relative advantages of these and similar variations are left for the reader to investigate.

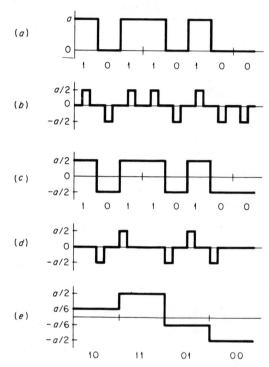

Fig. 9.1 *Digital waveforms. (a) Unipolar synchronous; (b) polar return-to-zero; (c) polar synchronous; (d) dicode; (e) quaternary.*

Regardless of the specific details, the system input signal can generally be described as a pulse train of the form

$$\sum_k a_k p\left(t - \frac{k}{s}\right)$$

where a_k is the amplitude level representing the kth message digit, $p(t)$ is the basic pulse shape with peak value $p(0) = 1$, the pulse-to-pulse spacing is $1/s$, and s is the signaling speed. For example, the polar synchronous signal of Fig. 9.1c has $a_k = \pm a/2$ and $p(t) = \Pi(t/\tau)$, where the pulse duration is $\tau = 1/s$. A return-to-zero signal would have $\tau < 1/s$, etc.

If the system is linear and distortionless over all frequencies, i.e., has infinite bandwidth, then $p(t)$ suffers no degradation in transmission, and an arbitrarily large signaling speed can be achieved by using very short pulses. But a practical system has finite bandwidth and less than ideal frequency response, causing the pulses to spread out and overlap. The engineer must therefore shape the output signal so as to *minimize intersymbol interference* due to overlapping and, at the same time, *maximize signaling rate*, objectives which are mutually contradictory. It is this problem we examine first.

Signal shaping—signaling speed

Assuming a linear system and superposition, the output waveform resulting from the input $\sum_k a_k p \left(t - \dfrac{k}{s} \right)$ can be written as

$$\sum_k A_k x \left(t - \frac{k}{s} - t_d \right)$$

where $A_k/a_k = K_0$ is the system voltage gain, t_d is the time delay, and $x(t)$ is the output pulse shape normalized to the peak value $x(0) = 1$. This formulation allows us to concentrate on the single pulse shape $x(t)$ and its spectrum $X(f)$, rather than complete messages.

Expressed in terms of the *system transfer function* $H(f)$ and the input pulse spectrum $P(f) = \mathcal{F}[p(t)]$, we have

$$K_0 x(t - t_d) = \mathcal{F}^{-1}[H(f)P(f)]$$

so

$$K_0 e^{-j\omega t_d} X(f) = H(f)P(f) \tag{9.1}$$

But the problems at hand, namely, intersymbol interference and signaling rate, are time-domain problems. It is therefore wise to observe that

$$x(t) = \frac{1}{K_0} \int_{-\infty}^{\infty} H(f)P(f)e^{j\omega(t+t_d)} \, df$$

and hence $x(t)$ is essentially the *impulse response* of a filter having transfer function $H(f)P(f)$. This viewpoint has important practical implications on the selection of $x(t)$, since $H(f)$ must be physically realizable.

Ignoring for simplicity the constant time delay t_d, consider the train of output pulses

$$\cdots + A_{-1} x \left(t + \frac{1}{s} \right) + A_0 x(t) + A_1 x \left(t - \frac{1}{s} \right) + \cdots$$

Since $x(t)$ has its peak at $t = 0$ and all other pulses peak at $t_k = k/s$, $k = \pm 1, \pm 2, \ldots$, a sure and direct cure for intersymbol interference is to require that $x(t) = 0$ for $|t| \geq 1/2s$. However, practical pulse shapes in finite-bandwidth systems generally have decaying tails and cannot be strictly timelimited. On the other hand, asymptotically timelimited functions, such as the gaussian pulse, require excessive spacing and therefore reduced signaling speed to suitably minimize the overlap. Accounting for these facts, elimination of intersymbol interference necessitates that $x(t_k) = 0$; in other words, $x(t)$ should have *periodic zero values* spaced by $1/s$, a condition known as *Nyquist's first criterion*.

One pulse shape having this property is the familiar sinc pulse $x(t) = \text{sinc } 2Bt$, with $B = s/2$. Recall that it was precisely this pulse that was used in Chap. 8 to demonstrate signaling at the maximum rate $s = 2B$. But sinc pulses are not the answer here. For one reason, the leading and trailing oscillations die off rather slowly; synchronization then becomes quite critical, and small variations of signaling speed result in substantial intersymbol interference. Moreover, the pulse spectrum $X(f) = (1/2B) \Pi(f/2B)$ requires that $H(f)$ have a sharp cutoff at $|f| = B$, which cannot be realized.

Taking a cue from the last remark, let us restrict $X(f)$ to resemble filter functions with a more gradual roll-off characteristic, as in Fig. 9.2.

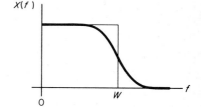

Fig. 9.2 *Pulse spectrum with vestigial symmetry.*

Nyquist's *vestigial-symmetry theorem* then states that if $X(f)$ is purely real (no imaginary part) and has odd symmetry about the nominal cutoff frequency W, then $x(t)$ is an even time function with a peak at $t = 0$ and periodic zero values spaced by $1/2W$. Therefore, it is possible to signal at $s = 2W$ without intersymbol interference. Proof of the theorem is left as an exercise for the reader.

Infinitely many functions have the vestigial symmetry required; some are practical, but most are not. One class of particular interest is the *sinusoidal* roll-off described by

$$X(f) = \begin{cases} \dfrac{1}{2W} & |f| < W - f_1 \\ \dfrac{1}{4W}\left[1 - \sin\dfrac{\pi(|f| - W)}{2f_1}\right] & W - f_1 \leq |f| \leq W + f_1 \\ 0 & |f| > W + f_1 \end{cases}$$

$$(9.2)$$

where $f_1 \leq W$ and the total bandwidth is $B = W + f_1$. [Note that this includes the rectangular case $X(f) = (1/2B)\Pi(f/2B)$ if $f_1 = 0$.] The resultant pulse shape found by transforming $X(f)$ is

$$x(t) = \operatorname{sinc} 2Wt \, \frac{\cos 2\pi f_1 t}{1 - (4f_1 t)^2} \tag{9.3}$$

which reduces to a sinc pulse when $f_1 = 0$.

Figure 9.3 shows $X(f)$ and $x(t)$ for $f_1 = 0$, $W/2$, and W. In all cases $x(t)$ has the desired zeros at $t_k = k/2W$, so one can signal synchronously at $s = 2W$. However, the leading and trailing oscillations decay

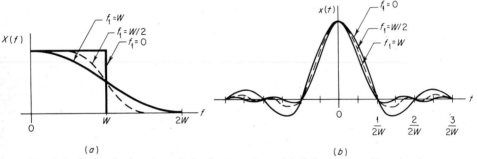

(a)   (b)

Fig. 9.3 *Pulses having sinusoidal roll-off spectra.* (a) *Pulse spectra;* (b) *pulse shapes.*

more rapidly if $f_1 \neq 0$, meaning that synchronization is less critical than with sinc pulses. Further inspection reveals two other convenient properties of $x(t)$ when $f_1 = W$: (1) the half-amplitude pulse width is exactly $1/2W$, which is the pulse-to-pulse spacing if $s = 2W$; (2) there are additional zero crossings at $t = \pm 3/4W$, $\pm 5/4W$, ... As a result, a digital waveform constructed from such pulses will have negligible intersymbol interference for any signaling speed $s \leq 2W$. Moreover, the corresponding system response $H(f)$ can be closely approximated.

Because of these advantages, practical data systems frequently use the 100 percent sinusoidal roll-off ($f_1 = W$) for which Eq. (9.2) reduces to

$$
\begin{aligned}
X(f) &= \frac{1}{4W}\left(1 + \cos\frac{\pi f}{2W}\right)\Pi\left(\frac{f}{2W}\right) \\
&= \frac{1}{2W}\cos^2\frac{\pi f}{4W}\,\Pi\left(\frac{f}{2W}\right)
\end{aligned} \tag{9.4}
$$

called the *raised-cosine* or *cosine-squared* pulse spectrum. The corresponding pulse shape is conveniently written in the form

$$x(t) = \frac{\operatorname{sinc} 4Wt}{1 - (4Wt)^2} \tag{9.5}$$

A binary message 10110100 as represented by these pulses is illustrated in Fig. 9.4 for $s = 2W$ and $A_k = \pm 1$.

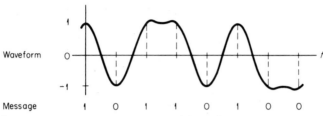

Fig. 9.4 *Baseband waveform for the binary message 10110100.*

Note, however, that the total bandwidth required for a raised-cosine spectrum is $B = W + f_1 = 2W$, and hence the maximum (synchronous) signaling speed is $s = B$, as compared with the theoretical maximum $s = 2B$ of an ideal rectangular response. With a 50 percent sinusoidal roll-off ($f_1 = W/2$), the signaling speed can be increased to $1.5B$, but the necessary system response is more difficult to obtain, and intersymbol interference may be a problem if synchronization is lost. Generally speaking, the maximum signaling speed for a given bandwidth B is

$$s_{\max} = \kappa B \qquad 1 \leq \kappa \leq 2 \tag{9.6}$$

where the larger values of κ call for more exacting system design and control.

Terminal filters

Having decided upon a suitable output pulse shape, whether that of Eq. (9.5) or any other, the system transfer function required to achieve pulse shaping is, from (9.1),

$$H(f) = K_0 e^{-j\omega t_d} \frac{X(f)}{P(f)}$$

where $P(f)$ is the input pulse spectrum. For example, taking $X(f)$ as given by (9.4) and assuming $p(t)$ is a rectangular pulse of duration $\tau \leq 1/s$, we have

$$H(f) = \frac{K_0}{2W\tau} \frac{\cos^2 \pi f/2W}{\operatorname{sinc} f\tau} \Pi\left(\frac{f}{2W}\right) e^{-j\omega t_d}$$

which is sometimes called a *modified raised cosine*. If $\tau \ll 1/2W$, then $H(f)$ is essentially a raised cosine.

Of course $H(f)$ is not a distinct physical entity, for it represents the fixed characteristics of the transmission medium, say $H_m(f)$, plus any terminal filters that might be included at transmitter and receiver for pulse-shaping purposes. Designating the latter as $H_t(f)$ and $H_r(f)$, respectively, the overall system frequency response is

$$H(f) = H_t(f)H_m(f)H_r(f)$$

the disposition of the various elements being indicated in Fig. 9.5.

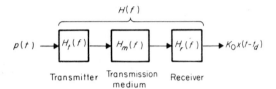

Fig. 9.5

Although we have been dealing with baseband transmission, this formulation is also applicable to modulation systems if the modulation-demodulation effects, referred to baseband, can be incorporated in $H_m(f)$. ·However, *nonlinear* modulation, e.g., FM or AM with envelope detection, entails a different approach, for which see Sunde (1959).

Given $X(f)$, $P(f)$, and $H_m(f)$, the terminal filters are designed according to the relationship

$$H_t(f)H_r(f) = \frac{K_0 e^{-j\omega t_d}}{H_m(f)} \frac{X(f)}{P(f)} \tag{9.7}$$

Obviously, any combination satisfying Eq. (9.7) is acceptable for the desired pulse shaping. The deciding factor as to how the individual terminal filters should be apportioned then comes from noise considerations. Specifically, since the receiving filter is located where the signal is contaminated by noise, and since there is some flexibility in the design of $H_r(f)$, it behooves the system engineer to choose a receiving filter that maximizes the output pulse amplitude relative to the noise.

The design of such *optimum receiving filters* is a rather difficult task and will not be covered here, the principles of optimization being deferred to Sec. 9.4. Suffice it to say that $H_r(f)$ should have a response that emphasizes those portions of the frequency band where the signal energy density is large and the noise power density is small and deemphasizes those portions where the reverse is true. Correspondingly, the transmitting filter $H_t(f)$ must compensate for the signal deemphasis of $H_r(f)$ by having an equalizing preemphasis response. Thus data terminal filters serve a function similar to that of FM preemphasis-deemphasis filters.

Noise and errors

The problem of noise and errors was briefly outlined in conjunction with our discussion of PCM decoding. It was there noted that error probabilities generally depend on the signaling waveform and the nature of the noise, as well as on the signal-to-noise ratio. Here we get into the details of error-probability calculations for several types of baseband signals, but we shall assume that the noise is always *gaussian*. This assumption is quite reasonable for *linear* baseband systems, since most electrical noise is gaussian-distributed and gaussian functions are invariant under linear operations. Furthermore, the analysis methods and results of this section will pave the way for more complex situations encountered later.

To begin with a simple case, consider a received polar binary signal having peak amplitudes $A_j = \pm A/2$ representing the binary digits 1 and 0. Let the signal be contaminated by additive gaussian noise $n(t)$ with zero mean and variance σ^2. A typical signal-plus-noise waveform $y(t)$ as it might appear at the input to the data demodulator is illustrated in Fig. 9.6. The function of the demodulator is then to convert the waveform $y(t)$ into suitable output message digits.

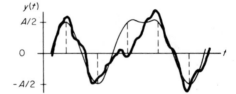

Fig. 9.6 *Polar binary waveform plus noise.*

A simple and direct demodulation technique is to decide, at the appropriate times, whether $y(t)$ is closer to $+A/2$ (a 1 presumably intended) or closer to $-A/2$ (a 0 presumably intended). Intuitively, the logical *decision rule* becomes: choose 1 if $y(t) > 0$, chose 0 if $y(t) < 0$, and flip a coin if $y(t) = 0$. (This last event, being rare, will receive no further attention.) The demodulator can therefore take the form of a synchronized decision circuit whose crossover or *threshold level* is set at zero volt, and conversion errors occur whenever the noise causes $y(t)$ to be on the "wrong" side of the threshold at the decision time.

Thus, insofar as error probabilities are concerned, we have two random variables

$$y_1 = \frac{A}{2} + n \qquad y_0 = -\frac{A}{2} + n$$

corresponding to the intended digits 1 and 0. Then, if a 1 was intended, the probability of conversion error is

$$P_{e_1} = P(y_1 < 0) = P\left(\frac{A}{2} + n < 0\right)$$

and similarly

$$P_{e_0} = P(y_0 > 0) = P\left(-\frac{A}{2} + n > 0\right)$$

Hence, the net error probability becomes

$$P_e = P_1 P_{e_1} + P_0 P_{e_0} \tag{9.8}$$

where P_1 and P_0 are the digit probabilities at the source, not necessarily equal but usually so. In any case $P_1 + P_0 = 1$, since one or the other must be transmitted.

Because n is gaussian with probability density function

$$p(n) = \frac{1}{\sqrt{2\pi}\,\sigma}\, e^{-n^2/2\sigma^2}$$

it follows that y_1 and y_0 are also gaussian-distributed with variance σ^2 but with mean values $\bar{y}_1 = A/2$ and $\bar{y}_0 = -A/2$. Figure 9.7 conveniently summarizes the situation by showing the density functions

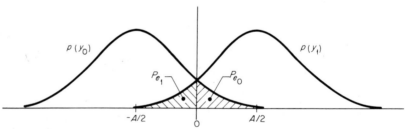

Fig. 9.7 *Signal-plus-noise probability density functions, polar binary waveform.*

$p(y_1)$ and $p(y_0)$. Recalling the area interpretation of probability densities gives

$$
\begin{aligned}
P_{e_1} = P(y_1 < 0) &= \int_{-\infty}^{0} p(y_1)\, dy_1 \\
&= \frac{1}{\sqrt{2\pi}\,\sigma} \int_{-\infty}^{0} \exp\left[-\frac{(y_1 - A/2)^2}{2\sigma^2}\right] dy_1 \\
P_{e_0} = P(y_0 > 0) &= \int_{0}^{\infty} p(y_0)\, dy_0 \\
&= \frac{1}{\sqrt{2\pi}\,\sigma} \int_{0}^{\infty} \exp\left[-\frac{(y_0 + A/2)^2}{2\sigma^2}\right] dy_0
\end{aligned}
\tag{9.9}
$$

which are indicated in the figure as shaded areas under the gaussian tails. From symmetry we see that the centered threshold level of zero volts yields $P_{e_1} = P_{e_0}$. Furthermore, this is the *optimum* threshold level if the digits are equiprobable ($P_1 = P_0$), since any other choice would increase P_{e_1} more than it reduces P_{e_0}, or vice versa.

A unique property of binary signals in gaussian noise is now also apparent in Fig. 9.7. If the intended amplitude is $+A/2$, superimposed *positive* noise excursions have no detrimental effect; and similarly for $-A/2$ with negative noise excursions. Hence, because $n(t)$ is equally likely to be positive or negative, the demodulator will be correct at least half the time, regardless of the signal-to-noise ratio. However, one should bear in mind that a binary message with 50 percent errors is 100 percent worthless.

Returning to the calculation, the change of variable $\sigma u = y_1 - A/2 = y_0 + A/2$ reduces (9.9) to

$$P_{e_1} = P_{e_0} = \frac{1}{\sqrt{2\pi}} \int_{A/2\sigma}^{\infty} e^{-u^2/2}\, du = \frac{1}{2} - \frac{1}{\sqrt{2\pi}} \int_0^{A/2\sigma} e^{-u^2/2}\, du$$

$$= \tfrac{1}{2} - \Phi\!\left(\frac{A}{2\sigma}\right)$$

where $\Phi(A/2\sigma)$ was defined by Eq. (4.36) in terms of the normal curve of error (see Fig. 4.6 or Table D for numerical values). Inserting this result into (9.8) gives for the net error probability[1]

$$P_e = (P_1 + P_0)\left[\tfrac{1}{2} - \Phi\!\left(\frac{A}{2\sigma}\right)\right]$$

$$= \tfrac{1}{2} - \Phi\!\left(\frac{A}{2\sigma}\right) \tag{9.10}$$

which is independent of the digit probabilities because $P_{e_1} = P_{e_0}$.

Finally, relating error probability and signal-to-noise ratio, we approximate the uncontaminated waveform as a string of full-width rectangular pulses of amplitude $\pm A/2$. The average signal power is then $S = (A/2)^2$, regardless of the digit probabilities, while $N = \sigma^2$ is the average noise power; hence $S/N = (A/2\sigma)^2$. But it is well to bear in mind that N depends upon the bandwidth, which, in turn, is proportional to signaling speed. Specifically, if the noise is white with positive-fre-

---

[1] Also written in the form $P_e = \tfrac{1}{2}[1 - \operatorname{erf} A/(2\sqrt{2}\,\sigma)] = \tfrac{1}{2}\operatorname{erfc} A/(2\sqrt{2}\,\sigma)$ where $\operatorname{erf} v = \dfrac{2}{\sqrt{\pi}} \displaystyle\int_0^v e^{-u^2}\, du$ is the *error function* and $\operatorname{erfc} v = 1 - \operatorname{erf} v$ is the *complementary error function*.

quency density η, then

$$N = \eta B = \frac{\eta s}{\kappa}$$

and

$$\left(\frac{A}{2\sigma}\right)^2 = \frac{S}{N} = 2\frac{\kappa}{2}\frac{S}{\eta s}$$

where we have used Eq. (9.6), κ being the pulse-shape factor.

Now S is the average signal power or the *energy* per unit time and s is the digit rate; therefore, $S/s = E$ is the *average energy per digit*. Similar to the normalized parameter $z = P_T/\eta W$ used for comparisons of analog modulation, we now define the digital-modulation parameter

$$\mathcal{E} = \frac{\kappa}{2}\frac{S}{\eta s} = \frac{\kappa}{2}\frac{E}{\eta} \tag{9.11}$$

which can be interpreted as the average digit energy relative to the noise power density, both being measured at the receiver, taking pulse shape into consideration via $\kappa/2$. Note that $\mathcal{E}_{max} = E/\eta$, since $\kappa/2 \le 1$.

For the case at hand $\mathcal{E} = \frac{1}{2}(S/N) = \frac{1}{2}(A/2\sigma)^2$, so

$$P_e = \frac{1}{2} - \Phi\left(\sqrt{\frac{S}{N}}\right)$$
$$= \frac{1}{2} - \Phi(\sqrt{2\mathcal{E}}) \qquad \text{polar binary} \tag{9.12}$$

It is a simple matter to modify our results for the case of a unipolar (on-off) binary signal having $A_j = A$ or 0. If the digits are again equiprobable, the optimum threshold level becomes $+A/2$, and P_e is correctly given by Eq. (9.10). But the average signal power is now

$$S = \sum_j P_j A_j^2 = \frac{1}{2} \times 0^2 + \frac{1}{2} \times A^2 = \frac{A^2}{2}$$

or double the power of a polar signal having the same amplitude spacing. (This of course merely reflects the wasted power in the redundant dc component of unipolar waveforms.) Therefore

$$P_e = \frac{1}{2} - \Phi\left(\sqrt{\frac{S}{2N}}\right)$$
$$= \frac{1}{2} - \Phi(\sqrt{\mathcal{E}}) \qquad \text{unipolar binary} \tag{9.13}$$

so the transmitted power must be twice that of before to achieve the same error probability. Illustrative of the significance of this 3-db power difference, suppose a polar system has $P_e = 10^{-6}$; the corresponding value for a unipolar system with the same transmitted power is $P_e = 4 \times 10^{-4}$, an error increase by a factor of 400.

Because we have taken due care in our examination of errors for binary signals, the extension to multilevel, or μ-ary, signals is quite straightforward, providing the noise is gaussian.   Consider, for instance, a polar trinary signal ($\mu = 3$) with output pulse amplitudes $A_j = A/2$, 0, or $-A/2$, representing the trinary digits 2, 1, and 0, respectively.   The equivalent to Fig. 9.7 has three gaussian density functions (Fig. 9.8), and

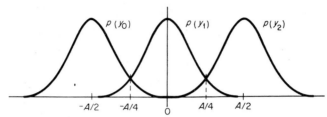

Fig. 9.8 *Signal-plus-noise probability density functions, polar trinary waveform.*

we see that two threshold levels are required.   Under the usual condition of equiprobable digits, the optimum threshold levels are easily shown to be $\pm A/4$, for which

$$P_{e_2} = P_{e_0} = \tfrac{1}{2} - \Phi\left(\frac{A}{4\sigma}\right)$$

whereas

$$P_{e_1} = 2\left[\tfrac{1}{2} - \Phi\left(\frac{A}{4\sigma}\right)\right]$$

because both positive and negative noise excursions cause errors when $A_j = 0$.   (By another choice of thresholds it is possible to equalize the per-digit error probabilities, but the cost is increased net probability P_e.) Hence

$$P_e = P_2 P_{e_2} + P_1 P_{e_1} + P_0 P_{e_0}$$
$$= \tfrac{4}{3}\left[\tfrac{1}{2} - \Phi\left(\frac{A}{4\sigma}\right)\right]$$

where we have inserted $P_2 = P_1 = P_0 = \tfrac{1}{3}$.

Generalizing to arbitrary μ with equal digit probabilities, similar reasoning gives

$$P_{e\mu} = \frac{2(\mu - 1)}{\mu}\left\{\tfrac{1}{2} - \Phi\left[\frac{A}{2(\mu - 1)\sigma}\right]\right\} \tag{9.14}$$

where the *spacing* between adjacent output pulse amplitudes is $A/(\mu - 1)$, that is, A is the peak-to-peak voltage, and the $\mu - 1$ threshold levels are centered between the pulse amplitudes. Equation (9.14) clearly reduces to (9.10) when $\mu = 2$. Nonetheless $P_{e\mu}$ does not tell the full story, for two reasons: (1) a μ-ary digit in general represents more information than a binary digit; (2) there are differing severities of error in μ-ary systems, depending on whether the noise shifts the apparent amplitude by one or more steps. Unfortunately an analytic assessment of these effects, say in terms of errors per bit, is quite difficult.

As to average power calculations, we proceed in the same fashion as was done for PCM in Sec. 8.4. Specifically, with $P_j = 1/\mu$ and amplitude spacing $A/(\mu - 1)$

$$S = \sum_{j=1}^{\mu} P_j A_j^2 = \begin{cases} \dfrac{\mu + 1}{12(\mu - 1)} A^2 & \text{polar} \\ \dfrac{2\mu - 1}{6(\mu - 1)} A^2 & \text{unipolar} \end{cases}$$

and therefore

$$\left[\frac{A}{2(\mu - 1)\sigma} \right]^2 = \frac{3}{\mu^2 - 1} \frac{S}{N} = \frac{6}{\mu^2 - 1} \varepsilon \qquad \text{polar} \qquad (9.15a)$$

or

$$\left[\frac{A}{2(\mu - 1)\sigma} \right]^2 = \frac{3}{2(\mu - 1)(2\mu - 1)} \frac{S}{N}$$

$$= \frac{3}{(\mu - 1)(2\mu - 1)} \varepsilon \qquad \text{unipolar} \qquad (9.15b)$$

which can be substituted into Eq. (9.14). Note that, for large μ, unipolar signaling requires 6 db more power.

Repeater systems—regeneration

Frequently in long-haul data transmission, *repeating amplifiers* must be inserted at regular intervals to maintain adequate signal levels. These repeaters are normally designed to have a power gain just compensating for the transmission loss between stations. And although long-haul systems are seldom baseband systems, the effects of repeaters on error probability can be discussed here without reference to any particular form of transmission.

If the repeaters are nothing more than amplifiers, they can do no better than amplify both signal and noise. But each repeater also adds noise of its own, which, as a rule, dominates all other sources of noise. Consequently, the signal-to-noise ratio progressively decreases with each station, so that after M links, the final signal-to-noise ratio is

$$\left(\frac{S}{N} \right)_M = \frac{1}{M} \frac{S}{N}$$

where S/N is the value after the first hop. Because error probability has a critical dependence on signal-to-noise ratio, the transmitted power *per station* must be increased linearly with M just to stay even, a factor not to be sneezed at, since, for example, there are about 100 repeaters in a transcontinental microwave system.

But, as mentioned in Chap. 7, there is an alternative method, namely, through the use of *regenerative* repeaters. A regenerative repeater consists of a complete data receiver and transmitter back to back in one package. The receiving portion converts incoming signals to message digits, making a few errors in the process; the digits are then delivered to the transmitting portion, which in turn generates a new data signal for transmission to the next station. The regenerated signal is thereby completely stripped of random noise but does contain some errors.

To analyze the system performance consider a binary signal with $p = P_{e_1} = P_{e_0}$ being the conversion error probability at each repeater. As a given digit passes from station to station, it may suffer cumulative conversion errors. If the number of erroneous conversions is *even*, they cancel out, and a correct digit is delivered to the destination. (Note that this is true only for *binary* data.) Now the probability of ν conversion errors in M links is given by the *binomial distribution* of Eq. (4.31), that is,

$$P_M(\nu) = \binom{M}{\nu} p^\nu (1 - p)^{M-\nu}$$

The net error probability is then the probability that ν is *odd;* specifically,

$$P_e = \sum_{\nu \text{ odd}} P_M(\nu) = \binom{M}{1} p(1 - p)^{M-1} + \binom{M}{3} p^3(1 - p)^{M-3} + \cdots$$
$$\approx Mp$$

where the approximation applies for $p \ll 1$ and M not too large. Hence the error probability increases linearly with the number of repeaters.

Taking p as given by Eq. (9.12), with S/N being the signal-to-noise ratio for one hop, the error probability for an M-link regenerative system is

$$P_e = M \left[\tfrac{1}{2} - \Phi \left(\sqrt{\frac{S}{N}} \right) \right]$$

whereas, without regeneration,

$$P_e = \tfrac{1}{2} - \Phi \left(\sqrt{\frac{S}{MN}} \right)$$

since the final signal-to-noise ratio is reduced by $1/M$. Figure 9.9 shows the power saving provided by regeneration, as a function of M, the error probability being fixed at $P_e = 10^{-4}$. Thus, for example, a 10-station nonregenerative baseband system requires about 8.5 db more transmitted power (per repeater) than a regenerative system.

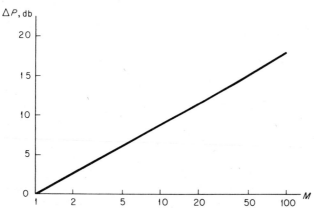

Fig. 9.9   *Power saving gained by regeneration, $P_e = 10^{-5}$.*

However, the design of regenerative repeaters is not trivial, and the cost and complexity of added equipment may counterbalance the power reduction in some instances. This is particularly true of modulated data transmission, for which regeneration entails complete demodulation at each repeater. Furthermore, nonregenerative repeaters can be used with any type of digital signal (or even analog signals for that matter), while regenerative systems are locked in on one particular signal type. Regeneration is therefore not a pat answer for all long-haul digital transmissions.

Other considerations

There are numerous other factors that, quite properly, might be called fundamental considerations in the design of practical baseband systems. These include: timing and synchronization, which is obviously critical in regeneration and data demodulation; optimum filter design for the suppression of both impulse noise and steady background noise; equalization[1] to correct for transmission imperfections and the consequent pulse distortion (a necessity if the channel characteristics are not fully known or

---

[1] A particularly attractive means of equalization for digital signals is through the use of *transversal filters*, which are essentially time-domain filters. See Kallman (1940) for a brief introduction.

are time-variable); time-division and frequency-division multiplexing techniques; etc.

Each of these is, however, a specialization in its own right and must be omitted here in deference to topics of more general interest. The reader wishing to pursue any of the aforementioned subjects will find abundant material in the professional literature, to which the selected readings at the end of the chapter are a guide and starting point.

9.2 DIGITAL-MODULATION SYSTEMS

Just as there are a multitude of modulation techniques for analog signals, so is it that digital information can be impressed upon a carrier wave in many ways. The three basic forms of digital modulation are known as *amplitude-shift keying, frequency-shift keying,* and *phase-shift keying,* which, as the names imply, are more or less equivalent to AM, FM, and PM, respectively.

This section covers the basic types and some of their variations, drawing upon the results of the previous section coupled with the modulation theory from Chaps. 5 and 6. Primary emphasis will be given to system performance in the presence of noise, i.e., error probabilities as a function of signal-to-noise ratio, and the relative merits of coherent detection. For the most part, we shall confine our attention to binary signals; the extension to μ-ary signals is not conceptually difficult but involves more arduous mathematics.

Amplitude-shift keying (ASK)

Given a digital message, the simplest modulation technique is ASK, wherein the c-w amplitude is switched between two or more values, usually *on* and *off* for binary signals. The resultant modulated wave then consists of RF pulses or *marks*, representing binary 1, and *spaces*, representing binary 0. Figure 9.10 shows an ASK waveform corresponding to the baseband signal of Fig. 9.4.

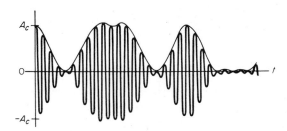

Fig. 9.10 *Binary ASK wave-*
      form.

Such waveforms can be generated by a conventional AM modulator having a polar input signal or by a balanced modulator with a unipolar input. In either case we have double-sideband linear modulation whose transmission bandwidth is always twice that of the baseband bandwidth. If s is the signaling speed in digits or pulses per second, then the necessary transmission bandwidth becomes

$$B_T = 2B = 2\frac{s}{\kappa}$$

where B is the baseband bandwidth and κ is the pulse-shape factor used in Eq. (9.6). In practice an even larger value of B_T may be desirable, so that, for purposes of minimizing noise, most of the pulse-shape filtering can be done after detection.

Alternately, to reduce transmission bandwidth or increase signaling speed for a given bandwidth, the modulated wave can be passed through a VSB filter. This results in VSB-ASK with $B_T \approx B = s/\kappa$ since one sideband is removed save for the vestigial portion. Complete sideband suppression, i.e., SSB, is not feasible for the reasons we have discussed in Sec. 5.4.

Turning to the receiving end—the most critical part of any data system—two methods of demodulating ASK are possible, *coherent* (synchronous) detection and *noncoherent* (envelope) detection. Both accomplish the same function insofar as the signal is concerned, with envelope detection being the easier to instrument. On the other hand, coherent detection is somewhat less vulnerable to noise. Since both methods are encountered in practice, and the analyses are quite different, we shall investigate each one in turn.

Coherent detection of ASK

For coherent detection the incoming signal is multiplied by a locally generated sinusoid, synchronized to the carrier, and the product is low-pass-filtered (Fig. 9.11). Taking the binary ASK signal

$$x_c(t) = A_j \cos \omega_c t \qquad A_j = A_c \text{ or } 0$$

and assuming the noise is *narrowband gaussian*, so that, per Eq. (4.57),

$$n(t) = n_c(t) \cos \omega_c t - n_s(t) \sin \omega_c t$$

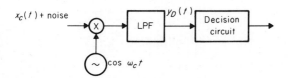

Fig. 9.11 *Coherent detection system for ASK.*

coherent detection of the signal plus noise results in

$$y_D(t) = K_D[A_j + n_c(t)]$$

where K_D is the detection constant. (The omitted details are identical to those for conventional synchronous detection, Sec. 5.6.)

Since $y_D(t)$ is the input to the data demodulator and $n_c(t)$ is known to be gaussian if $n(t)$ is gaussian, it follows that our results for unipolar baseband transmission are directly applicable here. In particular, the decision threshold should be set at $K_D A_c/2$ if the digits are equiprobable, and the error probability is

$$P_e = \tfrac{1}{2} - \Phi\left(\frac{A_c}{2\sigma}\right) \tag{9.16}$$

where $\sigma^2 = \overline{n_c{}^2} = \overline{n^2} = N$, N being the predetection noise power.

Putting P_e in terms of signal-to-noise ratio, or better yet in terms of the parameter $\mathcal{E}$ from Eq. (9.11), inspection of Fig. 9.10 shows the modulated signal power is approximately $S = \tfrac{1}{2}(A_c{}^2/2)$, while $\sigma^2 = N = \eta B_T = 2\eta s/\kappa$. Hence,

$$\left(\frac{A_c}{2\sigma}\right)^2 = \frac{S}{N} = \frac{\kappa S}{2\eta s} = \mathcal{E}$$

and (9.16) becomes

$$P_e = \tfrac{1}{2} - \Phi\left(\sqrt{\frac{S}{N}}\right) \tag{9.17a}$$

or

$$P_e = \tfrac{1}{2} - \Phi(\sqrt{\mathcal{E}}) \tag{9.17b}$$

Equation (9.17a) is somewhat deceptive, suggesting as it does that coherent ASK has the same error probability as polar baseband transmission. However, because the transmission bandwidth is double the baseband bandwidth, the predetection noise is twice the baseband noise. This fact is included in (9.17b), which, when compared with (9.12) and (9.13), reveals that coherent ASK is equivalent to *unipolar* baseband and requires 3 db more energy (or power) than polar baseband systems having the same signaling speed, noise density, etc.

As for μ-ary ASK, assuming equiprobable digits, it is readily shown that

$$P_{e\mu} = \tfrac{1}{2} - \Phi\left[\sqrt{\frac{3}{(\mu - 1)(2\mu - 1)}\frac{S}{N}}\right] \tag{9.18}$$

whose derivation follows from (9.14) and (9.15).

Noncoherent detection of ASK

An ideal envelope detector has as its output the envelope of the input wave. But envelope detection is a *nonlinear* operation, which, when performed on gaussian noise, generally produces a nongaussian output. Therefore, our previous conclusions are not applicable here, and we must give attention to the statistical properties of the input signal plus noise in order to evaluate error probabilities for noncoherent ASK. In addition, though the following development is somewhat involved, the results will be useful in other problems.

To begin the analysis the contaminated input signal is written in the form

$$y(t) = x_c(t) + n(t)$$
$$= A_j \cos \omega_c t + n_c(t) \cos \omega_c t - n_s(t) \sin \omega_c t$$
$$= \xi(t) \cos \omega_c t - n_s \sin \omega_c t$$

where we have introduced the new random variable

$$\xi(t) = A_j + n_c(t)$$

Converting $y(t)$ to the envelope-and-phase expression gives

$$y(t) = r(t) \cos [\omega_c t + \phi(t)]$$

where

$$r^2 = \xi^2 + n_s^2 \qquad \phi = \arctan \frac{n_s}{\xi}$$

and we are specifically concerned with the resultant envelope $r(t)$, which is the detected signal, and its probability density function $q(r)$.

Before plunging into the mathematics, let us speculate on the nature of $q(r)$ for certain extreme conditions, namely, when $A_j = 0$ and $A_j = A_c \gg \sqrt{N}$. In the former case $r(t)$ reduces to $r_n(t)$, the envelope of the noise alone; and this is known to have a *Rayleigh distribution*

$$q(r_n) = \frac{r_n}{N} e^{-r_n^2/2N} \qquad r_n \geq 0 \tag{9.19}$$

with mean value $\overline{r_n} = \sqrt{\pi N/2}$ (see Sec. 4.7). At the other extreme, if $A_c^2 \gg N$, corresponding to $S/N \gg 1$, then n_c and n_s are small compared to A_c most of the time. Hence

$$r = (A_c^2 + 2A_c n_c + n_c^2 + n_s^2)^{1/2}$$
$$= A_c \left[1 + 2\frac{n_c}{A_c} + \left(\frac{n_c}{A_c}\right)^2 + \left(\frac{n_s}{A_c}\right)^2 \right]^{1/2}$$
$$\approx A_c + n_c$$

for which $q(r)$ is approximately *gaussian* with mean $\bar{r} = A_c$ and variance N because n_c (and n_s) is a gaussian variate with $\bar{n}_c = 0$ and $\sigma^2 = N$. However, this conclusion must be qualified by the condition $r \geq 0$, envelopes being nonnegative.

As to the exact derivation of $q(r)$, we first note that $\xi = A_j + n_c$ is gaussian-distributed with $\bar{\xi} = A_j$ and variance N. Furthermore, since n_c and n_s are statistically independent, it follows that ξ and n_s are likewise independent, and their joint probability density function is the product of two gaussians

$$
\begin{aligned}
p(\xi, n_s) &= p(\xi)p(n_s) \\
&= \frac{1}{2\pi N} \exp - \frac{(\xi - A_j)^2 + n_s^2}{2N}
\end{aligned}
$$

Making the transformation

$$
q(r,\phi)\, dr\, d\phi = p(\xi, n_s)\, d\xi\, dn_s
$$

where $\xi = r \cos \phi$, $n_s = r \sin \phi$, etc., gives for the envelope and phase density function

$$
q(r,\phi) = \frac{r}{2\pi N} \exp - \frac{r^2 - 2A_j r \cos \phi + A_j^2}{2N} \tag{9.20}
$$

where $r \geq 0$ and $-\pi \leq \phi \leq \pi$ by definition.

Now, unlike the case of narrowband noise alone, the product term $r \cos \phi$ in (9.20) means that r and ϕ are not statistically independent, that is, $q(r,\phi) \neq q(r)q(\phi)$. Separation of $q(r)$ then requires the integration rule of Eq. (4.23)

$$
\begin{aligned}
q(r) &= \int_{-\pi}^{\pi} q(r,\phi)\, d\phi \\
&= \frac{r}{2\pi N} e^{-(A_j{}^2 + r^2)/2N} \int_{-\pi}^{\pi} \exp\left(\frac{A_j r}{N} \cos \phi\right) d\phi \\
&= \frac{r}{N} e^{-(A_j{}^2 + r^2)/2N} I_0\left(\frac{A_j r}{N}\right) \qquad r \geq 0 \tag{9.21}
\end{aligned}
$$

where $I_0(v)$ is the *modified Bessel function* of the first kind and zero order, defined in general as $I_0(v) = \dfrac{1}{2\pi} \displaystyle\int_0^{2\pi} \exp\,(v \cos u)\, du$.

The complexity of $q(r)$ as written above is perhaps discouraging, and indeed this is not a trivial function. However, for very large or very small arguments, $I_0(v)$ can be approximated by

$$
I_0(v) \approx \begin{cases} e^{v^2/4} & v \ll 1 \\[2mm] \dfrac{e^v}{\sqrt{2\pi v}} & v \gg 1 \end{cases}
$$

Then, with $A_j = 0$, Eq. (9.21) reduces to $q(r_n)$ of (9.19); whereas, with $A_j = A_c$ and $A_c^2/N \gg 1$,

$$q(r) \approx \sqrt{\frac{r}{2\pi A_c N}}\, e^{-(r-A_c)^2/2N} \qquad r \geq 0 \qquad (9.22)$$

which is essentially a gaussian distribution with $\bar{r} = A_c$ and $\sigma_r^2 = N$. This last conclusion stems from the fact that $\sqrt{r/2\pi A_c N} \approx 1/\sqrt{2\pi N}$ in the vicinity of $r = A_c$, where $q(r)$ has its peak. It should come as no surprise that the envelope-detector output is gaussian when $S/N \gg 1$, for we observed in Chap. 5 that an envelope detector acts just like a synchronous detector if the predetection signal-to-noise ratio is large.

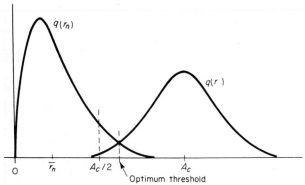

Fig. 9.12 *Probability density functions, noncoherent ASK detection.*

Turning to error probabilities, Fig. 9.12 shows the envelope density functions $q(r_n)$ and $q(r)$ that result when $A_j = 0$ or A_c, respectively. For simplicity the detection constant is taken as $K_D = 1$, and we assume $S/N \gg 1$, so (9.22) is applicable; we also assume the digits are equiprobable. Obviously, the decision threshold should be between 0 and A_c, but the best value is not necessarily $A_c/2$. In point of fact, there is no "best" threshold level in the sense of equalizing P_{e_1} and P_{e_0} and, at the same time, minimizing P_e. Moreover, the optimum level (minimum P_e) is a function of the signal-to-noise ratio;[1] it turns out to be the point where the two curves intersect, i.e., where $q(r_n) = q(r)$.

With $S/N \gg 1$ the optimum threshold is very close to $A_c/2$. Under

---

[1] The optimum threshold level is well approximated by $\tfrac{1}{2}A_c\sqrt{1 + 8N/A_c^2} = \tfrac{1}{2}A_c\sqrt{1 + 2/(S/N)}$.

this condition

$$P_{e_0} = \int_{A_c/2}^{\infty} q(r_n) \, dr_n = e^{-A_c^2/8N}$$

and

$$P_{e_1} = \int_0^{A_c/2} q(r) \, dr \approx \tfrac{1}{2} - \Phi\left(\frac{A_c}{2\sigma}\right)$$

where we have approximated P_{e_1} by the area under a gaussian tail from $-\infty$ to $A_c/2$, which is certainly not less than the actual area. Introducing the further approximation of Eq. (4.39), namely,

$$\tfrac{1}{2} - \Phi(v) \approx \frac{1}{\sqrt{2\pi}\,v} e^{-v^2/2} \qquad v \gg 1 \tag{9.23}$$

we obtain

$$P_{e_1} = \sqrt{\frac{2N}{\pi A_c^2}} \, e^{-A_c^2/8N}$$

bringing out the fact that P_{e_1} is much less than P_{e_0} when the threshold is chosen to minimize P_e and the signal-to-noise ratio is large.

Inserting $S/N = A_c^2/4N = \mathcal{E}$, the net error probability then becomes

$$\begin{aligned}
P_e &= P_1 P_{e_1} + P_0 P_{e_0} \\
&= \frac{1}{2} \frac{e^{-\frac{1}{2}(S/N)}}{\sqrt{2\pi(S/N)}} + \frac{1}{2} e^{-\frac{1}{2}(S/N)} \\
&\approx \tfrac{1}{2} e^{-\frac{1}{2}\mathcal{E}} \qquad \mathcal{E} \gg 1
\end{aligned} \tag{9.24}$$

assuming $P_1 = P_0 = \tfrac{1}{2}$. To compare this result with that of coherent detection under the same conditions, Eq. (9.17) is approximated per (9.23) by

$$P_{e\mathrm{CD}} \approx \frac{e^{-\frac{1}{2}\mathcal{E}}}{\sqrt{2\pi\mathcal{E}}} \qquad \mathcal{E} \gg 1$$

Hence, writing (9.24) in terms of $P_{e\mathrm{CD}}$,

$$P_{e\mathrm{NCD}} = \tfrac{1}{2} \sqrt{2\pi\mathcal{E}} \, P_{e\mathrm{CD}}$$

Figure 9.13 shows error probabilities for several binary systems as a function of $\mathcal{E}$. Curve A is polar baseband, curve C is coherent ASK (or unipolar baseband), and curve E is noncoherent ASK with fixed threshold at half amplitude. The latter is not optimum unless $S/N \gg 1$ but is more practical to instrument.

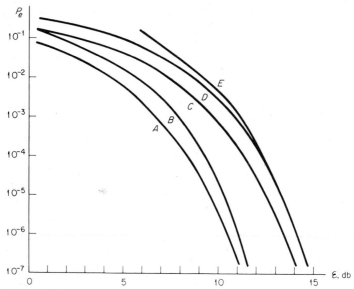

Fig. 9.13 *Error probabilities for various binary systems. A, polar baseband and coherent (phase-reversal keying); B, differentially coherent (phase-reversal keying); C, coherent ASK, VSB-ASK, and unipolar baseband; D, noncoherent FSK; E, noncoherent ASK.*

The figure supports our earlier assertion that noncoherent detection of ASK is more vulnerable to noise than coherent detection. However, for the low error probabilities usually required, say $P_e \leq 10^{-4}$, the difference is not very great. The slightly greater power requirement, typically 1 db, seems a fair price for the simplicity of envelope detection. On the other hand, if bandwidth conservation or increased signaling speed demands the use of VSB filtering, coherent detection would be a necessity.[1]

Frequency-shift keying (FSK)

An FSK waveform consists of constant-amplitude RF pulses having different frequencies, one frequency for each of the possible message symbols. Figure 9.14*a* is an idealized binary FSK signal, which, upon further examination, suggests that FSK can be treated as two (or more) interleaved on-off ASK signals of differing carrier frequencies, as in Fig. 9.14*b*. Pursuing this notion, FSK might be detected using ASK techniques, i.e., one synchronous or envelope detector for each frequency;

[1] A variation on VSB-ASK, roughly equivalent to AM with 50 percent modulation, does permit the use of envelope detection but requires about twice as much signal power.

the detector with the largest output is then presumed to indicate the transmitted frequency.

In point of fact, the conventional noncoherent detection system for binary FSK employs a pair of bandpass filters and envelope detectors,

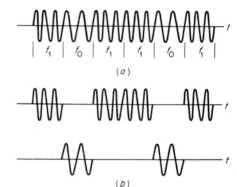

(a)

(b)

Fig. 9.14 *Binary FSK. (a) Idealized waveform; (b) decomposition into two ASK waves.*

as diagramed in Fig. 9.15. The detector outputs are subtracted to form the envelope difference $z = y_1 - y_0$, which is the input to the decision circuit. In the absence of noise $y_1 = K_D A_c$ and $y_0 = 0$ if the input frequency is f_1, and vice versa for f_0. Hence, the logical decision rule becomes: choose 1 if $z > 0$, choose 0 if $z < 0$. Correspondingly, the threshold level is set at zero volt.

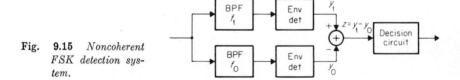

Fig. 9.15 *Noncoherent FSK detection system.*

As to error probabilities, one suspects that noncoherent FSK might be superior to noncoherent ASK; after all, conversion is made on the basis of two output signals. However, the transmitted power is twice the ASK power to achieve the same carrier amplitude (compare Figs. 9.10 and 9.14a), and the question naturally arises as to whether this method pays off in the long run.

For analysis purposes, suppose f_1 is sent and the signal is contaminated by additive gaussian noise. Drawing upon our study of envelope detection, we know that y_1 has the density function of Eq. (9.21) with $A_j = A_c$, while y_0 is the envelope of noise alone and therefore is Rayleigh distributed; specifically,

$$p(y_1) = \frac{y_1}{N_o} e^{-(A_c^2+y_1^2)/2N_o} I_0\left(\frac{A_c y_1}{N_o}\right) \qquad y_1 \geq 0$$

$$q(y_0) = \frac{y_0}{N_o} e^{-y_0^2/2N_o} \qquad y_0 \geq 0$$

assuming the same average noise power N_o out of each bandpass filter. The probability of a conversion error is then

$$P_{e_1} = P(z < 0) = P(y_1 < y_0)$$
$$= \int_0^\infty p(y_1) \left[\int_{y_1}^\infty q(y_0)\,dy_0\right] dy_1 \tag{9.25}$$

Performing the bracketed integral gives

$$\int_{y_1}^\infty q(y_0)\,dy_0 = e^{-y_1^2/2N_o}$$

so that

$$P_{e_1} = \frac{e^{-A_c^2/2N_o}}{N_o} \int_0^\infty y_1 e^{-y_1^2/N_o} I_0\left(\frac{A_c y_1}{N_o}\right) dy_1$$

which integrates to yield[1]

$$P_{e_1} = \tfrac{1}{2} e^{-A_c^2/4N_o}$$

It then follows from symmetry that $P_{e_2} = P_{e_1}$, and hence

$$P_e = \tfrac{1}{2} e^{-A_c^2/4N_o} \tag{9.26}$$

regardless of the digit probabilities.

Replacing the envelope detectors in Fig. 9.15 by synchronous detectors gives *coherent* FSK, for which

$$P_e = \tfrac{1}{2} - \Phi\left(\frac{A_c}{\sqrt{2N_o}}\right)$$
$$\approx \frac{e^{-A_c^2/4N_o}}{\sqrt{\pi A_c^2/N_o}} \qquad \frac{A_c^2}{N_o} \gg 1 \tag{9.27}$$

as is readily shown. A comparison of (9.26) and (9.27) for $P_e \leq 10^{-4}$ reveals, not unexpectedly, that the power margin provided by coherent FSK is inconsequential. Therefore, since coherent detection entails two synchronized oscillators, noncoherent FSK is used almost exclusively in practice.

For the comparison of FSK and ASK we must express (9.26) in terms of the parameter $\mathcal{E}$, for which we need the FSK transmission bandwidth as a function of signaling speed. Recalling that binary FSK can be decomposed into two ASK channels having carrier frequencies f_1 and

[1] See Schwartz, Bennett, and Stein (1966, app. A) for this and related integrations.

f_0, it follows that B_T should be about twice the ASK bandwidth.[1] Thus, as shown in Fig. 9.16,

$$B_T \approx 4B = 4\frac{s}{\kappa}$$

and the total predetection noise power is $N = \eta B_T = 4(\eta s/\kappa)$. But the

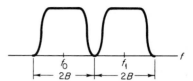

Fig. **9.16** *Representative FSK spectrum.*

separation filters are only half as wide as B_T, meaning that $N_o = N/2$, and therefore

$$\frac{A_c{}^2}{4N_o} = \frac{S}{N} = \frac{\kappa E}{4\eta} = \frac{1}{2}\varepsilon$$

where we have used $S = \frac{1}{2}A_c{}^2$. Substituting into (9.26) then gives

$$P_e = \frac{1}{2}e^{-\frac{1}{2}\varepsilon}$$

which is identical to noncoherent ASK when $\varepsilon \gg 1$, Eq. (9.24).

We now see that for most practical applications noncoherent FSK and ASK have the same error probability. What then are the advantages of FSK, if any? Upon further reflection, at least three are evident:

1. FSK has the constant-amplitude property whose merits were covered in conjunction with exponential modulation.
2. At smaller signal-to-noise ratios the ASK error probability becomes somewhat greater than that of FSK (see Fig. 9.13).
3. Referring back to Fig. 9.15, the FSK threshold level is zero volt, independent of the carrier amplitude. But for ASK the threshold level is a function of A_c and must be continually readjusted if the signal strength varies with time, not uncommon in radio transmission.

It is precisely for this last reason that FSK is preferred to ASK in applications where fading is expected and synchronous detection is not feasible.

One final point remains to be discussed here, namely, why binary FSK does not give the wideband noise reduction usually associated with

---

[1] At the risk of increased intersymbol interference, the bandwidth can be reduced to about $3s$ if $|f_1 - f_0| = s$. Further bandwidth reduction is possible using coherent detection or demodulation by means of analog frequency discrimination, the latter having much higher error rates than digital detection methods.

FM. For the system here described the reason is very simple: it is basically AM rather than FM. If one uses true FM, including detection by a limiter-discriminator,[1] it turns out that the transmission bandwidth required to control intersymbol interference is so large that the noise-reduction effect is essentially canceled by the increased predetection noise. However, for multilevel signals ($\mu > 2$) some advantage in the form of a bandwidth-power exchange is possible; after all, if $\mu \to \infty$, digital and analog signals are equivalent.

Phase-shift keying (PSK)

Phase-shift keying, as the name implies, is a digital modulation in which the carrier phase is switched between various discrete and equispaced values. For the particular case of binary PSK, the phase angles are chosen to be 0 and π rad, and the modulated wave becomes

$$
\begin{aligned}
x_c(t) &= A_c \cos (\omega_c t + \phi_j) & \phi_j &= 0 \text{ or } \pi \\
&= A_j \cos \omega_c t & A_j &= \pm A_c
\end{aligned}
$$

Thus, binary PSK is *phase-reversal keying* (PRK) and more akin to double-sideband suppressed-carrier AM than to phase modulation.

Illustrating this point, Fig. 9.17 shows the output of a DSB modulator when the input is the baseband signal of Fig. 9.4; note that there are

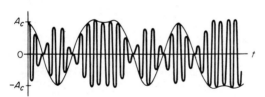

Fig. 9.17 *Binary PRK waveform.*

amplitude variations as well as phase reversals. The amplitude variations could, of course, be removed by a limiter. But limiting generally results in a larger transmission-bandwidth requirement or increased intersymbol interference. Consequently, practical phase-reversal systems use waveforms similar to Fig. 9.17, and $B_T = 2(s/\kappa)$, the same as ASK.[2]

It likewise follows that PRK can be detected in the same fashion as coherent ASK. Specifically, multiplying the signal plus noise $y(t) =$

---

[1] See Bennett and Davey (1965, chap. 9) for the error analysis.

[2] Note, however, that the constant-amplitude property of strictly exponential modulation is not preserved. If constant amplitude is required, the transmission bandwidth must be increased.

$x_c(t) + n(t)$ by a synchronous sinusoid $A_{LO} \cos \omega_c t$ and lowpass filtering yields

$$y_D(t) = K_D[\pm A_c + n_c(t)]$$

at the optimum conversion times.   The output signal therefore is a polar waveform with amplitude spacing $2A_c$, contaminated by gaussian noise of variance $\sigma^2 = \overline{n_c^2} = \overline{n^2} = N$.   Clearly, the decision threshold level should be set at zero volts, for which the conversion error probability is

$$P_e = \tfrac{1}{2} - \Phi\left(\frac{A_c}{\sigma}\right)$$

independent of the digit probabilities.   Then, since $S = \tfrac{1}{2}A_c^2$ and $N = \eta B_T = 2(\eta s/\kappa)$ if the input noise is white,

$$\left(\frac{A_c}{\sigma}\right)^2 = 2\frac{S}{N} = \frac{\kappa E}{\eta} = 2\mathcal{E}$$

and we have

$$P_e = \tfrac{1}{2} - \Phi(\sqrt{2\mathcal{E}}) \qquad \text{coherent PRK} \tag{9.28}$$

Equation (9.28) or Fig. 9.13 indicates that coherent PRK is equivalent to the best baseband transmission, insofar as noise performance is concerned, and requires 3 to 4 db less power than all other previously discussed types of digital modulation.

But this advantage is gained at the cost of synchronous detection, and noncoherent (envelope) detection cannot be employed because the information resides in phase reversals.   To get around synchronization problems a clever technique known as *phase-comparison* or *differentially coherent* PRK has been devised.   The detection system, diagramed in Fig. 9.18, uses the delayed previous digit (delayed by $1/s$) as a relative phase

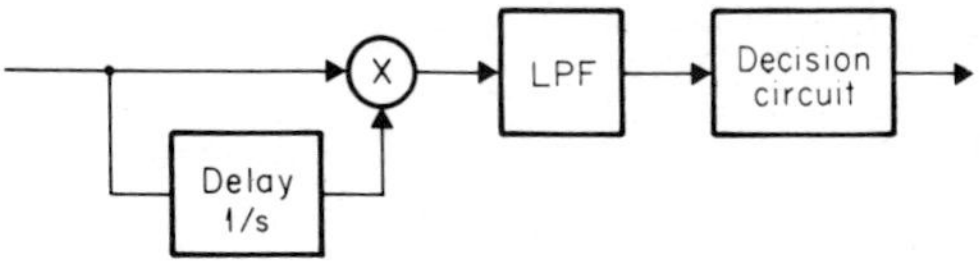

Fig. 9.18 *Phase-comparison (differential) PRK detection system.*

reference rather than having a fixed reference.   (Note the similarity to homodyne detection, Fig. 5.28.)   If adjacent digits are of like phase, their product results in a positive output (or binary 1); conversely, opposite phases result in a negative output (binary 0).   Thus, it is the *shift* or *no shift* between transmitted phase values that represents the message information, so appropriate coding, called *differential encoding*, is required at the transmitter.

Differential encoding starts with an arbitrary first digit and thereafter indicates the message digits by successive transition or no transition, essentially the reverse of the dicode wave in Fig. 9.1d. A transition stands for message 0, and no transition for message 1. The coding process is illustrated below. Differential encoding is most often used for PSK

Input message		1	0	1	1	0	1	0	0
Encoded message	1	1	0	0	0	1	1	0	1
Transmitted phase	0	0	π	π	π	0	0	π	0
Phase-comparison output		$+$	$-$	$+$	$+$	$-$	$+$	$-$	$-$
Output message		1	0	1	1	0	1	0	0

systems but is not restricted to such applications. In general, it is advantageous for systems having no sense of absolute polarity.

As to the noise performance of phase-comparison PRK, it might appear that differential detection requires twice as much power as coherent detection because the phase reference is itself contaminated by noise. However, the perturbations actually tend to cancel in the comparison process, somewhat like envelope comparison of FSK, so the degradation is not so great. Complete error analysis is rather messy, but yields the simple result[1]

$$P_e = \tfrac{1}{2}e^{-\varepsilon} \qquad \text{phase-comparison PRK} \qquad (9.29)$$

which is plotted as curve B in Fig. 9.13.

Relative to coherent PRK (curve A), the power penalty for differential detection is less than 1 db at $P_e \leq 10^{-4}$. Relative to noncoherent ASK or FSK (curves D and E), phase-comparison PRK has the nontrivial power margin of 3 db, by virtue of which this modulation method is gaining rapidly in popularity.

The only significant disadvantage is that because of the fixed delay time $1/s$ in the detector, the system is locked in on a specific signaling speed, thereby precluding variable-speed (asynchronous) transmission. Of minor annoyance is the fact that errors tend to occur in groups of two (why?).

Concluding this section, let us look briefly at multilevel phase-shift keying which, unlike PRK, is true phase modulation. A μ-ary PSK wave has the form

$$x_c(t) = A_c \cos (\omega_c t + \phi_j) \qquad \phi_j = 0, \frac{2\pi}{\mu}, \frac{4\pi}{\mu}, \ldots, \frac{2(\mu - 1)\pi}{\mu}$$

[1] See Downing (1964, chap. 8) for the derivation.

and is detected by phase discrimination with decision angles π/μ, $3\pi/\mu$, . . . centered between the expected carrier-phase values. Figure 9.19 shows a phasor diagram for quaternary PSK and the corresponding thresholds. The phase discriminator may employ either a coherent reference or differential phase comparison.

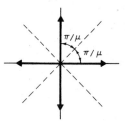

Fig. 9.19 *Phasors and thresholds for quaternary PSK, $\mu = 4$.*

Taking the case of coherent detection, suppose the transmitted phase ϕ_j is contaminated by narrowband gaussian noise. Because the absolute phase of the noise is arbitrary, we can write

$$n(t) = n_c(t) \cos (\omega_c t + \phi_j) - n_s(t) \sin (\omega_c t + \phi_j)$$

so the received signal plus noise becomes

$$\begin{aligned} y(t) &= x_c(t) + n(t) \\ &= r(t) \cos [\omega_c t + \phi_j + \phi_n(t)] \end{aligned}$$

where

$$\phi_n = \arctan \frac{n_s}{A_c + n_c}$$

If the discriminator has perfect limiting, only the noise-induced phase variation ϕ_n will cause conversion errors. Furthermore, by symmetry, the per-digit error probabilities are equal, and the net error probability is

$$P_{e\mu} = P \left(|\phi_n| > \pi/\mu \right) = 1 - \int_{-\pi/\mu}^{\pi/\mu} q(\phi_n) \, d\phi_n$$

since the thresholds are spaced by $2\pi/\mu$ rad.

The needed density function $q(\phi_n)$ can be found by integrating the joint density function of Eq. (9.20) over $r \geq 0$. The integration is straightforward, but tedious, and results in the formidable expression

$$q(\phi_n) = \frac{e^{-A_c{}^2/2N}}{2\pi} + \frac{A_c \cos \phi_n}{\sqrt{2\pi N}} e^{-(A_c \sin \phi_n)^2/2N} \left[\frac{1}{2} + \Phi \left(\frac{A_c \cos \phi_n}{\sqrt{N}} \right) \right]$$

To put this in more tractable form we make the reasonable assumption that $A_c{}^2/N \gg 1$, so the first term is negligible, $|\phi_n|$ is small most of the time, and $\frac{1}{2} + \Phi \approx 1$; hence

$$q(\phi_n) \approx \frac{A_c \cos \phi_n}{\sqrt{2\pi N}} e^{-(A_c^2/2N) \sin^2 \phi_n} \qquad |\phi_n| < \frac{\pi}{2}$$

The change of variable $u = (A_c/\sqrt{N}) \sin \phi_n$ then gives

$$\int_{-\pi/\mu}^{\pi/\mu} q(\phi_n) \, d\phi_n = \frac{2}{\sqrt{2\pi}} \int_0^{(A_c/\sqrt{N}) \sin \pi/\mu} e^{-u^2/2} \, du$$

$$= 2\Phi \left(\frac{A_c}{\sqrt{N}} \sin \frac{\pi}{\mu} \right)$$

Therefore, inserting $S/N = A_c^2/2N$,

$$P_{e\mu} = 2 \left[\frac{1}{2} + \Phi \left(\sqrt{\frac{2S}{N}} \sin \frac{\pi}{\mu} \right) \right]$$

$$\approx \frac{e^{-(S/N) \sin^2 \pi/\mu}}{\sqrt{\pi(S/N)} \sin (\pi/\mu)} \qquad \frac{S}{N} \gg 1 \tag{9.30}$$

where the approximation is quite good for $\mu \geq 4$.

Cahn (1959, 1960) has carried out the error analysis for μ-ary PSK with phase-comparison detection. The result is of the same form as Eq. (9.30), save that the signal-to-noise ratio must be increased by a factor of

$$\frac{\sin^2 (\pi/\mu)}{2 \sin^2 (\pi/2\mu)}$$

to achieve the same P_e as coherent detection. With $\mu = 4$ the penalty is about 2 db and increases asymptotically toward 3 db as $\mu \to \infty$. Thus, only for large μ does differential detection fully suffer from the noisy phase reference.

We have spent some time on multilevel PSK because it is of considerable practical interest for data transmission on voice telephone channels. Such channels are characterized by a fixed bandwidth and a relatively large signal-to-noise ratio. Since the bandwidth is fixed, the information rate in binary digits per second is most easily increased by increasing μ, the equivalent bit rate being $s \log_2 \mu$.

That multilevel PSK is better for this purpose than other types of digital modulation is attested to by comparison with coherent ASK. Assuming $S/N \gg 1$, Eq. (9.18) becomes

$$P_{e\mu_{ASK}} = \sqrt{\frac{(2\mu - 1)(\mu - 1)}{6\pi(S/N)}} \exp \left[-\frac{3(S/N)}{2(2\mu - 1)(\mu - 1)} \right]$$

Inserting $\mu = 4$, a representative value, and taking equal signal-to-noise ratios, we have

$$P_{e\mu_{\text{PSK}}} = \left(\frac{8}{49}\right)^{\frac{1}{4}} e^{-\frac{3}{4}(S/N)} P_{e\mu_{\text{ASK}}}$$

clearly indicating the superior noise immunity of PSK. For this example the power margin is about $\frac{7}{3}$, or 3.7 db. Even allowing for the 2-db loss entailed by phase-comparison detection, the PSK advantage is significant, particularly if the signal-to-noise ratio is fixed. And bear in mind that coherent ASK requires synchronous detection.

Summary

To summarize the results of this section, Table 9.1 lists the various types of binary modulation, their bit rates, and power requirements. The listing is approximately in order of increasing instrumentation complexity. Not indicated is the fact that differentially coherent PRK is operable only with synchronous signaling, while ASK has threshold-level difficulties when transmission is subject to fading.

Table 9.1   Comparison of binary modulation systems

Type	$\dfrac{s_{\max}}{(\kappa/2)B_T}$	ε, db, for $P_e = 10^{-4}$	Reference curve, Fig. 9.13
Polar baseband	2	8.4	A
Noncoherent ASK	1	12.3	E
Noncoherent FSK	$\frac{1}{2}$–$\frac{2}{3}$	12.3	D
Differentially coherent PRK	1	9.3	B
Coherent ASK or VSB-ASK	1–2	11.4	C
Coherent PRK	1	8.4	A

9.3 ERROR-CONTROL CODING

We have seen that error probability in digital transmission is a direct function of the signal-to-noise ratio. If, for a given system, the signal power is limited to some maximum value and errors are still unacceptably frequent, then some other means of improving reliability must be sought. Often, error-control coding provides the best solution.

In a nutshell, error-control coding is the calculated use of *redundancy*. Taking a hint from information theory, one systematically adds extra digits to the transmitted message, digits which themselves convey no information but make it possible for the receiver to detect or even correct errors in the information-bearing digits. Theoretically, near-errorless transmission is possible; more practically, there is the inevitable trade-off between transmission reliability, efficiency, and complexity of terminal equipment. Reflecting this factor, a multitude of error-detecting and error-correcting codes have been devised to suit various applications.

This section discusses the basic concepts of error-control coding and some of the commoner implementations. For simplicity, we shall deal only with binary codes. Furthermore, the treatment will be primarily qualitative since coding theory has progressed from a black art in its early days to a highly sophisticated mathematical discipline.

Error reduction versus bit rate

Because the essence of error-control coding is redundancy, it follows that the improved reliability entails a lower effective bit rate. Suppose, for example, that a code uses C check digits for every M message digits. If the gross signaling speed is s, then the actual message digits are delivered at a rate $s' = [M/(M + C)]s$, and the speed efficiency factor is

$$\alpha = \frac{s'}{s} = \frac{M}{M + C} \tag{9.31}$$

Generally speaking, codes that are both easily instrumented and effective in error control require a relatively large number of check digits, meaning small α. Thus, practical error control tends to go hand in hand with bit-rate reduction.

But there are other more simple-minded ways of decreasing errors at the expense of signaling speed, the signal power being fixed. And to properly evaluate the merits of error-control coding we should pause to consider these alternatives.

Recall that, in terms of signal-to-noise ratio, error probabilities are typically of order $e^{-(S/N)}$, where the noise power N is proportional to transmission bandwidth, which is proportional to signaling speed. Thus slower signaling and smaller bandwidth means larger signal-to-noise ratio. Specifically, if the speed is reduced to $s' = \alpha s$ and the transmission bandwidth accordingly narrowed, the new signal-to-noise ratio is $(s/s')(S/N) = (S/N)/\alpha$ and

$$P_e = e^{-(1/\alpha)(S/N)} = [e^{-(S/N)}]^{1/\alpha} = p^{1/\alpha} \tag{9.32}$$

where p is the original probability of error. With $p \ll 1$ and $1/\alpha \gg 1$, P_e can be substantially less than p.

An even more direct method, requiring no system modifications, is to merely repeat the entire message an *odd* number of times, say K, and use a majority-rule decision at the receiver for those digits which, because of errors, differ in repetition (recall Example 8.7).   This brute-force approach is essentially equivalent to bandwidth reduction, since the probability of the same message digit's being in error on all K transmissions is $p^K = p^{1/\alpha}$, where $\alpha = 1/K$ and $s' = \alpha s$ is the effective bit rate.

Clearly, both of these schemes are rather costly in terms of transmission time when p is not very small and yet one desires high reliability. It remains to be shown that error-control coding is a better approach. Unfortunately, the general analysis is quite complex, so we shall demonstrate its superiority by examples.

Error detection—parity-check codes

For many applications, errors can be rendered harmless if they are simply *detected* with no immediate attempt at correction.   This is true, for instance, in data telemetry when a large number of values are gathered for statistical analysis; erroneous values, if detected, are simply omitted from further processing, and the loss is negligible.   Similarly, given a *two-way* communication link, the fact that an error has been detected can be sent back to the transmitter for appropriate action, namely, retransmission.   Such *decision feedback* is especially advantageous if the system is subject to variable transmission conditions.   When conditions are good and errors infrequent, a low-redundancy code with its higher data rate is satisfactory; when conditions are unfavorable, as indicated by frequent error detection, the transmitter may switch to a code of higher redundancy or temporarily cease transmission.   But with or without feedback, simple error detection suffices only if the original error probability p is small to begin with and the probability of *undetected* errors is at a suitably low level.   The latter is, in fact, the equivalent error probability P_e.

Most error-detecting codes are based on the notion of *parity*.   The parity of a binary word is said to be even when the word includes an even number of 1s, while odd parity means an odd number of 1s.   Alternately, parity is the digit-by-digit *modulo* 2 sum[1] and can be easily measured by a flip-flop (initially in the 0 state) whose state is changed only by the 1s; at the end of the word the flip-flop reads 1 if the parity is odd, etc.

For error detection by *parity check*, we divide the message into groups of M digits and add one check digit to each group such that every $(M + 1)$-digit word has the same parity, say even.   Thus, the check digit is related to the message digits by

$$c = m_1 \oplus m_2 \oplus \cdots \oplus m_M$$

---

[1] Modulo-2 arithmetic, involving the binary digits 0 and 1, is the same as ordinary addition except that $1 \oplus 1 = 0$, $\oplus$ being the symbol for addition modulo 2. Moreover, there is no difference between addition and subtraction.

so that

$$m_1 \oplus m_2 \oplus \cdot \cdot \cdot \oplus m_M \oplus c = 0$$

The efficiency factor of Eq. (9.31) then becomes

$$\alpha = \frac{M}{M+1}$$

indicating reasonable efficiency if M is large.

Of the 2^{M+1} possible binary words having $M + 1$ digits, parity-check coding excludes precisely half, the half with odd parity. Therefore, if the parity of a received word is odd, we know there is an error—or three errors, or, in general, an odd number of errors. Error detection can then be implemented by checking the parity of each word as it arrives. Of course error correction is not possible, since we do not know where the errors are located within the word. Furthermore, an even number of errors preserves valid parity and hence goes undetected.

To calculate the various probabilities, let ϵ be the number of errors in the received word and let p be the per-digit error probability. An undetected error occurs when ϵ is even; according to the binomial distribution, this probability is

$$P_{ue} = \sum_{\epsilon \text{ even}} P_{M+1}(\epsilon) = \sum_{\epsilon \text{ even}} \binom{M+1}{\epsilon} p^\epsilon (1-p)^{M+1-\epsilon}$$

providing the errors are statistically independent. But when we accept a word having undetected errors, we get ϵ digit errors, of which the fraction $M/(M+1)$ are erroneous message digits. Thus the net error probability per message digit is

$$
\begin{aligned}
P_e &= \frac{M}{M+1} \sum_{\epsilon \text{ even}} \epsilon \binom{M+1}{\epsilon} p^\epsilon (1-p)^{M+1-\epsilon} \\
&= \frac{M}{M+1} \left[2 \frac{(M+1)M}{2 \cdot 1} p^2 (1-p)^{M-1} \right. \\
&\quad \left. + 4 \frac{(M+1)M(M-1)(M-2)}{4 \cdot 3 \cdot 2 \cdot 1} p^4 (1-p)^{M-3} + \cdot \cdot \cdot \right]
\end{aligned}
$$

Assuming $p \ll 1$, only the first term is significant, and if $Mp \ll 1$, then $(1-p)^{M-1} \approx 1$. With these conditions

$$P_e \approx \frac{M}{M+1} \left[2 \frac{(M+1)M}{2 \cdot 1} p^2 \right] = M^2 p^2 \tag{9.33}$$

equivalent to saying that more than two errors per word is unlikely.

The probability of a detected error is also of interest, for it indicates the amount of data that must be retransmitted or discarded. Since primarily single errors are detected, that is, $\epsilon = 1$, we have *per word*

$$P_{de} \approx \binom{M+1}{1} p(1-p)^M \approx (M+1)p$$

In a message of $N \gg 1$ total message digits, there are N/M words, of which $(N/M)(M+1)p$ have detected errors. If detected errors are discarded, the fractional number of message digits thrown away is

$$\frac{1}{N}\left[M \frac{N}{M}(M+1)p \right] = (M+1)p = P_{de}$$

As an illustration of the procedure and results let us take $M = 3$. The $\frac{1}{2} \times 2^4 = 8$ allowed 4-digit words are constructed as follows.

Message digits			Check digit
m_1	m_2	m_3	c
0	0	0	0
0	0	1	1
0	1	0	1
0	1	1	0
1	0	0	1
1	0	1	0
1	1	0	0
1	1	1	1

Then, if $p = 10^{-3}$, the error probability is

$$P_e = 3^2(10^{-3})^2 = 0.9 \times 10^{-5}$$

and errors have been reduced by a factor of about 100. Comparing this with bandwidth reduction or simple repetition schemes with the same effective data rate, we note that $\alpha = \frac{3}{4}$, for which (9.32) gives $P_e = (10^{-3})^{\frac{4}{3}} = 10^{-4}$. Parity checking is in this case 10 times more reliable, and its reliability advantage increases rapidly as p decreases.

Two final comments with respect to practical matters are in order here. First, it is preferable to use odd-word parity and an odd number of message digits per word; this ensures that every word has at least one transition, thereby aiding synchronization and preventing apparent loss of signal if the message contains an extended string of like digits. Second, as a result of impulse noise on switched circuits or short-duration fading on radio paths, errors may tend to occur in *bursts* of several successive digits; since multiple errors wreak havoc on parity checking, the check digits should be *interlaced* such that the digits checked are widely spaced. An example of interlacing is given in Fig. 9.20, where one parity word is indicated by lines connecting the digits.

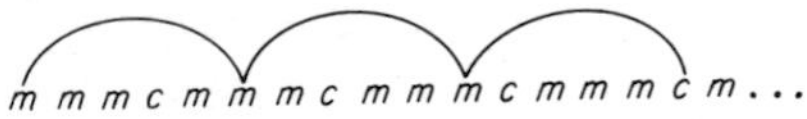

m m m c m m m c m m m c m m m c m . . .

Fig. 9.20 *Interlaced parity checking, m = message digit, c = check digit.*

Error correction by parity check

The idea of *error-correcting* codes is certainly far more appealing and exciting than mere error detection, suggesting as it does the one-way errorless transmission hypothesized by Shannon. And there are numerous applications, notably one-way links, for which error correction is a necessity. It also turns out that encoding at the transmitter is not much different for detection or correction; the receiving decoder is the problem in error-correcting systems. Consequently, when transmission equipment is constrained, *e.g.*, satellites, it may prove more practical to correct errors at the receiver than to provide for retransmission facilities, even though a two-way path is available.

Parity-check coding is readily extended to error correction by observing that correction requires the detection of an error and its *location* in the word. Thus, if two checks are made on the same word but in two different patterns, errors result in characteristic symptoms of invalid parity.

Probably the simplest error-correcting code is the square-array parity check, wherein M successive message digits are arranged in a square matrix whose rows and columns are each checked, as in Fig. 9.21,

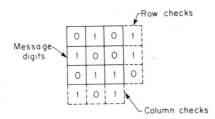

Fig. 9.21 *Square-array parity checking, $M = 9$.*

where $M = 9$. As in error detection, the check digits will be interlaced if errors bursts are expected. In general, the number of check digits per word is $C = 2\sqrt{M}$, so

$$\alpha = \frac{M}{M + 2\sqrt{M}}$$

and good efficiency requires large M.

Each complete word of $M + 2\sqrt{M}$ digits is then read out row by row for transmission and rematrixed at the receiver. A single message-digit error gives a row and column parity failure with the error at the intersection point, which can be corrected. Moreover, multiple errors not in the same row or column can be detected but not corrected. Because many errors so positioned are unlikely, the code is nominally single-error-correcting and double-error-detecting. Neglecting all but triple errors, the net error probability can be approximated as

$$P_e = \tfrac{1}{2}M(M + 2\sqrt{M} - 1)(M + 2\sqrt{M} - 2)p^3 \qquad (9.34)$$

providing $(M + 2\sqrt{M})\, p \ll 1$.

Inserting typical values for M and p, one finds this code not very attractive. For instance, taking $M = 36$ and $p = 10^{-3}$, we have

$$\alpha = \tfrac{3}{4}$$
$$P_e = 3.9 \times 10^{-5}$$

which is roughly equivalent to simple parity error detection with $M = 3$. Of course, most detected errors are also corrected by the square-array code, but at the cost of substantial decoding equipment.

A more effective error-correction technique is the Hamming code considered next.

Hamming code

Referring back to parity-check error detection, it can be seen that single errors are detected because each allowed word differs from the others in at least two digits; one digit alteration then produces an invalid word. The number of digits in which two binary words of the same length are different is called the *distance* (or *Hamming distance*), and, as a little further thought will verify, the correction of up to k errors per word requires a minimum distance of $2k + 1$. Thus, in principle, single-error correction (and double-error detection) is possible if we select coded words having minimum distance 3.

The trouble with the above approach is the practical difficulty in determining which valid word was intended when an invalid word is received. Hamming (1950) proposed a systematic method of adding check digits such that, upon receipt, an *error word* can be formed to spell out the single-error correction. The key to Hamming's method is as follows.

Consider a binary word of $M + C$ digits with at most one error. Error correction then requires $M + C + 1$ possible error words, namely, "no error," "first digit in error," "second digit in error," and so forth.

Now we are at liberty to use only the C check digits to generate the error word and, since there are 2^C combinations, we have the condition

$$2^C \geq M + C + 1$$

Full utilization of the code requires the equality to be realized, thereby restricting word length to $2^n - 1$, where $n = C$. For example, with $n = 3$ the word length is $2^3 - 1 = 7$, $C = 3$, and $M = 4$; it is this Hamming code we shall examine in detail.

To derive the encoding and correction process, let the transmitted word be $m_1 m_2 m_3 m_4 c_1 c_2 c_3$ and let the error word formed at the receiver be $e_1 e_2 e_3$ with the following meaning:

| Error word | | | Error |
e_1	e_2	e_3	location
0	0	0	No error
0	0	1	m_1
0	1	0	m_2
0	1	1	m_3
1	0	0	m_4
1	0	1	c_1
1	1	0	c_2
1	1	1	c_3

Examining the above shows that $e_1 = 1$ if the error is in m_4 or any of the check digits; hence e_1 is the parity check

$$e_1 = m'_4 \oplus c'_1 \oplus c'_2 \oplus c'_3$$

where the primes denote received digits. Similarly

$$e_2 = m'_2 \oplus m'_3 \oplus c'_2 \oplus c'_3$$
$$e_3 = m'_1 \oplus m'_3 \oplus c'_1 \oplus c'_3$$

Note that if there are two errors, both in the check digits, the error word is 000, which is all right since the message digits are correct. On the other hand, with one message-digit and one check-digit error, we may miss the message-digit error and erroneously "correct" another message digit, thereby compounding the felony. Fortunately, these and other multiple-error instances will be rare if p is small. (Alternately, at the cost of one more check digit, double errors can be detected.)

As to check-digit formation for this example, we have $e_1 e_2 e_3 = 000$ in absence of errors. Hence, at the transmitter, where there are no errors, the following must be true:

$$0 = m_4 \oplus c_1 \oplus c_2 \oplus c_3$$
$$0 = m_2 \oplus m_3 \oplus c_2 \oplus c_3$$
$$0 = m_1 \oplus m_3 \oplus c_1 \oplus c_3$$

Solving these modulo-2 simultaneous equations for the check digits gives

$$c_1 = m_2 \oplus m_3 \oplus m_4$$
$$c_2 = m_1 \oplus m_3 \oplus m_4$$
$$c_3 = m_1 \oplus m_2 \oplus m_4$$

The allowed words are then

m_1	m_2	m_3	m_4	c_1	c_2	c_3
0	0	0	0	0	0	0
0	0	0	1	1	1	1
0	0	1	0	1	1	0
.	.	.	.	.	.	.
.	.	.	.	.	.	.
.	.	.	.	.	.	.

assuming even parity.

Continuing the example, an uncorrected (and undetected) error occurs when $2 \leq \epsilon \leq 7$, hence

$$P_{ue} = 1 - [P(\epsilon = 0) + P(\epsilon = 1)]$$
$$= 1 - \left[\binom{7}{0} p^0 (1 - p)^7 + \binom{7}{1} p^1 (1 - p)^6 \right]$$

Expanding $(1 - p)^n$ in a binomial series and taking $7p \ll 1$, we then get

$$P_{ue} \approx 21p^2$$

Ignoring all but double errors, consistent with $p \gg 1$, each erroneous word has on the average $\frac{4}{7} \times 2$ erroneous message digits, so the net error probability per message digit is about

$$P_e = \frac{4}{7} \times 2 \times 21p^2 = 24p^2$$

and the speed-reduction factor is

$$\alpha = \frac{4}{7}$$

Going to longer word lengths improves efficiency but increases the chances of multiple errors.

Numerous variations on the basic Hamming code are possible, of course; the check digits can be interlaced, and the code can be expanded for multiple-error correction [for the latter, see Harman (1963, chap. 5)]. However, the instrumentation rapidly gets out of hand and nullifies the inherent simplicity of Hamming's procedure.

Other codes

Ideally an error-control code should combine reliability, efficiency, simple implementation, and invulnerability to error bursts. Not surprisingly, no known code entirely satisfies this tall order. However, there are several codes of practical interest that combine two or three of these properties, and we shall describe them briefly according to the general classifications of *cyclic* or *recurrent* codes. For detailed treatments see the references cited at the end of the chapter.

Cyclic codes are *block* codes in that a specific number of successive message digits are grouped together. The check digits are arranged such that a *shift register* with feedback can do the encoding. As illustration, Fig. 9.22 shows the shift-register configuration for producing a (7,3) cyclic

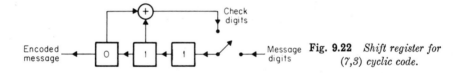

Fig. 9.22 *Shift register for (7,3) cyclic code.*

code, i.e., word length $M + C = 7$ digits, $M = 3$ message digits. The switch is first put in the lower position and 3 message digits, say 011, are fed in. The feedback loop is then closed, and the modulo-2 adder generates the first check digit, which, in this case, is $0 \oplus 1 = 1$. The first message digit is fed out, the first check digit fed in, and the cycle repeats until 4 check digits have been generated. (As the reader can verify, the complete word will be 0111001.) The register is then cleared, and 3 new message digits are inserted.

It can be shown that the (7,3) cyclic code has distance 4 and therefore is single-error-correcting, double-error-detecting. The encoding and decoding equipment is relatively simple, and reliability is high, but the efficiency is less than 50 percent ($\alpha = \frac{3}{7}$), and error bursts give problems. In retrospect, the Hamming code discussed above is a (7,4) cyclic code with distance 3. However, the checking pattern must be modified to permit shift-register encoding.

A type of cyclic code having greater reliability is the *Bose-Chaudhuri code*. With word length $2^n - 1$ and nk check digits this code can correct

k errors or detect (but not correct) $2k$ errors. In either mode the efficiency factor is $\alpha = (2^n - 1 - nk)/(2^n - 1)$. Many error patterns having more than $2k$ errors are also detected, so the code is reasonably invulnerable to error bursts. Encoding is done by a shift register of nk stages, and error-detection decoding is achieved by an identical shift register plus peripheral devices. However, error correction is more complicated and entails equipment tantamount to a small computer.

Recurrent codes, unlike all types discussed so far, do not divide the message digits into blocks but rather have a continuous encoding-decoding procedure. This eliminates the need for storage or buffering at the data terminals, thereby reducing equipment requirements. Usually, recurrent codes are designed for error-burst correction, in which capacity they have widespread interlacing and high redundancy. For example, a *Hagelbarger* code with 1 check digit per message digit ($\alpha = \frac{1}{2}$) can correct up to six successive errors if the preceding 19 digits are correct. Encoding and decoding are accomplished by simple shift registers. The code therefore is well suited to applications where error bursts are the main concern, terminal equipment must be minimized, and the bit-rate reduction can be tolerated.

Wozencraft and others have described random sequential codes which can be regarded as a form of recurrent coding. At very low rates, such codes offer great immunity to random errors.

9.4 INTRODUCTION TO OPTIMUM DETECTION

Stated rather crudely, optimization is the art of doing the best you can with what you have. Translating this definition to electrical communication, optimum detection can be viewed as the process that achieves maximum separation of a desired signal from undesired noise. But what constitutes maximum separation; what are the performance standards or measures of goodness by which a system is judged to be best? Obviously, criteria are variable and contingent upon the specific application. As a result, an extensive body of optimization theory containing many subtopics has been developed, starting from the basic works of Wiener (1949) and Kotel'nikov (1959). Here we propose to examine a small but important slice of the theory, namely, that part dealing with digital transmission.

Given a digital signal, the goal of optimum detection is quite clear; minimize the number of errors and maximize the signaling speed. Since error probability depends on signal-to-noise ratio, one approach to optimization is by way of *filter theory*, wherein one seeks a linear filter to maximize the signal relative to the noise. But optimum linear filtering is not necessarily an absolute optimization, so for a more general approach one must turn to *statistical decision theory*, having no built-in preconceptions about the nature of the system. Both viewpoints give valuable further insight to the study of digital communication and are fitting topics for the close of this chapter. Happily, as it turns out, both theories lead to the same system design if the noise is white gaussian.

Optimum (matched) filters

Recall that if $A/2$ is the peak signal amplitude at the input to a data demodulator and $N = \sigma^2$ is the noise power (or variance), then the probability of conversion error decreases rapidly with increasing $(A/2\sigma)^2 = A^2/4N$. Now suppose the received signal $x(t)$, representing one digital pulse arriving at $t = 0$, is contaminated by white noise having power spectral density $G_n(f)$. Presumably, a filter should be inserted ahead of the demodulator to limit the noise. The question is: What filter yields the largest value of $A^2/4N$?

To answer this question, and thereby determine the optimum receiving filter, let $H(f)$ be the filter transfer function and let $X(f) = \mathfrak{F}[x(t)]$ be the input-signal spectrum. The output signal is then the inverse transform

$$y(t) = \int_{-\infty}^{\infty} H(f)X(f)e^{j\omega t}\, df$$

which attains its peak amplitude $A/2$ at some particular time, say $t = t_0$, that is,

$$\frac{A}{2} = |y(t_0)| = \left| \int_{-\infty}^{\infty} H(f)X(f)e^{j\omega t_0}\, df \right|$$

Since the average output noise power is given by

$$N = \int_{-\infty}^{\infty} |H(f)|^2 G_n(f)\, df$$

the ratio we wish to maximize becomes

$$\frac{A^2}{4N} = \frac{\left| \int_{-\infty}^{\infty} H(f)X(f)e^{j\omega t_0}\, df \right|^2}{\int_{-\infty}^{\infty} |H(f)|^2 G_n(f)\, df} \tag{9.35}$$

where the only variable is $H(f)$.

Problems of this sort usually require the techniques of variational calculus. However, for this particular case, a well-known and useful relationship called *Schwarz's inequality* can be invoked as a laborsaving device. Derivation of the inequality is not difficult, and we shall take the time to outline it with the aid of signal-space concepts from Sec. 8.3.

Consider two time functions, say $u(t)$ and $v(t)$, and their vector representations **u** and **v** in D-dimensional signal space. By analogy with ordinary vectors, the scalar product $\mathbf{u} \cdot \mathbf{v}$ cannot exceed the product of the magnitudes $|\mathbf{u}|\,|\mathbf{v}|$; specifically,

$$(\mathbf{u} \cdot \mathbf{v})^2 = (|\mathbf{u}|\,|\mathbf{v}|\,\cos\theta)^2 \leq |\mathbf{u}|^2|\mathbf{v}|^2$$

where θ is the included angle. The equality applies if and only if $\cos\theta = \pm 1$, that is, if the vectors are *collinear*. Now, from Eq. (8.20), $|\mathbf{u}|^2 = 2BE_u$, where B is the signal bandwidth and $E_u = \int_{-\infty}^{\infty} u^2(t)\,dt = \int_{-\infty}^{\infty} |U(f)|^2\,df$ is the energy. Similarly, it can be shown that $|\mathbf{u} \cdot \mathbf{v}| = 2B\left|\int_{-\infty}^{\infty} u(t)v(t)\,dt\right| = 2B\left|\int_{-\infty}^{\infty} U(f)V^*(f)\,df\right|$. Therefore

$$\left[\int_{-\infty}^{\infty} u(t)v(t)\,dt\right]^2 \leq \int_{-\infty}^{\infty} u^2(t)\,dt \int_{-\infty}^{\infty} v^2(t)\,dt \tag{9.36a}$$

and

$$\left|\int_{-\infty}^{\infty} U(f)V^*(f)\,df\right|^2 \leq \int_{-\infty}^{\infty} |U(f)|^2\,df \int_{-\infty}^{\infty} |V(f)|^2\,df \tag{9.36b}$$

Consistent with **u** and **v** collinear, the equality condition is

$$u(t) = Kv(t) \quad\text{or}\quad U(f) = KV(f)$$

where K is any real constant. Finally, making the change $W(f) = V^*(f)$ in (9.36b), gives

$$\left|\int_{-\infty}^{\infty} U(f)W(f)\,df\right|^2 \leq \int_{-\infty}^{\infty} |U(f)|^2\,df \int_{-\infty}^{\infty} |W(f)|^2\,df \tag{9.36c}$$

with equality when $U(f) = KW^*(f)$. Equations (9.36a) to (9.36c) are all statements of Schwarz's inequality.

At this point a comparison of (9.35) and (9.36c) suggests the substitutions $U(f) = \sqrt{G_n(f)}\,H(f)$ and $W(f) = X(f)e^{j\omega t_0}/\sqrt{G_n(f)}$—recall that $G_n(f)$ is real and positive. Thus

$$\left|\int_{-\infty}^{\infty} H(f)X(f)e^{j\omega t_0}\,df\right|^2 \leq \int_{-\infty}^{\infty} |H(f)|^2 G_n(f)\,df \int_{-\infty}^{\infty} \frac{|X(f)|^2}{G_n(f)}\,df$$

and the equality applies only for

$$\sqrt{G_n(f)}\,H(f) = K\left[\frac{X(f)e^{j\omega t_0}}{\sqrt{G_n(f)}}\right]^* = K\frac{X^*(f)e^{-j\omega t_0}}{\sqrt{G_n(f)}}$$

Since $G_n(f)$ and $X(f)$ are invariant insofar as the maximization is concerned, it follows that

$$\left(\frac{A^2}{4N}\right)_{\max} = \int_{-\infty}^{\infty} \frac{|X(f)|^2}{G_n(f)}\,df \tag{9.37}$$

and the optimum filter required to achieve this value is

$$H(f) = K \frac{X^*(f)}{G_n(f)} e^{-j\omega t_0} \qquad (9.38)$$

Thus the optimum filter quite logically enhances those portions of the frequency band where the signal-to-noise spectral-density ratio is large, and vice versa.

Specializing to the case of *white* noise with $G_n(f) = \eta/2$, Eq. (9.37) becomes

$$\left(\frac{A^2}{4N}\right)_{\max} = \int_{-\infty}^{\infty} \frac{|X(f)|^2}{\eta/2} df = \frac{2E}{\eta} \qquad (9.39)$$

where $E = \int_{-\infty}^{\infty} |X(f)|^2 df = \int_{-\infty}^{\infty} x^2(t) dt$ is the received signal energy. The corresponding optimum filter is

$$H(f) = X^*(f)e^{-j\omega t_0} \qquad (9.40a)$$

with impulse response

$$h(t) = x(t_0 - t) \qquad (9.40b)$$

where, for simplicity, we have taken $K = \eta/2$.

Filters having the property of Eq. (9.40) are said to be *matched filters*, matched to $x(t)$ in that the impulse response has the shape of $x(t)$ delayed by t_0 and reversed in time. Matched filters were briefly discussed in Example 3.3, where it was noted that their effect is to maximize the output at $t = t_0$. Here we observe that matched filtering is a form of *correlation detection*. That is to say, in the absence of noise the filtered output is $x(t)$ correlated with itself, since

$$y(t) = x(t) * h(t) = \int_{-\infty}^{\infty} h(t')x(t - t') dt'$$
$$= \int_{-\infty}^{\infty} x(t_0 - t')x(t - t') dt'$$

At $t = t_0$, the peak output is then

$$y(t_0) = \int_{-\infty}^{\infty} x^2(t_0 - t') dt' = \int_{-\infty}^{\infty} x^2(t) dt = E$$

Relative to matched filters for digital transmission, recall in our analysis of polar binary baseband systems having white noise and pulse amplitudes $\pm A/2$ we developed the relationship

$$\left(\frac{A}{2\sigma}\right)^2 = \frac{A^2}{4N} = 2\frac{\kappa}{2}\frac{E}{\eta} = 2\mathcal{E}$$

where $\kappa/2 \leq 1$ reflects the pulse shape. Referring to Eq. (9.39) shows that $(A^2/4N)_{\max} = 2E/\eta = 2\mathcal{E}_{\max}$ *independent of the pulse shape* $x(t)$.

Therefore, with the help of matched filtering, it is possible to attain

$$\mathcal{E}_{\max} = \frac{E}{\eta}$$

without the design complications entailed by using $\kappa = 2$, that is, with other than sinc pulses.

However, there are two practical difficulties with matched filters: (1) usually they cannot be physically realized; (2) the output $y(t)$ has roughly twice the duration of $x(t)$, leading to problems of intersymbol interference. At the cost of increased transmission bandwidth, both of these problems can be overcome by using *rectangular* pulses (or the modulated equivalent). Then the matched filter's impulse response is also rectangular and, in absence of noise, the output waveform is *triangular* with twice the duration of the input pulse. To prevent overlapping outputs, the filter is discharged (shorted) every $1/s$ seconds, just after the output peaks. Since this process is equivalent to integrating over the input-pulse duration, such devices are known as *integrate-and-dump* filters. It is left to the reader to show that "dumping" does not affect $(A^2/4N)_{\max}$ given by Eq. (9.39).

As an alternative to the integrate-and-dump method, one can design the filter in reverse order, with $y(t)$ being a specified output pulse shape chosen to minimize intersymbol interference. The time delay t_0 is then relatively unimportant and can be as large as required, within reason, to facilitate filter approximation.

Optimum binary signaling

Let us now apply optimum filtering to binary data transmission in the presence of white gaussian noise. The purpose is to show, in rather general terms, the superiority of coherent detection methods and further prove that polar baseband and coherent phase-shift keying are indeed optimum systems.

We begin with the two digital pulses $x_1(t)$ and $x_0(t)$ assumed to be equiprobable. However, the shapes and energies may be different, so that

$$E_1 = \int_{-\infty}^{\infty} x_1{}^2(t)\, dt \qquad E_0 = \int_{-\infty}^{\infty} x_0{}^2(t)\, dt$$

and the average energy is

$$E = \tfrac{1}{2}(E_1 + E_0)$$

For notational convenience, we also define the *correlation coefficient* of $x_1(t)$ and $x_0(t)$ as

$$\rho = \frac{1}{E} \int_{-\infty}^{\infty} x_1(t)x_0(t)\, dt \tag{9.41}$$

which is bounded by

$$\rho^2 \leq 1$$

as can be easily proved with Schwarz's inequality, per Eq. (9.36a).

Because there are two pulses in question, optimum detection involves a pair of matched filters

$$h_1(t) = x_1(\tau - t) \qquad h_0(t) = x_0(\tau - t)$$

where, for clarity, $\tau = t_0$. In the absence of noise, each filter produces a maximum output only for its matched input. Therefore, we can subtract the outputs and use an appropriate decision threshold to determine which pulse actually arrived. Figure 9.23a diagrams the complete detector. Alternatively, since the filters are linear, the same effect is achieved more conveniently by a single filter with

$$h(t) = h_1(t) - h_0(t) = x_1(\tau - t) - x_0(\tau - t) \qquad (9.42)$$

giving the simplified system of Fig. 9.23b.

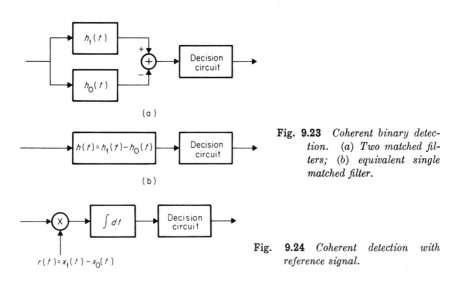

(a)

(b)

Fig. 9.23 *Coherent binary detection.* (a) *Two matched filters;* (b) *equivalent single matched filter.*

Fig. 9.24 *Coherent detection with reference signal.*

For either arrangement, the detection is properly classified as being *coherent* in the sense that the receiver has available stored copies of the uncontaminated pulse shapes. Emphasizing this point, Fig. 9.24 shows the equivalent product-type coherent detector with reference signal $r(t) = x_1(t) - x_0(t)$. The reader may wish to verify the equivalence of these three systems. The following analysis is couched in terms of the single-filter implementation.

Suppose that either $x_1(t)$ or $x_0(t)$ arrives at the receiver along with additive gaussian noise having $G_n(f) = \eta/2$. If $z(t)$ is the received signal plus noise, then the filtered output is

$$y(t) = z(t) * h(t)$$
$$= \int_{-\infty}^{\infty} z(t')[x_1(\tau - t + t') - x_0(\tau - t + t')] \, dt'$$

or, at the decision time $t = \tau$,

$$y(\tau) = \int_{-\infty}^{\infty} z(t')[x_1(t') - x_0(t')] \, dt' \qquad (9.43)$$

If $x_1(t)$ was actually transmitted, we get

$$y_1(\tau) = A_1 + n$$

where

$$A_1 = \int_{-\infty}^{\infty} x_1(t')[x_1(t') - x_0(t')] \, dt'$$
$$= \int_{-\infty}^{\infty} x_1{}^2(t') \, dt' - \int_{-\infty}^{\infty} x_1(t')x_0(t') \, dt'$$
$$= E_1 - \rho E$$

and we have introduced the correlation coefficient ρ from Eq. (9.41). Similarly, for $x_0(t)$ transmitted,

$$y_0(\tau) = A_0 + n$$

with

$$A_0 = \rho E - E_0$$

The peak-to-peak output amplitude spacing is thus

$$A_1 - A_0 = (E_1 - \rho E) - (\rho E - E_0)$$
$$= E_1 + E_0 - 2\rho E$$
$$= 2(1 - \rho)E \qquad (9.44)$$

which depends only on the average energy $E = \frac{1}{2}(E_1 + E_0)$.

Regardless of which signal was transmitted, the output noise is gaussian with $\bar{n} = 0$ and $\sigma^2 = \overline{n^2}$, since the input noise is presumed zero-mean gaussian and the filter is linear. To calculate the output noise variance we note that $\overline{n^2}$ is the average filtered noise power, that is,

$$\sigma^2 = \overline{n^2} = N = \int_{-\infty}^{\infty} |H(f)|^2 G_n(f) \, df$$
$$= \frac{\eta}{2} \int_{-\infty}^{\infty} |H(f)|^2 \, df$$

But $H(f) = \mathfrak{F}[h(t)]$, and, by analogy with Rayleigh's energy theorem,

$$\int_{-\infty}^{\infty} |H(f)|^2 \, df = \int_{-\infty}^{\infty} h^2(t) \, dt. \quad \text{Therefore}$$

$$
\begin{aligned}
\sigma^2 &= \frac{\eta}{2} \int_{-\infty}^{\infty} h^2(t) \, dt = \frac{\eta}{2} \int_{-\infty}^{\infty} h^2(\tau - t') \, dt' \\
&= \frac{\eta}{2} \int_{-\infty}^{\infty} [x_1{}^2(t') - 2x_1(t')x_0(t') + x_0{}^2(t')] \, dt' \\
&= \frac{\eta}{2} (E_1 - 2\rho E + E_0) \\
&= \eta(1 - \rho)E
\end{aligned}
\tag{9.45}
$$

The situation at the filter output can now be summarized as follows: the peak pulse amplitudes are $A_1 = E_1 - \rho E$ and $A_0 = \rho E - E_0$; the noise variations are gaussian-distributed with zero mean and variance $\sigma^2 = \eta(1 - \rho)E$; the digits are assumed equiprobable. It therefore follows that the decision level should be centered between A_1 and A_0, that is, at

$$D = \tfrac{1}{2}(A_1 - A_0) = \tfrac{1}{2}(E_1 - E_0)$$

and the logical decision rule becomes: choose 1 if $y > D$, choose 0 if $y < D$. The net error probability is thus

$$
\begin{aligned}
P_e &= P_1 P(A_1 + n < D) + P_0 P(A_0 + n > D) \\
&= \tfrac{1}{2} P[n < (1 - \rho)E] + \tfrac{1}{2} P[n > (1 - \rho)E]
\end{aligned}
$$

or

$$P_e = \tfrac{1}{2} - \Phi[\sqrt{(1 - \rho)\mathcal{E}_{\max}}] \tag{9.46}$$

where $\mathcal{E}_{\max} = E/\eta$.

Note that this result was obtained with a minimum of restrictions. In particular, nothing has been said about the shape of the pulses per se. But we now have a guideline for signal selection implicit in Eq. (9.46); i.e., to minimize P_e, $1 - \rho$ should be as large as possible. However, the correlation coefficient is bounded by $\rho^2 \leq 1$, so $1 - \rho \leq 2$. As is easily checked from (9.41), $\rho = -1$ when $x_1(t) = -x_0(t)$. Such signals are said to be *antipodal* and, with optimum coherent detection, yield the minimum error probability

$$P_{e\min} = \tfrac{1}{2} - \Phi(\sqrt{2\mathcal{E}_{\max}}) \tag{9.47}$$

Binary polar baseband and coherent PSK are two possible forms of antipodal signaling. On the other hand, if the signals are chosen to be *orthogonal* (or *uncorrelated*), meaning $\int_{-\infty}^{\infty} x_1(t)x_0(t) \, dt = 0$, then $\rho = 0$ and

$$P_e = \tfrac{1}{2} - \Phi(\sqrt{\mathcal{E}_{\max}})$$

with optimum coherent detection. Orthogonal signaling systems, of which unipolar baseband and coherent ASK are examples, pay a 3-db power penalty compared to antipodal signaling.

These conclusions are of course old hat, having been arrived at case by case in Secs. 9.1 and 9.2. What is new and significant is that Eq. (9.47) represents the *minimum* possible error probability for a given average energy (signal power) and noise density—or so we think. But one questionable condition has been imposed throughout, namely, that the system is linear. It therefore behooves us to undertake a broader investigation, dealing directly with the error probabilities and assuming nothing about the nature of the signals or the system. This is the viewpoint of statistical decision theory.

Elementary statistical decision theory

Statistical decision theory is concerned with extracting as much information as possible from a noise-corrupted signal by the application of *statistical inference.* There are two classes of problems amenable to this approach: (1) the *detection* problem, wherein one decides which of a set of possible signals is most likely to have caused the signal as received, and (2) the *estimation* problem, in which one estimates the most likely numerical value for some signal parameter. We shall here discuss only the detection problem and further restrict the discussion to binary signals. Obviously, we are barely scratching the surface of a fascinating subject; the reader whose interest is aroused thereby will find additional material in the references.

Consider the received signal plus noise

$$z(t) = x(t) + n(t)$$

where $x(t)$ is either $x_1(t)$ or $x_0(t)$, with a priori probabilities P_1 and P_0, respectively. We designate by H_1 the hypothesis that x_1 was the cause resulting in z, and by H_0 the hypothesis that x_0 was the cause. It is then desired to find a rule for deciding between H_1 and H_0 when z is received. This decision rule, or *hypothesis testing,* if it is to be reasonable, must include any knowledge we have about the cause-and-effect relationships. Since the observable effect z is a continuous random variable, such knowledge will be in the form of probability density functions.

Specifically, let the signal space Z contain all possible points at which the signal vector $\mathbf{z}$ might terminate. Now $\mathbf{z} = \mathbf{x} + \mathbf{n}$, and, presumably, the statistical properties of the noise are known. Thus, according to whether $\mathbf{x} = \mathbf{x}_1$ or $\mathbf{x}_0$, we have the probability density functions $p_1(z)$ and $p_0(z)$ defined on all points in Z; that is to say, $p_1(z)\, dz$ is the probability that the received signal vector terminates within dz of the point z when $\mathbf{x} = \mathbf{x}_1$, etc. Then, for each z we calculate the probability

$P(H_1|z)$ that H_1 is true, i.e., that x_1 was sent, and similarly $P(H_0|z)$. Clearly, the number of decision errors is minimized in the long run if we accept H_1 when $P(H_1|z) > P(H_0|z)$, and vice versa.   The decision rule therefore amounts to partitioning Z into the two disjoint volumes where $P(H_1|z) \gtrless P(H_0|z)$.   At the border line $P(H_1|z) = P(H_0|z)$ we flip a coin.

These conditional truth probabilities can be related to the density functions by invoking *Bayes' theorem*

$$P(B|A) = \frac{P(B)P(A|B)}{P(A)}$$

Thus, since $P(H_1|z) = p(x_1|z) \, dz$ and $P(z) = p(z) \, dz$, etc.,

$$P(H_1|z) = \frac{P_1 p_1(z) \, dz}{p(y) \, dz} \qquad P(H_0|z) = \frac{P_0 p_0(z) \, dz}{p(z) \, dz}$$

Correspondingly, the hypothesis test $P(H_1|z) \gtrless P(H_0|z)$ is expressed by the *likelihood ratio*

$$\lambda(z) = \frac{P_0}{P_1} \frac{P(H_1|z)}{P(H_0|z)} = \frac{p_1(z)}{p_0(z)} \tag{9.48}$$

such that we accept H_1 if $\lambda(z) > P_0/P_1$, etc.   When the digits are equiprobable ($P_1 = P_0$) the test becomes: compare $\lambda(z)$ to unity.

The statistical formulation at which we have arrived is a special case of *Bayes' solution* to the decision problem.   It is a special case because we have implicitly assumed that both types of error, i.e., mistaking 1 for 0 or 0 for 1, are equally serious.   In the radar detection problem, where missed pulses are far more costly than false alarms, appropriately unequal weighting factors would be incorporated in the likelihood-ratio test.[1]

Let us now calculate the likelihood ratio for binary signaling in the presence of white gaussian noise.   To do so we first note that $p_1(z) \, dz$, being the probability that $x_1 + n$ results in z, can be rewritten via

$$p_1(z) \, dz = P(z = x_1 + n) = P(n = z - x_1)$$

giving

$$p_1(z) = p(n = z - x_1)$$

Now assume for the moment that the channel is bandlimited in B; we consider a time interval T sec in length such that the various signals are described by their $2BT$ signal-space coordinates (or sample values), namely, $n(t_m)$, $x_1(t_m)$, and $z(t_m)$.   The probability that $n(t) = z(t) - x_1(t)$ is then the joint probability that each $n(t_m)$ equals the corresponding $z(t_m) - x_1(t_m)$.   But the coordinates $n(t_m)$ are statistically independent gaussian random variables with zero mean and variance $\sigma^2 = \eta B$, where

---

[1] See, for example, Davenport and Root (1958, chap. 14).

η is the noise power density. Therefore, $p(n = z - x)$ is the product of $2BT$ gaussian density functions, and

$$p_1(z) = \frac{1}{(\sqrt{2\pi\eta B})^{2BT}} \exp\left\{-\frac{\Sigma[z(t_m) - x_1(t_m)]^2}{2\eta B}\right\}$$

where the sum goes from $m = 1$ to $2BT$.

In similar fashion

$$p_0(z) = \frac{1}{(\sqrt{2\pi\eta B})^{2BT}} \exp\left\{-\frac{\Sigma[z(t_m) - x_0(t_m)]^2}{2\eta B}\right\}$$

so the likelihood ratio is

$$\lambda(z) = \frac{p_1(z)}{p_0(z)}$$

$$= \exp\left(-\frac{1}{2\eta B}\sum\{[z(t_m) - x_1(t_m)]^2 - [z(t_m) - x_0(t_m)]^2\}\right)$$

Passing to the limit $2BT \to \infty$, the sum becomes an integration over t with $1/2B \to dt$; hence

$$\lambda(z) = \exp\left\{-\frac{1}{\eta}\int_{-\infty}^{\infty}[z(t) - x_1(t)]^2 - [z(t) - x_0(t)]^2\, dt\right\}$$

and expanding the integral yields

$$-2\int_{-\infty}^{\infty} z(t)[x_1(t) - x_0(t)]\, dt + E_1 - E_0$$

where $E_1 = \int_{-\infty}^{\infty} x_1{}^2(t)\, dt$, etc. Thus we finally obtain

$$\lambda(z) = \exp\left\{\frac{2}{\eta}\int_{-\infty}^{\infty} z(t)[x_1(t) - x_0(t)]\, dt - \frac{E_1 - E_0}{\eta}\right\} \tag{9.49}$$

An examination of Eq. (9.49) reveals that the operation

$$\int_{-\infty}^{\infty} z(t)[x_1(t) - x_0(t)]\, dt$$

is identical to coherent binary detection, either by matched filtering or with a reference signal $r(t) = x_1(t) - x_0(t)$ [see Eq. (9.43)]. For coherent detection we found the decision level to be

$$D = \tfrac{1}{2}(E_1 - E_0)$$

when the digits are equally likely. Here the decision rule is compare $\lambda(z)$ to unity. But the test $\lambda(z) \gtrless 1$ is equivalently $\ln \lambda(z) \gtrless 0$; from (9.49)

$$\ln \lambda(z) = \frac{2}{\eta}\int_{-\infty}^{\infty} z(t)[x_1(t) - x_0(t)]\, dt - \frac{E_1 - E_0}{\eta}$$

so the test reduces to

$$\int_{-\infty}^{\infty} z(t)[x_1(t) - x_0(t)] \, dt \gtrless \tfrac{1}{2}(E_1 - E_0) = D$$

Therefore, without qualification, it can be said that optimum linear filtering does indeed yield the minimum error probability for the case under consideration.

Bear in mind, however, that we have been examining a particularly simple case. For more complex problems, e.g., unequal digit probabilities, multilevel signals, nongaussian noise, the results are not so clear-cut. Furthermore, one may be hard pressed to define what is meant by optimum.

PROBLEMS

9.1 Consider the pulse spectrum

$$X(f) = \begin{cases} \dfrac{1}{2W}[1 + X_1(f)] & |f| < W \\ \dfrac{1}{2W} X_1(f) & W < |f| < 2W \end{cases}$$

where $X_1(f)$ is real and has the symmetry property

$$X_1(W - u) = -X_1(W + u) \qquad 0 < u < W$$

Prove Nyquist's vestigial symmetry theorem by showing that

$$x(t) = \text{sinc } 2Wt + x_1(t)$$

where $x_1(k/2W) = 0$ for $k = 0, \pm 1, \pm 2, \ldots$.

9.2 Discuss digital transmission using gaussian pulse shapes of the form $x(t) = \exp(-\pi a^2 t^2)$. Specifically, to minimize intersymbol interference we might impose the condition $x(t = 1/s) = 0.01$, s being the signaling speed. Compare the resulting bandwidth requirement with that of the raised-cosine pulse spectrum.

9.3 Show, by writing the error probabilities for a decision level other than zero, that zero is indeed the best decision level for binary polar baseband transmission in mean-zero gaussian noise.

9.4 A trinary polar baseband system has pulse amplitudes $+A/2$, 0, and $-A/2$. Let the decision thresholds be at $\pm(1 - \epsilon)A/4$. If the noise is zero-mean gaussian, set up the equation which determines ϵ such that the *per-digit* error probabilities are equal. Also show that this results in a higher net error probability than when $\epsilon = 0$.

9.5 Suppose that binary data are transmitted in the form of *duration-modulated* pulses with $\tau_1 = 2/3s$ and $\tau_0 = 1/3s$. Select an appropriate decision rule and calculate the net error probability as a function of the signal-to-noise ratio.

9.6 Derive P_e for coherent FSK, Eq. (9.27).

9.7 When quaternary PSK is used to transmit binary data, there are two common encoding schemes as shown below:

Phase ϕ_j	Direct code	Gray code
0	00	00
$\dfrac{\pi}{2}$	01	10
π	10	11
$\dfrac{3\pi}{2}$	11	01

By referring to Fig. 9.19, explain why the Gray code results in fewer errors per binary digit.

9.8 An analog method of error detection, known as *null-zone detection*, uses a forbidden zone centered around each decision level such that if the signal-plus-noise voltage falls in this zone, no conversion is made—equivalent to a detected but uncorrected error. Obtain expressions for the probability of "detected errors" and undetected errors when null-zone detection is used in a polar binary baseband system, the null zone being $\pm\epsilon A/2$.

9.9 Repeat the Hamming code analysis of Sec. 9.3 with $C = 2$ check digits. In particular, construct the error words and parity-check relationships, and find P_e.

9.10 A (7,3) binary cyclic code is known to be single-error-correcting and double-error-detecting. Assuming $p \ll 1$, estimate the probabilities of corrected errors, detected errors, and undetected errors.

9.11 Figure P 9.1 shows the shift-register encoder for a Hagelbarger code with $\alpha = \frac{1}{2}$. The output switch alternately picks up the message digits and check digits. Neglecting the start-up transient, find the interlaced parity-check linkages and verify that six successive errors can be detected if 19 preceding digits are correct.

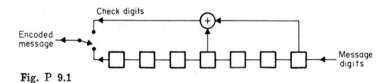

Fig. P 9.1

9.12 Prove the signal-vector statement of Schwarz's inequality

$$(\mathbf{u} \cdot \mathbf{v})^2 \leq |\mathbf{u}|^2 |\mathbf{v}|^2$$

by writing $\mathbf{u}$ and $\mathbf{v}$ in terms of their signal-space coordinates.

9.13 Find the optimum receiving filter $H(f)$ and the resulting $(A^2/4N)_{\max}$ when $x(t) = \text{sinc } 2Wt$ and $G_n(f) = (\eta/2)(1 + f^2)$.

9.14 Taking $x(t) = \Pi(st)$ and assuming white noise, show that an integrate-and-dump filter is equivalent to matched filtering.

9.15 Determine the optimum filter transfer function when the output pulse shape $y(t)$ is fixed and the input pulse shape $x(t)$ is variable, save that its energy $E_x = \int_{-\infty}^{\infty} x^2(t)\, dt$ is constant. Take the noise spectrum as $G_n(f)$. Relate $(A^2/4N)_{max}$ to E_x and η when the noise is white.

SELECTED SUPPLEMENTARY READING

Very few books are devoted primarily to digital data transmission. The most inclusive coverage is given by Bennett and Davey (1965), a useful introduction to both theoretical and practical matters. Other works containing appreciable treatment of the subject are, in order of increasing mathematical sophistication, Stein and Jones (1967), Schwartz, Bennett, and Stein (1966), and Wozencraft and Jacobs (1965).

However, with digital-system engineering being a rapidly developing field, the best source of up-to-date material is the professional literature, particularly the *Bell System Technical Journal* and the *IEEE Transactions on Communication Technology* (formerly the *Transactions on Communication Systems*). Of the many papers that could be cited here, the following have special merit: Arthurs and Dym (1962), a general discussion of digital-signal detection couched in the language of signal space; Salz (1965), for a description of FM systems using discriminators; and Jacobs (1967), which compares various μ-ary systems.

The reader interested in error-control coding will find Peterson (1961) a self-contained exposition. The original papers describing codes mentioned in Sec. 9.3 are Hamming (1950), Slepian (1956), Hagelbarger (1959), and Bose and Ray-Chaudhuri (1960).

Relative to coherent communication, Turin (1960) gives a tutorial treatment of matched filters, while Viterbi (1966) covers coherent detection of analog and digital signals. For an introduction to statistical decision theory, the concise monograph by Selin (1965) is recommended, as well as Harman (1963, chaps. 10 and 11) and Davenport and Root (1958, chaps. 11 and 14).

Appendix
NOISE IN
COMMUNICATION RECEIVERS

The study of noise in electrical communication is a vast and multifaceted subject, for noise stems from a variety of mechanisms and enters the system at every point. However, the effects are most serious where the signal level is lowest, namely, at the receiver. Hence, per Fig. A.1, the problem

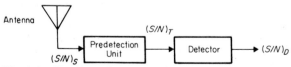

Fig. A.1 *Signal-to-noise ratios in a communication receiver.*

can be broken into three parts: (1) the input or source signal-to-noise ratio $(S/N)_S$ at the antenna terminals (transmission-line terminals in the case of wire systems); (2) the predetection signal-to-noise ratio $(S/N)_T$; and (3) the postdetection signal-to-noise ratio $(S/N)_D$.

To begin with the input, incoming noise is generally the result of random electromagnetic radiation intercepted by the antenna, this radiation being from both man-made and natural sources.   Examples of the former are ignition and commutator sparking, fluorescent lights, and x-ray machines, while the later includes atmospheric noise (static) and emissions from extraterrestrial bodies.[1]   In addition, the antenna itself may be a noise generator.   As a result, the incoming noise power depends on the type of antenna, where it is located and where it is looking, the operating frequency, time of day, and so forth.   Clearly, an investigation of all these factors is not possible here; moreover, although prediction formulas do exist, accurate determination of the input noise usually requires on-site measurements.   We shall therefore assume that $(S/N)_S$ is known, by one means or another.

At the other end of the receiver, the detector and any stages thereafter are usually relatively noiseless, so the relation between predetection and postdetection signal-to-noise ratios depends only on the type of modulation system.   These relationships were developed in Chaps. 5 to 7.

If the predetection unit is also free of noise, then $(S/N)_T$ equals $(S/N)_S$, and the job is done.   But this is seldom the case, since the RF and IF stages, the mixer, etc., often add internally generated noise whose power level is comparable to, or even greater than, the input noise.   An accurate assessment of system performance must then include the *receiver noise*.   And that is the task at hand.

This appendix discusses receiver noise in terms of the standard measures *noise figure* and *noise temperature*, measures which facilitate system design and evaluation by relating $(S/N)_T$ to $(S/N)_S$.   Starting with thermal noise and circuit-noise analysis, the description of noisy two-port networks will be developed.   It is then shown that the first few stages of a receiver are the critical elements in low-noise design.   The sources of receiver noise, e.g., noise from resistors, vacuum tubes, and transistors, are also considered, but very briefly.

A.1 THERMAL NOISE

Thermal noise, for our purposes, is the noise voltage due to the random motion of charged particles (usually electrons) in conducting media. From kinetic theory, the average energy of a particle at absolute temperature T is $\frac{1}{2}kT$ joules per degree of freedom, k being the Boltzmann con-

---

[1] Such emissions, known as *cosmic radio noise*, were first discovered by Karl Jansky in 1932, giving birth to the field of radio astronomy.   Much has been learned about our universe by studying this noise.   Nonetheless, it is a nuisance to the communication engineer.

stant.   We thus expect thermal-noise values to involve the product kT;
in fact, we shall develop a measure of noise power in terms of temperature.

Historically, Johnson (1928) and Nyquist (1928) first studied thermal
noise in metallic resistors, hence the designation *Johnson noise* or *resistance
noise*.   However, the same basic phenomenon is found in any linear passive media.

Resistance noise and passive networks

To a good approximation, resistance noise is *white noise* in that the noise
density is constant over all frequencies of interest in conventional electric
networks.   Specifically, expressed in *mean-square volts*, the power spectral
density at the open-circuited terminals of a resistance R at temperature
T is[1]

$$G_v(f) = 2RkT \qquad \text{volts}^2/\text{Hz} \tag{A.1}$$

where T is in degrees Kelvin.   Resistors to which Eq. (A.1) applies are
called *thermal resistors*.   Many practical resistors produce more noise than
thermal resistors; none produce less.

Relating (A.1) to the material and notation of Sec. 4.6, we can write
$G_v(f) = \eta/2$, where

$$\eta = 4RkT \tag{A.2}$$

is the positive-frequency noise density in mean-square volts per unit bandwidth.   Thus, the open-circuit mean-square noise voltage in a bandwidth
B is

$$\overline{v_n^2} = \eta B = 4RkTB \tag{A.3}$$

and the *rms* voltage is $\sqrt{4RkTB}$.

Of course we cannot speak of *total* voltage since, in theory, white
noise has infinite power.   The problem is avoided by observing that any
real two-port device or measuring instrument has a *noise equivalent
bandwidth* B_N, defined in Eq. (4.53) as

$$B_N = \frac{1}{H_0^2} \int_0^\infty |H(f)|^2 \, df$$

where $H_0 = |H(f)|_{\max}$ is the midband voltage gain.   Therefore, the filtered
noise has a finite mean-square value given by

$$\overline{v_n^2} = \int_{-\infty}^\infty G_v(f)|H(f)|^2 \, df = 2 \int_0^\infty \frac{\eta}{2} |H(f)|^2 \, df$$
$$= H_0^2 \eta B_N = H_0^2 4RkTB_N$$

[1] See Lawson and Uhlenbeck (1950, chap. 4) for the thermodynamics derivation.

For convenience, the standard temperature, or *room temperature*, is taken to be $T_0 = 290°K$ (63°F).   This is rather chilly for room temperature by American standards (probably realistic for a European laboratory lacking central heat), but it does simplify numerical evaluation since, with $k = 1.37 \times 10^{-23}$ joule/deg,

$$kT_0 \approx 4 \times 10^{-21} \text{ watt-sec}$$

Hence, a 10-kilohm thermal resistance at room temperature produces $\eta = 4 \times 10^4 \times 4 \times 10^{-21} = 16 \times 10^{-17}$ volt^2/Hz.   An rms voltmeter with $B_N = 100$ kHz would read the open-circuit value as $\sqrt{4RkT_0B_N} = (16 \times 10^{-17} \times 10^5)^{1/2} = 4$ μv.

When analyzing networks having resistance noise, we can replace the noisy resistor by a fictitious *noiseless* resistance of the same value, together with an equivalent *noise generator*.   If there are two or more sources of noise, one must bear in mind that *noise powers add*, assuming statistical independence.   (The assumption of statistical independence is quite valid for physically distinct thermal resistors.)   It is therefore common practice to specify noise sources in terms of power spectral density or mean-square values, though one must carefully distinguish between such sources and conventional signal generators.

Using Eq. (A.3), the Thévenin equivalent model of a thermal resistor is as shown in Fig. A.2a, where the mean-square voltage generator is

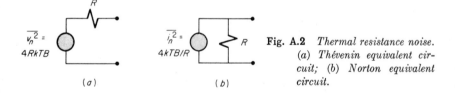

$$\overline{v_n^2} = \\ 4RkTB$$

$$\overline{i_n^2} = \\ 4kTB/R$$

(a)   (b)

Fig. A.2   *Thermal resistance noise. (a) Thévenin equivalent circuit; (b) Norton equivalent circuit.*

shaded to indicate its special nature.   The short-circuit mean-square current is obviously

$$\overline{i_n^2} = \frac{\overline{v_n^2}}{R^2} = \frac{4kT}{R} B \tag{A.4}$$

which leads to the Norton model of Fig. A.2b.   Both models have identical terminal properties, as the reader can easily verify.   Note, however, that it is the temperature (thermal agitation) which causes $\overline{i_n^2}$, not the reverse; hence, there is no additional ohmic heating or power dissipation in R due to self-noise current.   (If this were the case, a short across the terminals in Fig. A.2a would cause the resistor to spontaneously and continually increase in temperature.)   On the other hand, resistance noise can cause power dissipation in an *external* load circuit.

Now consider a passive one-port (two-terminal) network containing only inductors, capacitors, and resistors (Fig. A.3). If all resistors are at

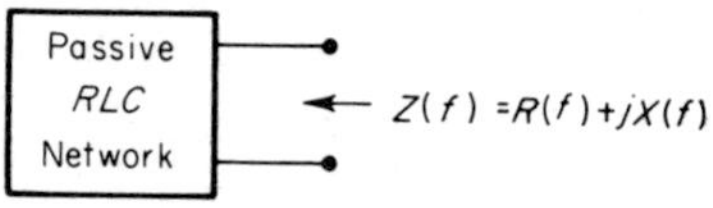

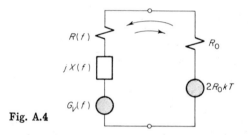

Fig. A.3 — Passive RLC Network — $Z(f) = R(f) + jX(f)$

temperature T (thermal equilibrium) and the complex terminal impedance is $Z(f) = R(f) + jX(f)$, then the output power spectrum is

$$G_v(f) = 2R(f)kT \tag{A.5}$$

and

$$\overline{v_n^2} = \int_{-\infty}^{\infty} G_v(f)\, df = 2kT \int_{-\infty}^{\infty} R(f)\, df$$

The noise thus depends on the *apparent* resistance, and the reactive elements have introduced a frequency dependence so that it is no longer white noise.

Equation (A.5), called *Nyquist's formula*, can be proved by invoking the second law of thermodynamics, as follows. Consider the circuit in Fig. A.4, which shows the Thévenin equivalent of Fig. A.3, including a

Fig. A.4

noise generator having $G_v(f)$, whose value is unknown, and a loading resistor R_0. The latter has its own noise generator $G_{v0}(f) = 2R_0 kT$. If the system is in *thermal equilibrium* at temperature T, the noise power delivered from $Z(f)$ to R_0 exactly equals the power flow from R_0 to $Z(f)$—otherwise there would be a self-sustaining temperature difference, and one could build a perpetual motion machine. Furthermore, this power balance must hold over any frequency band such as $f_1 < f < f_1 + df$.

Now the mean-square current in df flowing to the left is

$$\overline{i_L^2} = \frac{2R_0 kT\, df}{[R_0 + R(f_1)]^2 + [X(f_1)]^2}$$

while to the right

$$\overline{i_R^2} = \frac{G_v(f_1)\, df}{[R_0 + R(f_1)]^2 + [X(f_1)]^2}$$

But an equal power exchange requires

$$\overline{i_L^2}R(f_1) = \overline{i_R^2}R_0$$

or

$$2R_0 kT \, df \, R(f_1) = G_v(f_1) \, df \, R_0$$

for any value of f_1. Hence

$$G_v(f) = 2R(f)kT$$

which completes the proof.

As a special case of Nyquist's formula, suppose a network consists entirely of resistors in thermal equilibrium at temperature T. Then

$$G_v(f) = 2R'kT$$

where R' is the equivalent resistance seen looking into the terminals. This result is most gratifying when compared with (A.1) since the network could be replaced by a single resistor R' for all other purposes. If, however, the temperatures are not equal, the noise equivalent circuit must be analyzed in detail (see Example A.2 for an illustration).

Example A.1

An interesting application of several of the above ideas is the noise-voltage calculation at the terminals of a parallel RC circuit (Fig. A.5a).

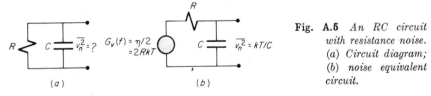

Fig. A.5 *An RC circuit with resistance noise. (a) Circuit diagram; (b) noise equivalent circuit.*

Since

$$Z(f) = \frac{R}{1 + j2\pi fRC} = R\frac{1 - j(f/f_3)}{1 + (f/f_3)^2} \qquad f_3 = \frac{1}{2\pi RC}$$

then $R(f) = R/[1 + (f/f_3)^2]$, and

$$\overline{v_n^2} = 2kT \int_{-\infty}^{\infty} \frac{R \, df}{1 + (f/f_3)^2} = 2\pi RkTf_3 = \frac{kT}{C}$$

It may seem surprising that $\overline{v_n^2}$ depends on C but not on R—after all, the resistor is the source of the noise. We shall return to this in a moment; first, let us check the result using the Thévenin model.

Separating the resistor and its noise, as in Fig. A.5b, the mean-square voltage in df across the capacitor is

$$\overline{dv_n{}^2} = \left| \frac{1/j\omega C}{R + 1/j\omega C} \right|^2 2RkT \, df$$

Hence

$$\overline{v_n{}^2} = \int_{-\infty}^{\infty} \frac{2RkT \, df}{1 + (2\pi fRC)^2} = \frac{kT}{C}$$

as before.

But Fig. A.5b suggests another interpretation. We can think of an input white-noise generator having $\eta = 2G_v(f) = 4RkT$ followed by a noiseless RC filter whose noise equivalent bandwidth is $B_N = (\pi/2)f_3$ and whose midband gain is $H_0 = |H(0)| = 1$. Thus $\overline{v_n{}^2} = \eta B_N = kT/C$. The reason why $\overline{v_n{}^2}$ is independent of R is now more obvious: increasing R increases the noise density η but decreases the noise bandwidth B_N. The effects precisely cancel each other.

Available power and noise temperature

Instead of dealing with mean-square voltage or current, describing thermal noise by its *available power* cleans up the notation and speeds calculations. Available power is the *maximum* power that can be delivered to a load from a source having fixed nonzero source resistance. The familiar maximum-power-transfer theorem states that this power is delivered only when the load impedance is the complex conjugate of the source impedance. The load is then said to be *matched* to the source, a condition which may or may not be desired in a communication receiver.

Let a sinusoidal signal source have impedance $Z_S = R_S + jX_S$, and let the open-circuit *rms* voltage be v_S (Fig. A.6). If the load is matched so that $Z_L = Z_S^* = R_S - jX_S$, then the terminal voltage is $v_S/2$, and the available power is

$$P_a = \frac{(v_S/2)^2}{R_S} = \frac{v_S{}^2}{4R_S} \tag{A.6}$$

Using the Thévenin model, we can extend this concept to a thermal resistor treated as a *noise source* (Fig. A.7). The *available noise power* is

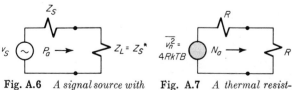

Fig. A.6 *A signal source with matched load.*

Fig. A.7 *A thermal resistance with matched load.*

by comparison

$$N_a = \frac{\overline{v_n^2}}{4R} = \frac{4RkTB}{4R} = kTB \tag{A.7}$$

which depends only on the temperature and bandwidth. A thermal resistor therefore delivers a maximum of kT watts per unit bandwidth, regardless of the value of R!

Clearly, temperature is the fundamental parameter of thermal noise. But there are other white-noise sources which are *nonthermal* in the sense that the noise power is unrelated to a physical temperature. Nonetheless, we can speak of the *noise temperature* T_S of any white-noise source, thermal or nonthermal, by defining

$$T_S = \frac{N_a}{kB} \tag{A.8}$$

where N_a is the maximum noise power the source can deliver in bandwidth B.

For thermal sources T_S is a physical temperature; for nonthermal sources it is a measure of the available noise power. This means that a nonthermal white-noise source having equivalent internal resistance R_S and noise temperature T_S can be replaced by a thermal resistor $R = R_S$ at physical temperature $T = T_S$ without altering the terminal properties.

Example A.2

Two resistors R_1 and R_2 at temperatures T_1 and T_2 are connected in series to form a white-noise source. Applying the Thévenin model (and recalling that mean-square values add) the open-circuit terminal voltage is $\overline{v_n^2} = 4(R_1 T_1 + R_2 T_2)kB$, and the equivalent resistance is $R' = R_1 + R_2$. Thus, the available noise power is

$$N_a = \frac{\overline{v_n^2}}{4R'} = \frac{4(R_1 T_1 + R_2 T_2)kB}{4(R_1 + R_2)}$$

and hence

$$T_S = \frac{N_a}{kB} = \frac{R_1 T_1 + R_2 T_2}{R_1 + R_2}$$

Note that if $R_1 = R_2$, the noise temperature is the arithmetic average of the physical temperatures, that is, $T_S = \frac{1}{2}(T_1 + T_2)$.

Since T_S is not a physical temperature in this example, the series connection is a nonthermal source, as should be expected since the resistors are not in thermal equilibrium. However, viewed from the terminals, this nonthermal source looks identical to a thermal resistance R' at temperature T_S.

A.2 THE DESCRIPTION OF NOISY TWO–PORT NETWORKS

The predetection portion of a communication receiver consists of many two-port networks—amplifiers, mixers, attenuators, etc.—connected in cascade. Each two-port has internally generated noise that tends to degrade system performance. Thus, S/N at the output may be substantially less than at the input. This section deals with the description of noisy two-ports and the relationship of input and output signal-to-noise ratios.

A general treatment of linear two-ports with internal noise can be quite complicated, as indicated by Fig. A.8, which assumes all signals

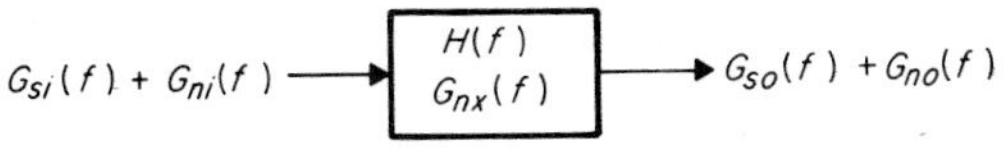

$$G_{si}(f) + G_{ni}(f) \longrightarrow \boxed{\begin{array}{c} H(f) \\ G_{nx}(f) \end{array}} \longrightarrow G_{so}(f) + G_{no}(f)$$

Fig. A.8

and noise are additive. The output-signal power spectrum is

$$G_{so}(f) = |H(f)|^2 G_{si}(f)$$

and the output-noise power spectrum is

$$G_{no}(f) = |H(f)|^2 G_{ni}(f) + G_{nx}(f)$$

where $G_{nx}(f)$ is the power spectrum of the *excess noise* introduced by the two-port itself. Therefore, the output signal-to-noise ratio is

$$\left(\frac{S}{N}\right)_o = \frac{\int_{-\infty}^{\infty} |H(f)|^2 G_{si}(f)\, df}{\int_{-\infty}^{\infty} [|H(f)|^2 G_{ni}(f) + G_{nx}(f)]\, df}$$

This formidable expression is greatly simplified if (1) the input noise is white, or at least has uniform density η over the passband of the device, and (2) the amplitude response is essentially constant at $|H(f)| = H_0$ over the frequency range of the input signal. Under these conditions

$$\left(\frac{S}{N}\right)_o = \frac{H_0^2 \int_{-\infty}^{\infty} G_{si}(f)\, df}{\eta \int_0^{\infty} |H(f)|^2\, df + \int_{-\infty}^{\infty} G_{nx}(f)\, df}$$

$$= \frac{H_0^2 P_S}{H_0^2 \eta B_N + N_x} \tag{A.9}$$

where P_S is the input-signal power and N_x is the total excess noise power at the output. Equation (A.9) brings out the fact that a two-port amplifies (or attenuates) the input signal and noise by the same amount and then adds noise of its own. Noting that $P_S/\eta B_N$ is the input signal-to-noise ratio (referred to the bandwidth of the two-port), we see that the signal-to-noise ratios are equal only when the two-port is *noiseless*, that is, when $N_x = 0$.

Effective noise temperature

Despite the simplifications, Eq. (A.9) is at best a cumbersome description of a two-port's noisiness. A more tractable measure is obtained if we assume, for the moment, that all impedances are matched; i.e., looking into the two-port, the source sees a matched impedance, and the output of the two-port sees a matched load (Fig. A.9). Then the source delivers

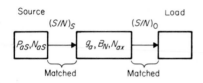

Fig. A.9 *A noisy two-port with matched source and load.*

its *available* signal power P_{aS}, and the output signal power is

$$P_{aO} = g_a P_{aS}$$

where g_a is the *available power gain* of the device. Similarly, the output noise power is

$$N_{aO} = g_a N_{aS} + N_{ax}$$

N_{ax} being the available excess noise power at the output. If the input noise is white and represented by a noise temperature T_S, then

$$N_{aS} = kT_S B_N$$

is the available source noise in the equivalent bandwidth B_N.

Combining these terms gives the output signal-to-signal ratio as

$$\left(\frac{S}{N}\right)_o = \frac{P_{aO}}{N_{aO}} = \frac{g_a P_{aS}}{g_a k T_S B_N + N_{ax}} = \frac{P_{aS}}{k T_S B_N + N_{ax}/g_a}$$

and taking the source signal-to-noise ratio as $(S/N)_S = P_{aS}/N_{aS} = P_{aS}/kT_S B_N$, we have

$$\left(\frac{S}{N}\right)_o = \frac{(S/N)_S}{1 + (N_{ax}/g_a k T_S B_N)} \leq \left(\frac{S}{N}\right)_S \tag{A.10}$$

It was noted before that $(S/N)_o = (S/N)_S$ only when the two-port is noiseless; but (A.10) shows that the signal-to-noise ratios can be nearly equal, despite excess noise, providing that $N_{ax} \ll g_a k T_S B_N$. Obviously what counts is not the absolute value of the excess noise power but its value relative to the source noise, a conclusion that certainly makes sense.

Pursuing this thought further, we observe that $N_{ax}/g_a k B_N$ depends only on the parameters of the two-port and has the dimensions of temperature. Therefore, we define the *effective input noise temperature* (also called the *amplifier temperature*) by

$$T_e = \frac{N_{ax}}{g_a k B_N} \tag{A.11}$$

The meaning of T_e is illustrated in Fig. A.10, which shows that a noisy

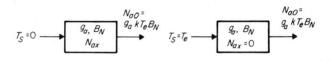

(a)   (b)

Fig. A.10   *A noisy amplifier with $T_S = 0$ and its noiseless equivalent with $T_S = T_e$.*

amplifier with $T_S = 0$ can be replaced by a noiseless amplifier with $T_S = T_e$. In short, T_e is a measure of noisiness *referred to the input*. If the device is noiseless, $T_e = 0$.

Substituting (A.11) into (A.10) gives

$$\left(\frac{S}{N}\right)_o = \frac{(S/N)_S}{1 + T_e/T_S} \tag{A.12a}$$

$$\left(\frac{S}{N}\right)_o = \frac{g_a P_{aS}}{g_a k (T_S + T_e) B_N} \tag{A.12b}$$

Equation (A.12b) is particularly informative, for it says that under matched conditions, the available output noise is

$$N_{aO} = g_a k (T_S + T_e) B_N \tag{A.13}$$

and the output noise temperature is $g_a(T_S + T_e)$.

But what if impedances are not matched? In that case, all powers will be *less* than the available powers, being reduced by a mismatch factor. Nonetheless, Eqs. (A.10) and (A.12) are still valid because they are power *ratios* taken at specific points, and the mismatch factor cancels out. Thus, although an amplifier may not actually be operated with matched impedances, its available power gain g_a and effective input noise temperature T_e are significant parameters. As a bonus, they are relatively easy to measure.

Attenuators

Although we have been thinking of two-ports as amplifiers, the above results apply equally well to attenuators, the principal difference being that the available power gain of an attenuator is less than 1. Consequently it is common to refer to the attenuation or *loss* L, defined as

$$L = \frac{1}{g_a}$$

so that $P_{aO} = P_{aS}/L$.

Now consider an attenuator composed entirely of *resistive elements* in thermal equilibrium at room temperature T_0. (This is a good approximation for lossy transmission line and waveguide, as well as for actual networks of resistors.) Assuming the source temperature is also T_0 and that all impedances are matched, looking back into the output terminals, we see nothing but thermal resistance and a noise bandwidth. Therefore, the available output noise power is simply $N_{aO} = kT_0 B_N$. But, from (A.13), $N_{aO} = g_a k(T_S + T_e)B_N$ for *any* two-port with matched impedances. The attentuator in question has $g_a = 1/L$ and $T_S = T_0$, so

$$T_e = (L - 1)T_0 \tag{A.14}$$

independent of the bandwidth or source temperature.

Noise figure

The effective input noise temperature is most useful in describing low-noise amplifiers, devices with $T_e \ll T_S$. But when the excess noise is large, the *integrated noise figure F* proves more convenient.

Noise figure is defined as the actual output noise power divided by the output noise power if the two-port were noiseless, the source being at room temperature T_0. The definition is a power ratio and can be written in terms of available powers as

$$F = \frac{N_{aO}}{g_a k T_0 B_N} \tag{A.15a}$$

Now with $T_S = T_0$, $N_{aO} = g_a k(T_0 + T_e)B_N$, and hence

$$F = \frac{T_0 + T_e}{T_0} = 1 + \frac{T_e}{T_0} \tag{A.15b}$$

or, from (A.11),

$$F = 1 + \frac{N_{ax}}{g_a k T_0 B_N} \tag{A.15c}$$

It should be apparent that in general $F \geq 1$ (the actual output noise cannot be less than the amplified source noise) and $F = 1$ for a *noiseless* amplifier.

As to the relationship of input and output signal-to-noise ratios, writing (A.15*b*) as

$$T_e = (F - 1)T_0 \tag{A.16}$$

and substituting in (A.12*a*) yields

$$\left(\frac{S}{N}\right)_o = \frac{(S/N)_s}{1 + (F - 1)(T_0/T_s)} \tag{A.17}$$

If, *and only if*, the source temperature is room temperature, Eq. (A.17) reduces to the simple result

$$\left(\frac{S}{N}\right)_o = \frac{1}{F}\left(\frac{S}{N}\right)_s$$

Prior to the advent of low-noise designs, most source temperatures were essentially T_0 (the source noise came from thermal resistance), so the latter expression could be used indiscriminantly.   For modern systems, in which noise is a crucial factor, Eq. (A.17) must be used.

One final remark here.   Comparing (A.16) with (A.14) shows that the noise figure of a resistive attenuator is just

$$F = L = \frac{1}{g_a} \tag{A.18}$$

if the attenuator is at room temperature.

Example A.3

A simple laboratory technique for measuring T_e or F requires a "hot" thermal resistance and a relative power meter, both impedance-matched to the device in question, plus a thermometer.   The procedure is as follows:

1. Connect the resistor, at room temperature, to the input of the two-port and record the output-power indication N_1.   Let the meter calibration constant be C_m, so $N_1 = C_m N_{aO} = C_m g_a k(T_0 + T_e)B_N$.

2. Heat the resistor until the output-power reading has *doubled* and note the resistor temperature T_R.   The new reading is

$$N_2 = C_m g_a k(T_R + T_e)B_N$$

Now $N_2 = 2N_1$, so $C_m g_a k(T_R + T_e)B_N = 2C_m g_a k(T_0 + T_e)B_N$, and hence

$$T_e = T_R - 2T_0 \qquad F = 1 + \frac{T_e}{T_0} = \frac{T_R}{T_0} - 1$$

Note that we do not have to find g_a or B_N, nor do we need the calibration constant C_m. However, the resistor will be on the hot side since it must reach a minimum of $2T_0 = 585°F$!

A.3   RECEIVER–NOISE CALCULATIONS

Let us now put together a number of two-ports representing the entire predetection unit, thereby relating the predetection signal-to-noise ratio to the input signal-to-noise ratio at the antenna terminals. For this purpose we shall find the *overall* receiver noise temperature or noise figure in terms of the properties of the individual stages.

Consider two amplifiers in cascade, as in Fig. A.11, again assuming

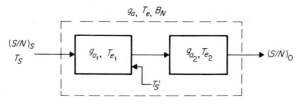

Fig. A.11   *Two noisy amplifiers in cascade.*

matched impedances. Clearly the available power gain of the combination is $g_a = g_{a_1} g_{a_2}$. However, the overall noise temperature T_e requires further thought.

One approach is to examine the situation at the input to the second amplifier, treating all that comes before as a new source with $T'_S = g_{a_1}(T_S + T_e)$. We could then plug into (A.12b) and turn the crank.

But a more instructive method is to follow through the various noise sources and determine their contribution at the output. In particular, N_{aO} will be the sum of three terms:

1. Amplified source noise: $g_{a_1} g_{a_2} N_{aS} = g_{a_1} g_{a_2} k T_S B_N$
2. Excess noise from the first stage, amplified by the second stage:
   $g_{a_2} N_{ax_1} = g_{a_2}(g_{a_1} k T_{e_1} B_N)$
3. Excess noise from the second stage: $N_{ax_2} = g_{a_2} k T_{e_2} B_N$

It has been implicitly assumed that both stages see white noise over their passband and that B_N is the overall noise bandwidth. Actually, good design suggests that the final stage should have the smallest possible bandwidth to minimize unnecessary noise at the demodulator. Thus, as a rule, B_N can be taken as the noise bandwidth of the last stage.

Returning to the calculation, we sum the three terms and factor to yield

$$N_{aO} = g_{a_1}g_{a_2}k\left[T_S + T_{e_1} + \frac{T_{e_2}}{g_{a_1}}\right]B_N$$

But, by definition of g_a and T_e for the cascade

$$N_{aO} = g_a k(T_S + T_e)B_N$$

Hence

$$T_e = T_{e_1} + \frac{T_{e_2}}{g_{a_1}}$$

and therefore

$$F = F_1 + \frac{F_2 - 1}{g_{a_1}}$$

Iterating this procedure with three or more cascaded two-ports, one obtains for the overall effective input temperature and noise figure

$$T_e = T_{e_1} + \frac{T_{e_2}}{g_{a_1}} + \frac{T_{e_3}}{g_{a_1}g_{a_2}} + \cdots \tag{A.19}$$

$$F = F_1 + \frac{F_2 - 1}{g_{a_1}} + \frac{F_3 - 1}{g_{a_1}g_{a_2}} + \cdots \tag{A.20}$$

The latter is known as *Friis' formula.*

Equations (A.19) and (A.20) indicate that the receiver noise may be dominated by the first stage. Suppose, for example, that $g_{a_1} \gg 1$; then the overall noise temperature is essentially that of the first stage itself, $T_e \approx T_{e_1}$. The physical reason is that with large first-stage gain, the amplified source noise (plus N_{ax_1}) will be much greater than any noise added by later stages. On the other hand, if the first stage is an *attenuator* of loss $L = 1/g_{a_1} > 1$—such as a lossy transmission line—the effect is twice cursed, since $T_{e_1} = (L - 1)T_0$ and

$$T_e = (L - 1)T_0 + LT_{e_2} + \cdots$$

This situation should be avoided if at all possible.

Clearly, the first few stages are of prime concern in the design of low-noise communication receivers. In particular, the first stage should have a small noise temperature and reasonably large gain. With a good "front end" or preamplifier, the remaining stages merely serve as additional amplification and filtering, amplifying both signal and noise without appreciably changing the ratio.

To complete our calculations, suppose the predetection unit has an overall effective input noise temperature T_e and a noise bandwidth approximately equal to the modulated signal bandwidth, $B_N \approx B_T$. At the antenna terminals let the received signal power be P_S, and let the noise temperature (*antenna temperature*) be T_S (Fig. A.12). (Receiving antennas are usually impedance-matched at their terminals, so P_S

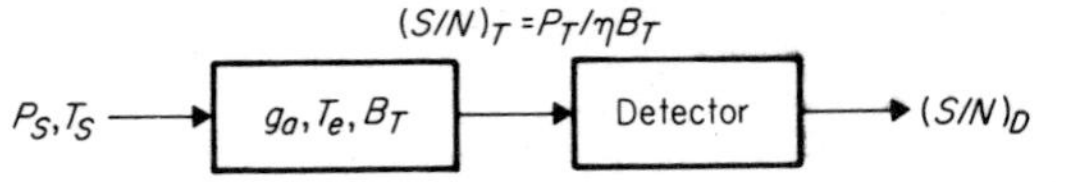

$$(S/N)_T = P_T/\eta B_T$$

$$P_S, T_S \longrightarrow \boxed{g_a, T_e, B_T} \longrightarrow \boxed{\text{Detector}} \longrightarrow (S/N)_D$$

Fig. A.12

is the *available* power.) The predetection signal-to-noise ratio is thus

$$\left(\frac{S}{N}\right)_T = \frac{g_a P_S}{g_a k (T_S + T_e) B_T} = \frac{P_S}{k T_N B_T} \tag{A.21}$$

where $T_N = T_S + T_e$ is the *system temperature* referred to the antenna terminals. In the past we have written $(S/N)_T = P_T/\eta B_T$. By comparison with (A.21) it follows that

$$P_T = g P_S \qquad \eta = g k T_N$$

where g is the *actual* power gain of the predetection unit, that is, $g = g_a$ if impedances are matched.

Example A.4

In adverse locations it is often necessary to put a television receiving antenna on a tall mast. A long and therefore lossy cable connects antenna and receiver. To overcome the effects of the cable, a preamplifier can be mounted at the antenna, as in Fig. A.13. The system parameters are given in decibels, which is standard practice, but must be converted before use in (A.19) or (A.20).

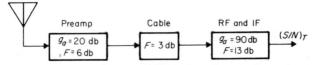

Fig. **A.13** *A receiver with preamplification at the antenna terminals.*

Inserting values in (A.20), with $F_2 = L = 2$, yields for the overall noise figure

$$F = 4 + \frac{2 - 1}{100} + \frac{20 - 1}{100 \times \frac{1}{2}} = 4.39 = 6.4 \text{ db}$$

which shows that the noise performance is essentially that of the pre-amplifier alone.   (Note that the RF and IF gain does not enter these calculations.)   Thus, if the source temperature is $T_S = T_0$, $(S/N)_T$ is 6.4 db less than the signal-to-noise ratio at the antenna terminals.

If the preamplifier is omitted, $F = 2 + 2(20 - 1) = 40$, and the output noise is about 10 times as large as before.

Example A.5

By space-age standards, the previous example is a very noisy receiver. Low-noise systems, such as the satellite ground station of Fig. A.14, have

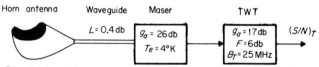

Fig. A.14   *A low-noise receiver of the type used in satellite ground stations.*

sufficiently small excess noise for the noise figure to be numerically awkward to handle.   Thus, the calculations are usually carried out in terms of temperatures.

The lossy waveguide has $L = 0.4$ db $= 1.10$, so $T_{e_1} = (1.10 - 1)T_0 = 29°$K.   For the traveling-wave tube (TWT), $T_{e_3} = (F_3 - 1)T_0 = 870°$K; of course the *physical* temperature of the TWT is much less. Thus

$$T_e = 29 + 1.1 \times 4 + \frac{1.1 \times 870}{400} \approx 36°\text{K}$$

Note that, despite the small waveguide loss, its noise contribution dominates the system.   Interchanging the positions of maser and waveguide would be highly desirable but difficult, since the guide is the flexible connection permitting the horn antenna to be steered.   (With a parabolic dish antenna the maser can be mounted directly at the feed, eliminating waveguide loss and noise.   However, dish antennas tend to have larger noise temperatures, which cancel out the saving.)

Under typical conditions, the antenna temperature may be $T_S = 22°$K, giving a system noise temperature of $T_N = T_S + T_e = 58°$K $= 0.2T_0$.   Hence, from (A.21),

$$\left(\frac{S}{N}\right)_T = \frac{P_S}{4 \times 10^{-21}(T_N/T_0)B_T} = \frac{P_S}{2 \times 10^{-14}}$$

A 20-db predetection signal-to-noise ratio requires an available signal power at the antenna terminals of $P_S = 10^2 \times 2 \times 10^{-14} = 2 \ \mu\mu\text{w}$.

A.4 THE SOURCES OF RECEIVER NOISE

We have discussed noise from the system viewpoint, as described by noise temperatures and noise figure. To conclude the study, let us briefly examine the specific sources of noise in receiver components and relate them to the system description.   Needless to say, a detailed treatment of this broad and fascinating subject is beyond the scope of this book and belongs more properly in an electronics text.   The interested reader will find abundant supplementary material in the literature.

Tube noise

Current in a vacuum tube is carried by the electron stream.   Because each electron has a discrete charge and a random arrival time at the plate, the collected current consists of small and random fluctuations about the average, or dc, value.   This situation is roughly analogous to dropping a stream of buckshot on a kettledrum, hence the descriptive name *shot noise*.

Shot noise in a temperature-limited vacuum diode was examined in Example 4.9, where we found that the power spectrum was relatively flat up to frequencies of order $1/\tau_e$, τ_e being the electron transit time from cathode to plate.   Specifically, the shot-noise power spectrum is given by the *Schottky formula* as

$$G_i(f) = eI_b \qquad |f| \ll \frac{1}{\tau_e}$$

where e is the electronic charge and I_b is the dc current.   It then follows that the short-circuit mean-square current in a bandwidth B is simply

$$\overline{i_b{}^2} = 2eI_bB$$

In conventional diode operation, i.e., space-charge-limited, the presence of space charge in the virtual cathode tends to smooth the electron-stream variations, thereby reducing the noise power.   This effect is reflected by writing

$$\overline{i_b{}^2} = \Gamma^2 2eI_bB \tag{A.22}$$

where Γ^2 is the *space-charge reduction factor* and depends on the electrode geometry, applied voltage, cathode temperature, etc.[1]   Typically, Γ^2 is in the range of 0.01 to 0.1.   Note that temperature-limited diodes ($\Gamma^2 = 1$) are much noisier than conventional diodes, a property used to advantage in certain laboratory noise generators discussed shortly.

[1] See Bennett (1960, chap. 4) or Davenport and Root (1958, chap. 7) for details.

A negative-grid triode is a space-charge-limited diode with added control grid.   Hence, triode noise is primarily shot noise, and the mean-square plate current is given by Eq. (A.22) with appropriate reduction factor.   However, it is often more convenient to treat the triode as being *noiseless* and refer the tube noise to the grid circuit.   This is done by adding a fictitious thermal resistance R_{eq} at temperature T_0 in series with the grid, resulting in the small-signal model of Fig. A.15, where the plate

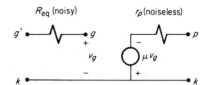

Fig. A.15   *Simplified triode equivalent circuit with tube noise represented by* $R_{eq}.$

resistance r_p is noiseless.   Since the grid normally draws no current, R_{eq} has no other effect than noise generation.

Combined theoretical and experimental studies have shown that for a wide variety of triodes with oxide-coated cathodes, the equivalent grid resistance is approximately

$$R_{eq} \approx \frac{2.5}{g_m} = 2.5\,\frac{r_p}{\mu} \tag{A.23}$$

where g_m is the transconductance and μ the amplification factor.   For example, the mean-square open-circuit output voltage in Fig. A.15 would be

$$\overline{v_n^2} = \mu^2 \overline{v_g^2} = \mu^2 4 R_{eq} k T_0 B$$
$$\approx 10 \mu r_p k T_0 B$$

for the tube noise alone.

Multigrid tubes (tetrodes and pentodes) have shot noise plus an additional effect due to fluctuating division of the electron stream among the various collecting electrodes, called *partition noise*.   Both shot and partition noise can be represented in the grid circuit of pentodes by a modified equivalent resistance

$$R_{eq} = \frac{I_b}{I_b + I_s}\left(\frac{2.5}{g_m} + \frac{20 I_s}{g_m^2}\right) \tag{A.24}$$

where I_b and I_s are the dc plate and screen currents, respectively.   Comparing (A.24) with (A.23), one would expect pentodes to be noisier than triodes of equal transconductance.   As a rule that conclusion is correct; however, some high-quality pentodes do have relatively low R_{eq}.

A few representative values of g_m and R_{eq} are given in Table A.1. Note the extremely high R_{eq} for the 6SA7, a *pentagrid converter* having three collecting electrodes and hence substantial partition noise.   Obviously, such tubes should be avoided in low-noise designs.

Table A.1   Noise equivalent grid resistances

Tube	Application	g_m, μmhos	R_{eq}, ohms
6CW4	Triode amplifier	12,500	200
6J5	Triode amplifier	2,600	1,250
6AC7/1852	Pentode amplifier	9,000	600–760
6SK7	Pentode amplifier	2,000	9,400–11,500
6AC7/1852	Pentode mixer	3,400	3,000
6SA7	Pentagrid converter	450	210,000

Besides shot and partition noise, two other phenomena may occur in vacuum tubes, namely, *flicker noise* (or *1/f noise*) and *induced grid noise*.   Flicker noise is a low-frequency effect traced to gradual deterioration of the cathode surface.   Its power spectrum decreases with increasing f (hence the alternate designation) and is usually negligible above a few kilohertz.   At the other extreme, induced grid noise first becomes noticeable around 10 MHz and increases as frequency squared until it dominates all other tube noise at about 100 MHz and up.   As the name suggests, it is due to random control-grid voltage induced by the electron stream. This voltage is in turn amplified and appears at the plate terminals.

Diode noise generators

A noise generator incorporating a temperature-limited diode is shown in Fig. A.16a.   The choke coil and blocking capacitor serve to separate the direct current and the shot-noise current so that $\overline{i_b^2}$ flows only through

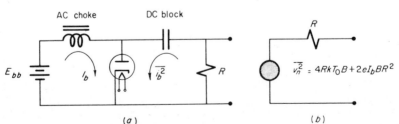

Fig. A.16   *Diode noise generator.*   *(a) Circuit diagram; (b) noise equivalent circuit.*

the diode and the shunt resistor R. The value of I_b is controlled by the cathode temperature via the filament voltage.

With the terminals open-circuited, the mean-square noise voltage is the sum of the resistance noise plus shot noise, namely,

$$\overline{v_n{}^2} = 4RkT_0B + 2eI_bBR^2$$

if R is at room temperature. Because a temperature-limited diode is a current source (infinite impedance), the output impedance is just R, and the noise equivalent circuit is as shown in Fig. A.16b. The available noise power is then

$$N_{aS} = kT_0B + \tfrac{1}{2}eI_bRB = kT_0\left(1 + \frac{eI_bR}{2kT_0}\right)B$$

Thus, if $eI_bR \gg 2kT_0$, a noise temperature much greater than T_0 can be attained. Moreover, the temperature is accurately determined by metering the direct current I_b.

Such generators are usually used as the "hot resistor" for noise-figure measurements as in Example A.3. The plate current is switched from off to on and adjusted so that the output power reading doubles. The equivalent resistance temperature is then $T_R = T_0(1 + eI_bR/2kT_0)$, so

$$F = \frac{T_R}{T_0} - 1 = \frac{eI_bR}{2kT_0}$$
$$\approx 20I_bR$$

If $R = 50$ ohms, a common value, then I_b in milliamperes is very nearly equal to the noise figure.

Diodes specifically designed for this application have very small cathode-plate spacing and are operated at relatively high plate voltages. The transit time can then be about 10^{-9} sec or less, giving a white-noise spectrum up to 1 GHz. At microwave frequencies, or for higher noise powers, gas-filled diodes or fluorescent tubes must be employed.

Triode-amplifier noise figures

A typical grounded-cathode triode amplifier is given in schematic form in Fig. A.17a. To simplify the analysis we shall assume a midrange operating frequency such that flicker noise and induced grid noise are negligible, and all bypass capacitors are effectively short circuits. This leads to the equivalent circuit of Fig. A.17b, where R_{eq} represents the tube shot noise, r_p is noiseless, and R_g is a conventional thermal resistor. The source noise is, as usual, represented by R_S at an appropriate temperature T_S. The load noise from R_L is treated as external to the amplifier.

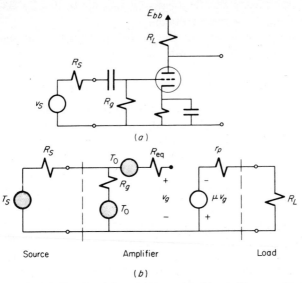

Fig. A.17 *Simple triode amplifier.* (a) *Circuit diagram;* (b) *noise equivalent circuit.*

Looking back from the grid terminal, we see three noisy resistors, R_S, R_g, and R_{eq}. The latter two are at room temperature, and, taking $T_S = T_0$ to facilitate calculations, the mean-square grid voltage becomes

$$\overline{v_g^2} = 4R'kT_0B \qquad \text{where} \qquad R' = R_{eq} + \frac{R_g R_S}{R_g + R_S}$$

Since the plate circuit adds no other noise (that is the advantage of the R_{eq} representation), the available noise power is

$$N_{aO} = \frac{\mu^2 \overline{v_g^2}}{4r_p} = \frac{\mu^2 R' k T_0 B}{r_p}$$

The available power gain g_a is readily found by considering available *signal* powers, that is, $P_{aS} = v_S^2/4R_S$, $P_{aO} = (\mu v_g)^2/4r_p$, and $v_g = [R_g/(R_g + R_S)]v_S$, hence

$$g_a = \frac{P_{aO}}{P_{aS}} = \frac{\mu^2 R_S}{r_p} \left(\frac{R_g}{R_g + R_S} \right)^2$$

Note that g_a is quite different from the square of the *actual* voltage gain

$$\left(\frac{v_O}{v_S} \right)^2 = \left[\frac{\mu R_L R_g}{(R_L + r_p)(R_g + R_S)} \right]^2$$

With $T_S = T_0$ the noise figure is $F = N_{aO}/g_a k T_0 B$ by definition, therefore

$$F = \frac{R'}{R_S} \left(\frac{R_g + R_S}{R_g} \right)^2$$
$$= 1 + \frac{R_S}{R_g} + \frac{R_{eq}}{R_S} \left(\frac{R_g + R_S}{R_g} \right)^2 \tag{A.25}$$

But $R_g \gg R_S$ for most triode amplifiers,[1] so

$$F \approx 1 + \frac{R_{eq}}{R_S} \tag{A.26}$$

which is also a valid approximation for grounded-plate (cathode follower) and grounded-grid configurations with $R_g \gg R_S$.

Equation (A.25) shows that F depends on R_g and R_S as well as R_{eq}. Now R_{eq} and R_g are usually fixed for a given tube, so low-noise design requires optimizing the *source impedance*. Differentiating (A.25) with respect to R_S and again assuming $R_g \gg R_S$ yields

$$R_{S_{opt}} = \sqrt{R_{eq} R_g}$$

and

$$F_{min} = 1 + 2 \sqrt{\frac{R_{eq}}{R_g}} \tag{A.27}$$

Normally R_S is also fixed, but we can approach F_{min} by inserting a step-up transformer between source and amplifier. Of course the ohmic losses in the transformer must be small to minimize added thermal noise.

It is interesting to compare (A.27) with (A.25) when the source impedance *matches* the amplifier input. For $R_S = R_g$, Eq. (A.25) becomes $F = 2 + 4(R_{eq}/R_g)$, obviously much greater than F_{min}. This clearly indicates that impedance matching is not necessarily the best thing in a communications receiver.

With reasonable care it is quite possible to build triode amplifiers having noise figures less than 1.5 (0.2 db) and power gains of 30 db or so, providing the operating frequency is below about 10 MHz. From 10 to 200 MHz, low-noise amplifiers call for the high-frequency stability of a pentode and the noise characteristics of a triode (no partition noise). An ingenious solution to this dilemma is the *cascode* circuit devised by Wallman et al. (1948). A cascode consists of two triodes, the first having grounded cathode to provide voltage gain and the second having grounded

---

[1] At high frequencies a grid-to-cathode shunt conductance due to transit-time loading may invalidate this approximation and becomes an important design consideration.

grid to provide stability. The overall noise figure is approximately that of the first stage alone, while the overall gain is somewhat less.

Certain premium triodes, called *nuvistors*, have been designed especially for high-frequency applications, notably in cascode amplifiers. Nuvistors such as the 6CW4 have relatively small shunt capacitance, very small R_{eq}, and can be operated with rather sizable values of R_g—see Eq. (A.27). Many nuvistors also have minimal flicker noise, making them desirable for sensitive low-frequency instrumentation as well.

Transistor noise and noise figures

While tube noise is rather easily described, at least to a first approximation, noise in semiconductors is a good bit more involved and less well understood. There are, however, certain similarities to tube noise.

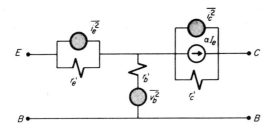

Fig. A.18 *Transistor noise equiv-*
alent circuit.

A simplified noise model for the junction transistor is given in Fig. A.18, which shows three different noise generators:

1. *Thermal noise*, due to the ohmic base resistance r_b' and appearing as the random base voltage

$$\overline{v_b^2} = 4r_b' kTB$$

2. *Shot noise*, due to fluctuations in the minority-carrier stream crossing the junction and appearing as the random emitter current

$$\overline{i_e^2} = 2eI_E B$$

where I_E is the dc emitter current.

3. *Partition noise*, due to base-recombination variations appearing as the random collector current

$$\overline{i_c^2} = 2eI_C \left(1 - \frac{|\alpha|}{\alpha_0}\right) B$$

where

$$\alpha = \frac{\alpha_0}{1 + j(f/f_\alpha)}$$

and α_0 is the common-base direct-current gain and f_α is the alpha-cutoff frequency.

Further complicating the matter is the fact that $\overline{i_e{}^2}$ and $\overline{i_c{}^2}$ are not statistically independent, and at low frequencies a surface-leakage effect contributes $1/f$ noise similar to vacuum-tube flicker effect.   These have been omitted in the circuit diagram.

With the aid of Fig. A.18 and the usual assumption that r_c' is very large compared to all other resistances, the noise figure of a common-base or common-emitter amplifier can be calculated.   After some manipulation one finds that[1]

$$F = 1 + \frac{r_e'}{2R_S} + \frac{r_b'}{R_S}$$
$$+ \frac{(1 - \alpha_0)(R_S + r_e' + r_b')^2}{2\alpha_0 r_e' R_S}\left[1 + \frac{1}{1 - \alpha_0}\left(\frac{f}{f_\alpha}\right)^2\right] \quad (A.28)$$

which is a function of operating frequency.   Figure A.19 shows F versus

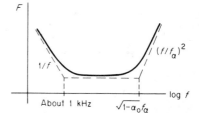

Fig. A.19 *Typical curve of transistor noise figure versus frequency.*

frequency on logarithmic coordinates and includes $1/f$ noise.   Note that the noise figure rapidly increases above the beta-cutoff frequency $f = \sqrt{1 - \alpha_0}\, f_\alpha$.

Inserting values in the above expression, typical transistor noise figures are in the range of 2 to 5 db at frequencies below the beta cutoff. Special low-noise transistors for RF amplifiers can achieve $F = 3$ db with a power gain of 10 db or more.

Some comments about the relative merits of tubes and transistors for low-noise applications would seem in order at this point.   However, because of the rapidly evolving state of the art in semiconductors, the best one can say is that further reduction of tube noise is unlikely, whereas continued progress is expected in low-noise transistor design.

[1] See, for example, Pettit and McWhorter (1961, chap. 13).

PROBLEMS

A.1 Two resistors R_1 and R_2 at temperatures T_1 and T_2 are connected in parallel. Draw the noise equivalent circuit and write expressions for N_a and T_S.

A.2 Referring to the above problem, under what conditions will the power transfer from R_1 to R_2 equal that from R_2 to R_1?

A.3 Find and sketch $G_v(f)$ for the circuit of Fig. P A.1. The resistors are equal and at the same temperature.

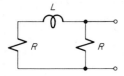

Fig. P A.1

A.4 A certain oscillator produces a 10-kHz sine wave and has an internal resistance of 600 ohms. To improve the ratio of open-circuit signal voltage to rms noise voltage, a capacitor is placed across the output terminals. What is the optimum value for the capacitance?

A.5 When the source temperature at the input to a certain amplifier is changed from T_0 to $2T_0$, the available output noise increases by one-third. Find the amplifier's T_e and F.

A.6 The two-port network of Fig. P A.2 is sometimes used to approximately match 300-ohm antennas to 50-ohm amplifiers. Calculate F and T_e.

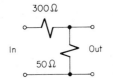

Fig. P A.2

A.7 A sine-wave generator can be used in place of the hot resistor in Example A.3. In step 1 the generator is off but connected to the input, so its source resistance delivers noise power. For step 2 the generator is turned on and adjusted so that its signal power P_S causes the output indication to double. Show that the noise figure is given by $F = P_S/kT_0B_N$ and discuss the disadvantages of this method.

A.8 A resistive two-port network having $g_a < 1$ is at temperature $T \neq T_0$. A matched thermal resistor at temperature $T_S \neq T$ is connected to the input. What is the output noise temperature?

A.9 Obtain the cascade relationship $F = F_1 + (F_2 - 1)/g_{a_1}$ by starting with Eq. (A.12b) and noting that the source temperature seen by the second amplifier is $T'_S = g_{a_1}(T_S + T_{e_1})$.

A.10 Two cascaded amplifiers have the following specifications: $T_{e_1} = 3T_0$, $g_{a_1} = 10$ db, $F_2 = 13.2$ db, $g_{a_2} = 50$ db. If $B_N = 100$ kHz and $T_S = 10T_0$, what input signal power is required to achieve an output S/N of 30 db?

A.11 A cable repeater system consists of k identical amplifiers each having $F \gg 1$ and g_a. The amplifiers are connected by lengths of cable having loss $L \gg 1$.

Usually, the amplifier gain just compensates for the cable loss, that is, $g_a = 1/L$. Write an expression for the overall noise figure and justify the saying that "doubling the number of repeaters increases the overall noise figure by 3 db."

A.12 The *Haus-Adler noise measure* is defined as

$$M = \frac{F - 1}{1 - 1/g_a}$$

Show that a cascade of two amplifiers has the lowest overall noise figure if the first amplifier has the lowest M. *Hint:* Calculate the difference $F_{12} - F_{21}$ for the two possible configurations.

A.13 A space-charge-limited diode has $\Gamma^2 = 0.02$ and a dynamic plate conductance of 0.005 mho. It is operated at $I_b = 100$ ma in series with a 400-ohm resistor. Draw the noise equivalent circuit and calculate the total rms noise voltage across the resistor in $B = 1$ MHz.

A.14 The triode amplifier of Fig. A.17a has $\mu = 100$, $r_p = 20$ kilohms, $R_L = 20$ kilohms, $R_g = 10$ kilohms, and $R_S = 50$ ohms. Calculate the actual voltage gain, the available power gain, F, and $F_{\min}$.

A.15 Write an expression for the noise figure of the cathode-follower amplifier shown in Fig. P A.3, including the noise from R_L.

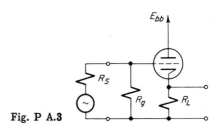

Fig. P A.3

A.16 Figure P A.4 shows a simplified *cascode* amplifier. Draw the noise equivalent circuit and write expressions for the available power gain and noise figure. Compare with a conventional triode amplifier.

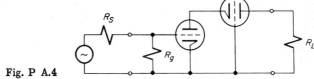

Fig. P A.4

A.17 Examining Eq. (A.28), we see that a low-noise transistor should have small r'_b, small $1 - \alpha_0$, and large f_α. Are these consistent with desired transistor parameters for applications where noise is not crucial?

A.18 Starting with Eq. (A.28) and the assumptions $f^2 \ll (1 - \alpha_0)f_\alpha^2$ and $r'_e \ll r'_b$, show that the noise figure is minimized when $R_S^2 = r'^2_b + 2\alpha_0 r'_b r'_e/(1 - \alpha_0)$. Taking $r'_e = 20$ ohms, $r'_b = 100$ ohms, and $1 - \alpha_0 = 0.02$, evaluate the resulting $F_{\min}$.

SELECTED SUPPLEMENTARY READING

Numerous books cover the material of this Appendix in varying degrees of depth and emphasis. One of the best general references is Bennett (1960), which, in addition to receiver noise, discusses noise in electromagnetic radiation, semiconductor noise, measurement techniques, and the design of low-noise equipment. Other useful works include Lawson and Uhlenbeck (1950, chaps. 4 and 5), van der Ziel (1954), and Baghdady (1960, chaps. 15 to 17). The latter contains a concise but comprehensive treatment of antenna noise.

As to some of the particular topics, Pierce (1956) and van der Ziel (1958) are excellent survey papers on the physical sources of noise, while Pettit and McWhorter (1961) have further details on noise in vacuum-tube and transistor amplifiers. Noise in microwave systems and the role of lasers and masers is summarized by Siegman (1961) using the informative transmission-line approach. For a qualitative yet highly illuminating discussion of low-noise electronics, see Jolly (1967).

Table A
Fourier Transform Pairs

Listed below are some of the common Fourier transform theorems and pairs based on the definition

$$X(f) = \int_{-\infty}^{\infty} x(t)e^{-j\omega t}\, dt \qquad \omega = 2\pi f$$

providing $X(f)$ exists. For more extensive tabulations, see Campbell and Foster (1948) or Erdélyi (1954).

Theorems

Operation	Function	Transform
Superposition	$a_1 x_1(t) + a_2 x_2(t)$	$a_1 X_1(f) + a_2 X_2(f)$
Time delay	$x(t - t_0)$	$X(f)e^{-j\omega t_0}$
Scale change	$x(at)$	$\dfrac{1}{|a|} X\left(\dfrac{f}{a}\right)$
Frequency translation	$x(t)e^{j\omega_c t}$	$X(f - f_c)$
Modulation	$x(t) \cos \omega_c t$	$\frac{1}{2}[X(f - f_c) + X(f + f_c)]$
Differentiation	$\dfrac{d^n x(t)}{dt^n}$	$(j\omega)^n X(f)$
Integration	$\displaystyle\int_{-\infty}^{t} \cdots \int_{-\infty}^{t_2} x(t_1)\, dt_1 \cdots dt_n$	$(j\omega)^{-n} X(f)$
Convolution	$x_1(t) * x_2(t)$	$X_1(f) X_2(f)$
Multiplication	$x_1(t) x_2(t)$	$X_1(f) * X_2(f)$

Transforms

Function	$x(t)$	$X(f)$
Rectangular	$\Pi\left(\dfrac{t}{\tau}\right)$	$\tau \operatorname{sinc} f\tau$
Triangular	$\Lambda\left(\dfrac{t}{\tau}\right)$	$\tau \operatorname{sinc}^2 f\tau$
Gaussian	$e^{-\pi(t/\tau)^2}$	$\tau e^{-\pi(f\tau)^2}$
Exponential	$e^{-|t|/\tau}$	$\dfrac{2\tau}{1 + (2\pi f\tau)^2}$
Sinc	$\operatorname{sinc} 2Wt$	$\dfrac{1}{2W} \Pi\left(\dfrac{f}{2W}\right)$
Constant	1	$\delta(f)$
Phasor	$e^{j(\omega_c t + \phi)}$	$e^{j\phi}\delta(f - f_c)$
Sinusoid	$\cos(\omega_c t + \phi)$	$\frac{1}{2}[e^{j\phi}\delta(f - f_c) + e^{-j\phi}\delta(f + f_c)]$
Fourier series	$\displaystyle\sum_{n=-\infty}^{\infty} c_n e^{jn\omega_0 t}$	$\displaystyle\sum_{n=-\infty}^{\infty} c_n \delta(f - nf_0)$
Impulse	$\delta(t - t_0)$	$e^{-j\omega t_0}$
Ideal sampling wave	$\displaystyle\sum_{m=-\infty}^{\infty} \delta(t - mT_s)$	$\dfrac{1}{T_s} \displaystyle\sum_{n=-\infty}^{\infty} \delta\left(f - \dfrac{n}{T_s}\right)$
Signum	$\operatorname{sgn} t$	$\dfrac{-j}{\pi f}$

Table B
Useful Mathematical Relations

Certain of the mathematical relations encountered in this text are listed below for convenient reference.   However, this table is not intended as a substitute for more comprehensive handbooks.

Trigonometric identities

$e^{\pm j\theta} = \cos\theta \pm j\sin\theta$

$\cos\theta = \dfrac{1}{2}\left(e^{j\theta} + e^{-j\theta}\right)$

$\sin\theta = \dfrac{1}{2j}\left(e^{j\theta} - e^{-j\theta}\right)$

$\sin^2\theta + \cos^2\theta = 1$
$\cos^2\theta - \sin^2\theta = \cos 2\theta$

$\cos^2\theta = \frac{1}{2}(1 + \cos 2\theta)$
$\cos^3\theta = \frac{1}{4}(3\cos\theta + \cos 3\theta)$

$\sin^2 \theta = \frac{1}{2}(1 - \cos 2\theta)$

$\sin^3 \theta = \frac{1}{4}(3 \sin \theta - \sin 3\theta)$

$\sin (\alpha \pm \beta) = \sin \alpha \cos \beta \pm \cos \alpha \sin \beta$

$\cos (\alpha \pm \beta) = \cos \alpha \cos \beta \mp \sin \alpha \sin \beta$

$\tan (\alpha \pm \beta) = \dfrac{\tan \alpha \pm \tan \beta}{1 \mp \tan \alpha \tan \beta}$

$\sin \alpha \sin \beta = \frac{1}{2} \cos (\alpha - \beta) - \frac{1}{2} \cos (\alpha + \beta)$

$\cos \alpha \cos \beta = \frac{1}{2} \cos (\alpha - \beta) + \frac{1}{2} \cos (\alpha + \beta)$

$\sin \alpha \cos \beta = \frac{1}{2} \sin (\alpha - \beta) + \frac{1}{2} \sin (\alpha + \beta)$

$\sin (\omega t + \phi) = \cos \left(\omega t + \phi - \dfrac{\pi}{2} \right)$

$A \cos (\omega t + \alpha) + B \cos (\omega t + \beta) = C \cos \omega t - S \sin \omega t$
$$= R \cos (\omega t + \phi)$$

where

$C = A \cos \alpha + B \cos \beta$

$S = A \sin \alpha + B \sin \beta$

$R = \sqrt{C^2 + S^2} = \sqrt{A^2 + B^2 + 2AB \cos (\alpha - \beta)}$

$\phi = \arctan \dfrac{S}{C} = \arctan \dfrac{A \sin \alpha + B \sin \beta}{A \cos \alpha + B \cos \beta}$

Series expansions and approximations

The first one or two terms of the following series are often useful approximations.   Unless otherwise indicated, the approximation condition is $|x| \ll 1$.

$(1 + x)^n = 1 + nx + \dfrac{n(n - 1)}{2!} x^2 + \cdots \qquad |nx| \ll 1$

$e^x = 1 + x + \dfrac{1}{2!} x^2 + \cdots$

$a^x = 1 + x \ln a + \dfrac{1}{2!} (x \ln a)^2 + \cdots$

$\ln (1 + x) = x - \frac{1}{2}x^2 + \frac{1}{3}x^3 + \cdots$

$\sin x = x - \dfrac{1}{3!} x^3 + \dfrac{1}{5!} x^5 - \cdots$

$\cos x = 1 - \dfrac{1}{2!} x^2 + \dfrac{1}{4!} x^4 - \cdots$

$\tan x = x + \frac{1}{3}x^3 + \frac{2}{15}x^5 + \cdots$

$\arcsin x = x + \frac{1}{6}x^3 + \frac{3}{40}x^5 + \cdots$

$\arctan x = \begin{cases} x - \frac{1}{3}x^3 + \frac{1}{5}x^5 - \cdots & |x| \ll 1 \\[2mm] \dfrac{\pi}{2} - \dfrac{1}{x} + \dfrac{1}{3x^3} - \cdots & x \gg 1 \end{cases}$

$\operatorname{sinc} x = 1 - \dfrac{1}{3!} (\pi x)^2 + \dfrac{1}{5!} (\pi x)^4 - \cdots$

$$J_n(x) = \begin{cases} \dfrac{x^n}{2^n n!}\left[1 - \dfrac{x^2}{2^2(n+1)} + \dfrac{x^4}{2 \cdot 2^4(n+1)(n+2)} - \cdots\right] & 0 \le x \ll 1 \\[3mm] \sqrt{\dfrac{2}{\pi x}}\, \cos\left(x - \dfrac{n\pi}{2} - \dfrac{\pi}{4}\right) & x \gg 1 \end{cases}$$

$$I_0(x) = \begin{cases} 1 + \dfrac{x^2}{2^2} + \dfrac{x^4}{2^4 4^2} + \cdots \approx e^{x^2/4} & 0 \le x \ll 1 \\[3mm] \dfrac{e^x}{\sqrt{2\pi x}} & x \gg 1 \end{cases}$$

Definite integrals

For all the following integrals the integration limits are 0 to ∞.

$$\int \frac{x^{m-1}}{1+x^n}\, dx = \frac{\pi/n}{\sin\,(m\pi/n)} \qquad n > m > 0$$

$$\int \frac{\sin x}{x}\, dx = \int \frac{\tan x}{x}\, dx = \frac{\pi}{2}$$

$$\int \frac{\sin x \cos nx}{x}\, dx = \begin{cases} \dfrac{\pi}{2} & n^2 < 1 \\[2mm] \dfrac{\pi}{4} & n^2 = 1 \\[2mm] 0 & n^2 > 1 \end{cases}$$

$$\int \frac{\sin^2 x}{x^2}\, dx = \frac{\pi}{2}$$

$$\int \frac{\cos nx}{1+x^2}\, dx = \frac{\pi}{2} e^{-|n|}$$

$$\int \text{sinc}\, x \, dx = \int \text{sinc}^2 x \, dx = \tfrac{1}{2}$$

$$\int \sin x^2 \, dx = \int \cos x^2 \, dx = \frac{1}{2}\sqrt{\frac{\pi}{2}}$$

$$\int x^n e^{-ax}\, dx = \frac{n!}{a^{n+1}} \qquad n \ge 1,\ a > 0$$

$$\int e^{-a^2 x^2}\, dx = \frac{1}{2a}\sqrt{\pi} \qquad a > 0$$

$$\int x^2 e^{-x^2}\, dx = \tfrac{1}{4}\sqrt{\pi}$$

$$\int e^{-ax} \cos x \, dx = \frac{a}{1+a^2} \qquad a > 0$$

$$\int e^{-ax} \sin x \, dx = \frac{1}{1+a^2} \qquad a > 0$$

$$\int e^{-a^2 x^2} \cos bx \, dx = \frac{1}{2a}\sqrt{\pi}\, e^{-(b/2a)^2}$$

Table C
The Sinc Function

Numerical values of the sinc function, sinc $x = (\sin \pi x)/\pi x$, and its square are tabulated below for x from 0 to 4 in increments of 0.25.

x	sinc x	sinc^2 x
0.00	1.000	1.000
0.25	0.900	0.811
0.50	0.637	0.405
0.75	0.300	0.090
1.00	0.000	0.000
1.25	−0.180	0.032
1.50	−0.212	0.045
1.75	−0.129	0.017
2.00	0.000	0.000
2.25	0.100	0.010
2.50	0.127	0.016
2.75	0.082	0.007
3.00	0.000	0.000
3.25	−0.069	0.005
3.50	−0.091	0.008
3.75	−0.060	0.004
4.00	0.000	0.000

Table D
Gaussian Probabilities

The function

$$\Phi(k) = \frac{1}{\sqrt{2\pi}} \int_0^k e^{-u^2/2} \, du$$

is the probability a gaussian variate of mean m and variance σ^2 will be observed between m and $m + k\sigma$. Specifically, from symmetry and the fact that $\Phi(\infty) = \frac{1}{2}$, we have

$$P(m < X < m + k\sigma) = P(m - k\sigma < X < m) = \Phi(k)$$
$$P(|X - m| < k\sigma) = 2\Phi(k)$$
$$P(X > m + k\sigma) = P(X < m - k\sigma) = \frac{1}{2} - \Phi(k)$$
$$P(|X - m| > k\sigma) = 1 - 2\Phi(k)$$

For large k the following approximation is useful

$$\frac{1}{2} - \Phi(k) \approx \frac{1}{\sqrt{2\pi}\, k} e^{-k^2/2} \qquad k \gg 1$$

Numerical values of $\Phi(k)$ and $\frac{1}{2} - \Phi(k)$ are listed below for selected k.

k	$\Phi(k)$	$\frac{1}{2} - \Phi(k)$
0.0	0.000	0.500
0.1	0.040	0.460
0.2	0.079	0.421
0.3	0.118	0.382
0.4	0.156	0.344
0.5	0.191	0.309
0.6	0.226	0.274
0.7	0.258	0.242
0.8	0.288	0.212
0.9	0.316	0.184
1.0	0.341	0.159
1.25	0.394	0.106
1.5	0.433	0.067
1.75	0.460	0.040
2.0	0.477	0.023
2.25	0.488	0.012
2.5	0.494	6.2×10^{-3}
2.75	0.497	3.0×10^{-3}
3.0	0.499	1.4×10^{-3}
3.5	0.500	2.3×10^{-4}
4.0	0.500	3.3×10^{-5}
4.5	0.500	3.4×10^{-6}
5.0	0.500	2.9×10^{-7}

Table E
Glossary of Symbolic Notation

Operations

Operation	Definition	Reference section
Fourier transform	$\mathcal{F}[x(u)] = \int_{-\infty}^{\infty} x(u)e^{-i2\pi uv}\, du = X(v)$	2.3
Inverse Fourier transform	$\mathcal{F}^{-1}[X(v)] = \int_{-\infty}^{\infty} X(v)e^{i2\pi uv}\, dv = x(u)$	2.3
Hilbert transform	$\hat{x}(u) = x(u) * \dfrac{1}{\pi u} = \dfrac{1}{\pi}\int_{-\infty}^{\infty} \dfrac{x(u')}{u - u'}\, du$	3.5
Convolution	$x(u) * y(u) = \int_{-\infty}^{\infty} x(u')y(u - u')\, du'$	
	$\quad\quad\quad\;\; = \int_{-\infty}^{\infty} y(u')x(u - u')\, du'$	2.4
Time average	$\langle x(t) \rangle = \lim_{T\to\infty} \dfrac{1}{T}\int_{-T/2}^{T/2} x(t)\, dt$	2.2, 2.3
Statistical or ensemble average	$\bar{x} = \int_{-\infty}^{\infty} xp(x)\, dx$	4.3
Autocorrelation	$R_x(\tau) = \langle x(t)x(t + \tau) \rangle$	2.6
	$\quad\quad\;\; = \overline{x(t)x(t + \tau)}$	4.5

Functions

Function	*Definition*	*Reference section*
Sinc	$\operatorname{sinc} u = \dfrac{\sin \pi u}{\pi u}$	2.2
Rectangular	$\Pi(u) = \begin{cases} 1 & \|u\| \leq \frac{1}{2} \\ 0 & \|u\| > \frac{1}{2} \end{cases}$	2.3
Triangular	$\Lambda(u) = \begin{cases} 1 - \|u\| & \|u\| \leq 1 \\ 0 & \|u\| > 1 \end{cases}$	2.4
Impulse	$\delta(u) = \lim\limits_{a \to \infty} \begin{cases} ae^{-(au)^2} \\ a\Pi(au) \\ a \operatorname{sinc} au \\ a \operatorname{sinc}^2 au \end{cases}$	2.5
Sign	$\operatorname{sgn} u = \begin{cases} 1 & u > 0 \\ -1 & u < 0 \end{cases}$	3.5
Gaussian probability	$\Phi(k) = \dfrac{1}{\sqrt{2\pi}} \displaystyle\int_0^k e^{-u^2/2}\, du$	4.4

Miscellaneous symbols

$n! = 1 \cdot 2 \cdots n$	n factorial
$\dbinom{m}{n} = \dfrac{m!}{n!(m-n)!}$	Binomial coefficient
$\approx$	Approximately equals
$\sim$	Of the same numerical order
$\displaystyle\int_T$	$\displaystyle\int_{-T/2}^{T/2}$
$[a,b]$	The interval $a \leq u \leq b$
$\leftrightarrow$	Denoting a Fourier transform pair
$\oplus$	Addition modulo 2
$\bigstar$	Material which may be omitted

References

Abbott, E.
    "Flatland," 6th ed., rev., Blackwell, Oxford, 1950.
Abramson, N.
    "Information Theory and Coding," McGraw-Hill, New York, 1963.
Albert, A. L.
    "Electrical Communication," 2d ed., Wiley, New York, 1940.
Armstrong, E. H.
    A Method of Reducing Disturbances in Radio Signaling by a System of Fre-
    quency Modulation, *Proc. IRE*, vol. 24, pp. 689–740, May, 1936.
Arthurs, E. and H. Dym
    On the Optimum Detection of Digital Signals in the Presence of White Gaussian
    Noise—A Geometric Interpretation and a Study of Three Basic Data Trans-
    mission Systems, *IRE Trans. Commun. Systems*, vol. CS-10, pp. 336–372,
    December, 1962.
Aseltine, J. A.
    "Transform Method in Linear System Analysis," McGraw-Hill, New York,
    1958.
Ash, R.
    "Information Theory," Wiley, New York, 1965.
Baghdady, E. J. (ed.)
    "Lectures on Communication System Theory," McGraw-Hill, New York, 1960.

Beckmann, P.
"Probability in Communication Engineering," Harcourt, Brace & World, New York, 1967.

Bendat, J. S.
"Principles and Applications of Random Noise Theory," Wiley, New York, 1958.

Bennett, W. R.
Spectra of Quantized Signals, *Bell System Tech. J.*, vol. 27, pp. 446–472, July, 1948.

Methods of Solving Noise Problems, *Proc. IRE*, vol. 44, pp. 609–638, May, 1956.

"Electrical Noise," McGraw-Hill, New York, 1960.

Bennett, W. R., and J. R. Davey
"Data Transmission," McGraw-Hill, New York, 1965.

Black, H. S.
"Modulation Theory," Van Nostrand, Princeton, N.J., 1953.

Bode, H. W.
"Network Analysis and Feedback Amplifier Design," Van Nostrand, Princeton, N.J., 1945.

Bose, R. C., and D. K. Ray-Chaudhuri
On a Class of Error-correcting Binary Group Codes, *Inform. Control*, vol. 3, pp. 68–79, March, 1960.

Bowers, F. K.
What Use Is Delta Modulation to the Transmission Engineer?, *Trans. AIEE*, pt. I, vol. 76, pp. 142–147, May, 1957.

Bracewell, R.
"The Fourier Transform and Its Applications," McGraw-Hill, New York, 1965.

Brillouin, L.
"Science and Information Theory," Academic, New York, 1956.

Cahn, C. R.
Performance of Digital Phase-modulation Communication Systems, *IRE Trans. Commun. Systems*, vol. CS-7, pp. 3–6, May, 1959.

Combined Digital Phase and Amplitude Modulation Communication Systems, *IRE Trans. Commun. Systems*, vol. CS-8, pp. 150–155, September, 1960.

Campbell, G. A., and R. M. Foster
"Fourier Integrals for Practical Applications," Van Nostrand, Princeton, N.J., 1948.

Carson, J. R.
Notes on the Theory of Modulation, *Proc. IRE*, vol. 10, pp. 57–64, February, 1922 (reprinted in *Proc. IEEE*, vol. 51, pp. 893–896, June, 1963).

Chaffee, J. G.
The Application of Negative Feedback to Frequency-modulation Systems, *Proc. IRE*, vol. 27, pp. 317–331, May, 1939.

Christian, E., and E. Eisenmann
"Filter Design Tables and Graphs," Wiley, New York, 1966.

Churchill, R. V.
"Fourier Series and Boundary Value Problems," 2d ed., McGraw-Hill, New York, 1963.

Close, C. M.
"The Analysis of Linear Circuits," Harcourt, Brace & World, New York, 1966.

Cooper, G. R., and C. D. McGillem
"Methods of Signal and System Analysis," Holt, New York, 1967.

Costas, J. P.
Synchronous Communication, *Proc. IRE*, vol. 44, pp. 1713–1718, December, 1956.

Craig, E. J.
"Laplace and Fourier Transforms for Electrical Engineers," Holt, New York, 1964.

Cramér, H.
"The Elements of Probability Theory," Wiley, New York, 1955.

Cuccia, C. L.
"Harmonics, Sidebands, and Transients in Communication Engineering," McGraw-Hill, New York, 1952.

Davenport, W. B., Jr., and W. L. Root
"Introduction to Random Signals and Noise," McGraw-Hill, New York, 1958.

D'Azzo, J. J., and C. H. Houpis
"Feedback Control System Analysis and Synthesis," 2d ed., McGraw-Hill, New York, 1965.

DeRusso, P. M., R. J. Roy, and C. M. Close
"State Variables for Engineers," Wiley, New York, 1965.

Downing, J. J.
"Modulation Systems and Noise," Prentice-Hall, Englewood Cliffs, N.J., 1964.

Drake, A. W.
"Fundamentals of Applied Probability Theory," McGraw-Hill, New York, 1967.

Erdélyi, A. (ed.)
"Tables of Integral Transforms," 2 vols., McGraw-Hill, New York, 1954.

Everitt, W. L., and G. E. Anner
"Communication Engineering," 3d ed., McGraw-Hill, New York, 1956.

Franklin, R. H., and H. B. Law
Trends in Digital Communication by Wire, *IEEE Spectrum*, vol. 3, pp. 52–58, November, 1966.

Friedman, B.
"Principles and Techniques of Applied Mathematics," Wiley, New York, 1956.

Frutiger, P.
Noise in FM Receivers with Negative Frequency Feedback, *Proc. IEEE*, vol. 54, pp. 1506–1520, November, 1966.

Gibson, J. E.
"Nonlinear Automatic Control," McGraw-Hill, New York, 1963.

Giger, A. J., and J. G. Chaffee
The FM Demodulator with Negative Feedback, *Bell System Tech. J.*, vol. 42, pp. 1109–1135, July, 1963.

Goldman, S.
"Frequency Analysis, Modulation and Noise," McGraw-Hill, New York, 1948.

"Information Theory," Prentice-Hall, Englewood Cliffs, N.J., 1953.

Guillemin, E. A.
"Communication Networks," vol. II, Wiley, New York, 1935.

"The Mathematics of Circuit Analysis," Wiley, New York, 1949.

"Synthesis of Passive Networks," Wiley, New York, 1957.

Hagelbarger, D. W.
Recurrent Codes: Easily Mechanized Burst-correcting Binary Codes, *Bell System Tech. J.*, vol. 38, pp. 969–984, July, 1959.

Hamming, R. W.
Error Detecting and Error Correcting Codes, *Bell System Tech. J.*, vol. 26, pp. 147–160, April, 1950.

Hancock, J. C.
"An Introduction to the Principles of Communication Theory," McGraw-Hill, New York, 1961.

Harman, W. W.
   "Principles of the Statistical Theory of Communication," McGraw-Hill, New York, 1963.
Hartley, R. V.
   Transmission of Information, *Bell System Tech. J.*, vol. 7, pp. 535–563, July, 1928.
Henney, K. (ed.)
   "Radio Engineering Handbook," 5th ed., McGraw-Hill, New York, 1959.
Jacobs, I.
   Comparison of *M*-ary Modulation Systems, *Bell System Tech. J.*, vol. 46, pp. 843–863, May–June, 1967.
Jahnke, E., and F. Emde
   "Tables of Functions," 4th ed., Dover, New York, 1945.
Javid, M., and E. Brenner
   "Analysis, Transmission, and Filtering of Signals," McGraw-Hill, New York, 1963.
Johnson, J. B.
   Thermal Agitation of Electricity in Conductors, *Phys. Rev.*, vol. 32, pp. 97–109, July, 1928.
Jolly, W. P.
   "Low Noise Electronics," American Elsevier, New York, 1967.
Kallman, H. E.
   Transversal Filters, *Proc. IRE*, vol. 28, pp. 302–310, July, 1940.
Kotel'nikov, V. A. (trans. by R. A. Silverman)
   "The Theory of Optimum Noise Immunity," McGraw-Hill, New York, 1959.
Lathi, B. P.
   "Signals, Systems and Communications," Wiley, New York, 1965.
Lawson, J. L., and G. E. Uhlenbeck
   "Threshold Signals," McGraw-Hill, New York, 1950.
Lee, Y. W.
   "Statistical Theory of Communication," Wiley, New York, 1960.
Lighthill, M. J.
   "An Introduction to Fourier Analysis and Generalized Functions," Cambridge, New York, 1958.
Linden, D. A.
   A Discussion of Sampling Theorems, *Proc. IRE*, vol. 47, pp. 1219–1226, July, 1959.
Marshall, J. L.
   "Introduction to Signal Theory," International Textbook, Scranton, Pa., 1965.
Nichols, M. H., and L. L. Rauch
   "Radio Telemetry," 2d ed., Wiley, New York, 1956.
Nyquist, H.
   Certain Factors Affecting Telegraph Speed, *Bell System Tech. J.*, vol. 3, pp. 324–346, April, 1924.

   Certain Topics in Telegraph Transmission Theory, *Trans. AIEE*, vol. 47, pp. 617–644, April, 1928.

   Thermal Agitation of Electric Charge in Conductors, *Phys. Rev.*, vol. 32, pp. 110–113, July, 1928.
Oliver, B. M.
   Efficient Coding, *Bell System Tech. J.*, pp. 724–750, July, 1952.
Oliver, B. M., J. R. Pierce, and C. E. Shannon
   The Philosophy of PCM, *Proc. IRE*, vol. 36, pp. 1324–1332, November, 1948.
Panter, P. F.
   "Modulation, Noise, and Spectral Analysis," McGraw-Hill, New York, 1965.

Papoulis, A.
"The Fourier Integral and Its Applications," McGraw-Hill, New York, 1962.

"Probability, Random Variables, and Stochastic Processes," McGraw-Hill, New York, 1965.

Pappenfus, E. W., W. B. Bruene, and E O. Schoenike
"Single Sideband Principles and Circuits," McGraw-Hill, New York, 1964.

Peterson, W. W.
"Error Correcting Codes," Wiley, New York, 1961.

Pettit, J. M., and M. M. McWhorter
"Electronic Amplifier Circuits," McGraw-Hill, New York, 1961.

Pfeiffer, P. E.
"Concepts of Probability Theory," McGraw-Hill, New York, 1965.

Pierce, J. R.
Physical Sources of Noise, *Proc. IRE*, vol. 44, pp. 601–608, May, 1956.

"Symbols, Signals and Noise," Harper & Row, New York, 1961.

Pratt, F.
"Secret and Urgent," Bobbs-Merrill, New York, 1942.

Reeves, A. H.
The Past, Present, and Future of PCM, *IEEE Spectrum*, vol. 2, pp. 58–63, May, 1965.

Rice, S. O.
Mathematical Analysis of Random Noise, *Bell System Tech. J.*, vol. 23, pp. 282–332, 1944, and vol. 24, pp. 46–156, 1945.

Statistical Properties of a Sine-wave plus Random Noise, *Bell System Tech. J.*, vol. 27, pp. 109–157, January, 1948.

Communication in the Presence of Noise—Probability of Error for Two Encoding Schemes, *Bell System Tech. J.*, vol. 29, pp. 60–93, January, 1950.

Rowe, H. E.
"Signals and Noise in Communication Systems," Van Nostrand, Princeton, N.J., 1965.

Salz, J.
Performance of Multilevel Narrow-band FM Digital Communication Systems, *IEEE Trans. Commun. Technology*, vol. COM-13, pp. 420–424, December, 1965.

Sanders, R. W.
Communication Efficiency Comparisons of Several Communication Systems, *Proc. IRE*, pp. 575–588, April, 1960.

Schwartz, M.
"Information Transmission, Modulation, and Noise," McGraw-Hill, New York, 1959.

Schwartz, M., W. R. Bennett, and S. Stein
"Communication Systems and Techniques," McGraw-Hill, New York, 1966.

Selin, I.
"Detection Theory," Princeton, Princeton, N.J., 1965.

Shannon, C. E.
A Mathematical Theory of Communication, *Bell System Tech. J.*, vol. 27, pp. 379–423, July, 1948, and vol. 27, pp. 623–656, October, 1948.

Communication in the Presence of Noise, *Proc. IRE*, vol. 37, pp. 10–21, January, 1949.

Prediction and Entropy of Printed English, *Bell System Tech. J.*, vol. 30, pp. 50–64, January, 1951.

Siegman, A. E.
    Thermal Noise in Microwave Systems, *Microwave J.*, vol. 4, pp. 81–90, March, 1961, vol. 4, pp. 66–73, April, 1961, and vol. 4, pp. 93–104, May, 1961.
Slepian, D.
    A Class of Binary Signaling Alphabets, *Bell System Tech. J.*, vol. 35, pp. 203–234, January, 1956.
Slepian, D., H. J. Landau, and H. O. Pollak
    Prolate Spheroidal Wave Functions, Fourier Analysis and Uncertainty, *Bell System Tech. J.*, vol. 40, pp. 43–84, January, 1961.
Sommerville, D. M.
    "An Introduction to the Geometry of N Dimensions," Dutton, New York, 1929.
Stein, S., and J. J. Jones
    "Modern Communication Principles with Application to Digital Signaling," McGraw-Hill, New York, 1967.
Still, A.
    "Communication through the Ages," Holt, New York, 1946.
Stuart, R. D.
    "An Introduction to Fourier Analysis," Methuen, London, 1966.
Stumpers, F. L.
    Theory of Frequency-modulation Noise, *Proc. IRE*, vol. 36, pp. 1081–1902, September, 1948.
Sunde, E. D.
    Ideal Binary Transmission by AM and FM, *Bell System Tech. J.*, vol. 38, pp. 1357–1425, November, 1959.
Terman, F. E.
    "Electronic and Radio Engineering," 4th ed., McGraw-Hill, New York, 1955.
Turin, G. L.
    An Introduction to Matched Filters, *IRE Trans. Inform. Theory*, vol. IT-6, pp. 311–329, June, 1960.
Tuttle, D. F., Jr.
    "Network Synthesis," vol. I, Wiley, New York, 1958.
Van der Ziel, A.
    "Noise," Prentice-Hall, Englewood Cliffs, N.J., 1954.

    Noise in Junction Transistors, *Proc. IRE*, vol. 46, pp. 1019–1038, June, 1958.
Viterbi, A. J.
    "Principles of Coherent Communication," McGraw-Hill, New York, 1966.
Wallman, H., A. B. Macnee, and C. P. Gadsden
    A Low-noise Amplifier, *Proc. IRE*, vol. 36, pp. 700–708, June, 1948.
Wheeler, H. A.
    The Interpretation of Amplitude and Phase Distortion in Terms of Paired Echos, *Proc. IRE*, vol. 27, pp. 359–385, June, 1939.
Wiener, N.
    "Cybernetics," Wiley, New York, 1948.

    "Extrapolation, Interpolation, and Smoothing of Stationary Time Series," Wiley, New York, 1949.
Woodward, P. M.
    "Probability and Information Theory with Applications to Radar," Pergamon Press, New York, 1953.
Wozencraft, J. M., and I. M. Jacobs
    "Principles of Communication Engineering," Wiley, New York, 1965.
Wozencraft, J. M., and B. Reiffen
    "Sequential Decoding," M.I.T., Cambridge, Mass., 1961.

Index

Index